Modules in Applied Mathematics: Volume 1

Edited by William F. Lucas

Modules in Applied Mathematics

Volume 1
Differential Equation Models
Martin Braun, Courtney S. Coleman, and Donald A. Drew, *Editors*

Volume 2
Political and Related Models
Steven J. Brams, William F. Lucas, and Philip D. Straffin, Jr., *Editors*

Volume 3
Discrete and System Models
William F. Lucas, Fred S. Roberts, and Robert M. Thrall, *Editors*

Volume 4
Life Science Models
Helen Marcus-Roberts and Maynard Thompson, *Editors*

Differential Equation Models

Edited by
Martin Braun, Courtney S. Coleman,
and Donald A. Drew

With 166 Illustrations

Springer-Verlag
New York Heidelberg Berlin

Martin Braun

Department of Mathematics
Queens College
Flushing, NY 11367
USA

Courtney S. Coleman

Department of Mathematics
Harvey Mudd College
Claremont, CA 91711
USA

Donald A. Drew

Department of Mathematical
  Sciences
Rensselaer Polytechnic
  Institute
Troy, NY 12181
USA

AMS Subject Classifications: 00A69, 34A01, 35A02

Library of Congress Cataloging in Publication Data

Modules in applied mathematics.
    Includes bibliographies.
    Contents: v. 1. Differential equation models / edited by Martin Braun, Courtney
S. Coleman, and Donald A. Drew. — v. 2. Political and related models / edited by
Steven J. Brams, William F. Lucas, and Philip D. Straffin, Jr.
    1. Mathematics—1961– . 2. Mathematical models. I. Lucas, William F., 1933– .
QA37.2.M6 1982   510   82-10439

This book was prepared with the support of NSF grants Nos. SED77-07482, SED75-
00713, and SED72-07370. However, any opinions, findings, conclusions, and/or
recommendations herein are those of the authors and do not necessarily reflect the
views of NSF.

Typeset by Asco Trade Typesetting Ltd., Hong Kong.
Printed and bound by R. R. Donnelley & Sons, Harrisonburg, VA.
Printed in the United States of America.

9 8 7 6 5 4 3 2 1

ISBN 0-387-90695-9 Springer-Verlag New York Heidelberg Berlin
ISBN 3-540-90695-9 Springer-Verlag Berlin Heidelberg New York

Preface

The purpose of this four volume series is to make available for college teachers and students samples of important and realistic applications of mathematics which can be covered in undergraduate programs. The goal is to provide illustrations of how modern mathematics is actually employed to solve relevant contemporary problems. Although these independent chapters were prepared primarily for teachers in the general mathematical sciences, they should prove valuable to students, teachers, and research scientists in many of the fields of application as well. Prerequisites for each chapter and suggestions for the teacher are provided. Several of these chapters have been tested in a variety of classroom settings, and all have undergone extensive peer review and revision. Illustrations and exercises are included in most chapters. Some units can be covered in one class, whereas others provide sufficient material for a few weeks of class time.

Volume 1 contains 23 chapters and deals with differential equations and, in the last four chapters, problems leading to partial differential equations. Applications are taken from medicine, biology, traffic systems and several other fields. The 14 chapters in Volume 2 are devoted mostly to problems arising in political science, but they also address questions appearing in sociology and ecology. Topics covered include voting systems, weighted voting, proportional representation, coalitional values, and committees. The 14 chapters in Volume 3 emphasize discrete mathematical methods such as those which arise in graph theory, combinatorics, and networks. These techniques are used to study problems in economics, traffic theory, operations research, decision theory, and other fields. Volume 4 has 12 chapters concerned with mathematical models in the life sciences. These include aspects of population growth and behavior, biomedicine (epidemics, genetics and bio-engineering), and ecology.

These four volumes are the result of two educational projects sponsored by The Mathematical Association of America (MAA) and supported in part by the National Science Foundation (NSF). The objective was to produce needed material for the undergraduate curriculum. The first project was undertaken by the MAA's Committee on the Undergraduate Program in Mathematics (CUPM). It was entitled Case Studies and Resource Materials for the Teaching of Applied Mathematics at the Advanced Undergraduate Level, and it received financial support from NSF grant SED72-07370 between September 1, 1972 and May 31, 1977. This project was completed under the direction of Donald Bushaw. Bushaw and William Lucas served as chairmen of CUPM during this effort, and George Pedrick was involved as the executive director of CUPM. The resulting report, which appeared in late 1976, was entitled *Case Studies in Applied Mathematics*, and it was edited by Maynard Thompson. It contained nine chapters by eleven authors, plus an introductory chapter and a report on classroom trials of the material.

The second project was initiated by the MAA's Committee on Institutes and Workshops (CIW). It was a summer workshop of four weeks duration entitled Modules in Applied Mathematics which was held at Cornell University in 1976. It was funded in part by NSF grant SED75-00713 and a small supplemental grant SED77-07482 between May 1, 1975 and September 30, 1978. William F. Lucas served as chairman of CIW at the time of the workshop and as director of this project. This activity lead to the production of 60 educational modules by 37 authors.

These four volumes contain revised versions of 9 of the 11 chapters from the report *Case Studies in Applied Mathematics*, 52 of the 60 modules from the workshop Modules in Applied Mathematics, plus two contributions which were added later (Volume 2, Chapters 7 and 14), for a total of 63 chapters. A preliminary version of the chapter by Steven Brams (Volume 2, Chapter 3), entitled "One Man, N Votes," was written in connection with the 1976 MAA Workshop. The expanded version presented here was prepared in conjunction with the American Political Science Association's project Innovation in Instructional Materials which was supported by NSF grant SED77-18486 under the direction of Sheilah K. Mann. The unit was published originally as a monograph entitled *Comparison Voting*, and was distributed to teachers and students for classroom field tests. This chapter was copyrighted by the APSA in 1978 and has been reproduced here with its permission.

An ad hoc committee of the MAA consisting of Edwin Beckenbach, Leonard Gillman, William Lucas, David Roselle, and Alfred Willcox was responsible for supervising the arrangements for publication and some of the extensive efforts that were necessary to obtain NSF approval of publication in this format. The significant contribution of Dr. Willcox throughout should be noted. George Springer also intervened in a crucial way at one point. It should be stressed, however, that any opinions or recommendations are those of the particular authors, and do not necessarily reflect the views of NSF, MAA, the editors, or any others involved in these project activities.

There are many other individuals who contributed in some way to the realization of these four volumes, and it is impossible to acknowledge all of them here. However, there are two individuals in addition to the authors, editors and people named above who should receive substantial credit for the ultimate appearance of this publication. Katherine B. Magann, who had provided many years of dedicated service to CUPM prior to the closing of the CUPM office, accomplished the production of the report *Case Studies in Applied Mathematics*. Carolyn D. Lucas assisted in the running of the 1976 MAA Workshop, supervised the production of the resulting sixty modules, and served as managing editor for the publication of these four volumes. Without her efforts and perseverance the final product of this major project might not have been realized.

July 1982   W. F. Lucas

Preface for Volume 1

Volume 1 consists of twenty-three chapters concerned with mathematical modeling and problem solving using differential equations. The chapters in Part I deal with the very beginning, and often the most important part of the modeling process: how to translate the given problem into a mathematical problem. The first chapter by Henderson West shows how to translate various word problems into differential equations, while Chapter 3 by Frauenthal deals with the special case of population growth models, a subject of much current interest. The second chapter, also by Henderson West, describes how to analyze a differential equation, and how to draw qualitative conclusions from it. These three chapters were written with a clarity and painstaking attention to detail that is not often found in textbooks, and thus are "must reading" for the beginning student of modeling.

The three chapters by Braun in Part II deal with three diverse and important problems that can be modeled, and completely solved, by first order differential equations. It is interesting to note that the work described in these units (and indeed, many of the modules in this volume) was originally done not by mathematicians, but by chemists, biologists and sociologists.

Part III is essentially a continuation of Part II, the difference being that the problems in this section are modeled by higher order linear equations and by solvable systems of first order equations. Systems of differential equations can be used to model very complex and even esoteric problems, and the results obtained are often very exciting, as seen in the three modules by Braun, Coleman, and Powers.

The five chapters by Baker and Drew in Part IV describe some applications of mathematics to problems in traffic theory, another popular source of interesting modeling problems. The results obtained are not as powerful and spectacular as the results obtained in the previous chapters. Nevertheless, they are still extremely important as they illustrate how mathematical modeling can often aid in our understanding of complex and even uncontrollable and unsolvable problems.

The five chapters by Braun and Coleman in Part V deal with systems of non-linear equations and their application to important problems in biology and ecology. Powerful results are obtained via the methods of the qualitative theory of differential equations. It is interesting to note that the qualitative theory of differential equations evolved, originally, from problems in physics and astronomy. These same techniques have had important applications to problems in the biological sciences. Indeed, the existence of such a powerful and polished theory has motivated many mathematicians to undertake the study of several outstanding problems in biology and ecology.

The chapters in Part VI deal with systems that can be modelled by partial differential equations, one of the more difficult areas of mathematical analysis. The unit by Borrelli is basic in that it carefully describes the theory and the actual modeling. The remaining three chapters by Drew, Meyer and Porsching deal with concrete and important real life applications.

July 1982

MARTIN BRAUN
COURTNEY S. COLEMAN
DONALD A. DREW

Contents

Contents for the Companion Volumes

PART I

DIFFERENTIAL EQUATIONS, MODELS, AND WHAT TO DO WITH THEM

CHAPTER 1

Setting Up First-Order Differential Equations from Word Problems

Beverly Henderson West*

1. Introduction

"Word problems" are sometimes troublesome; but you have learned that most noncalculus applied problems can be conquered with careful translating and attention to the kinds of units involved. A trivial illustration of this type is as follows.

EXAMPLE 1. One Sunday a man in a car leaves A at noon and arrives at B at 3:20 p.m. If he drove steadily at 55 mi/h, how far is B from A?

Solution: • distance = rate × time

$$= (55 \text{ mi/h})(3\tfrac{1}{3} \text{ h})$$

$$= (55)(\tfrac{10}{3}) \text{ mi}$$

$$\bullet = 183\tfrac{1}{3} \text{ mi.}$$

Note: In this and the other examples of this chapter, the *key* mathematical statements (equations, solutions, initial conditions, answers, etc.) are preceded by bullets (•) to stand out among the calculations. An even better way to emphasize the key statements, especially in handwritten work, would be to draw boxes around them or to use color highlighting.

Word problems involving differential equations may be more difficult than the applied problems you have dealt with heretofore. Contrast Example 1 with Example 2.

* Department of Mathematics, Cornell University, Ithaca, NY 14853.

EXAMPLE 2. One Sunday a man in a car leaves A at noon and arrives at B at 3:20 p.m. He started from rest and steadily increased his speed, as indicated on his speedometer, to the extent that when he reached B he was driving at 60 mi/h. How far is B from A?

Solution: An inexperienced student might suspect that not enough information is provided. However, the steadily increasing speedometer reading means that the man's speed or velocity is a *linear function* of time, and velocity is the derivative of distance S as a function of time. So,

$$\bullet \frac{dS}{dt} = at + b \quad (\text{mi/h})$$

and, by integration,

$$\bullet S = \tfrac{1}{2}at^2 + bt + c \quad (\text{mi}).$$

If t is measured in hours, the remaining information in the problem tells us that

$$\bullet \; \text{①} \; S(0) = 0; \quad \text{②} \; \frac{dS}{dt}(0) = 0; \quad \text{③} \; \frac{dS}{dt}(3\tfrac{1}{3}) = 60; \quad \text{④} \; S(3\tfrac{1}{3}) = ?.$$

The first three conditions are enough to evaluate the three constants a, b, c, and the fourth will then give us the answer to the question in the problem:

$$c = 0 \quad (\text{from ①})$$
$$b = 0 \quad (\text{from ②})$$
$$a = 18 \quad (\text{from ③})$$

so

$$S(3\tfrac{1}{3}) = 9t^2 = 9(\tfrac{10}{3})^2 = \bullet \; 100 \text{ mi, the distance from } A \text{ to } B.$$

Now, you may well have solved this problem differently, but all these ingredients must have been implicit in your solution. For instance, you might have realized that starting from rest would immediately give $dS/dt = at$, but you would still have been using information from the problem (condition (2)) to evaluate a constant which would otherwise have been there.

Consider another simple differential equation word problem which is commonly encountered.

EXAMPLE 3. The growth rate of a population of bacteria is in direct proportion to the population. If the number of bacteria in a culture grew from 100 to 400 in 24 h, what was the population after the first 12 h?

Solution: The first sentence tells what is true at *any* instant; the second gives information on specific instants. If we denote the population by $y(t)$, the first tells us that

$$\bullet \frac{dy}{dt} = ky,$$

which has a general solution

$$\bullet\, y = A\, e^{kt}.$$

The two constants A and k may be evaluated by the information in the second sentence of the problem, that

$$\bullet\, ① \; y(0) = 100 \quad \text{and} \quad \bullet\, ② \; y(24) = 400,$$

where we have chosen hours as units for t. ① implies $y(0) = A\, e^0 = A = 100$. ② implies $y(24) = 100\, e^{24k} = 400$, which gives $k = \ln 4/24$. So

$$\bullet\, y(t) = 100\, e^{t \ln 4/24}.$$

We are asked to find $\bullet\, y(12) = 100\, e^{(12/24)\ln 4} = 200$ bacteria. (All the details of this solution are considered in Exercise 1.)

Note: We chose the units for t to be *hours*. The problem could be solved equally well with the *day* as the unit for t (then 24 h gives $t = 1$ and 12 h gives $t = 1/2$). As long as you are *consistent* throughout the problem, the choice is yours. In Exercise 1 you can confirm that you will get the same final answer of population $= 200$ by measuring time in days instead of hours, although you will encounter different values for some of the constants en route.

You can already see that word problems involving differential equations tend to require even more translating than Example 1 and that you cannot plug all the numbers at once into one equation. Let us summarize the pattern we have followed in Examples 2 and 3.

For a situation involving a quantity y which is going to depend on time t, we set up an equation that gives the relation between y', y, and t at *any particular moment* t. From this equation, integration gives a new equation, involving y and t, not y'. This new equation contains constants of integration and is still true for *any* specific t. *Now* information given in the problem that is true only at *specific* times (e.g., "the train's brakes failed at 11:22 a.m.") is used to evaluate the constants of integration and any other parameters (linearity, proportionality, etc.). Then at last we have a function $y(t)$ which can be evaluated immediately for any additional values of t. We shall expand upon this pattern in the next section.

2. Guidelines

Given the infinite number of guises in which word problems occur, we cannot learn a foolproof method of setting up and solving all of them, but we can list crucial areas to which you must pay careful attention.

Keep in mind that you are trying to find a *solution curve* for a differential equation. The idea behind differential equations is that if you know the derivative at each point on a curve *and* where the curve begins, then you can reconstruct this solution curve.

1) *Translation.* Many common words occur which indicate "derivative", such as "rate," "growth" (in biology and population studies), "decay" (in radioactivity), and "marginal" (in economics). Words like "change," "varies," "increase," and "decrease" are signals to look carefully at *what* is changing—again, a derivative may be called for. (See Example 5 in the next section.)

Ask yourself whether any principles or physical laws govern the problem under consideration. Are you expected to apply a known law, or must you derive what is appropriate for the problem? Answers to these questions are important guides on how to proceed. It can be particularly difficult to see how to begin the solution when you first read a problem in some field with which you are not familiar. We pay special attention to this in the examples that follow.

Many problems fall into the following pattern:

$$\frac{\text{net rate}}{\text{of change}} = \frac{\text{rate of}}{\text{input}} - \frac{\text{rate of}}{\text{outgo.}}$$

If you can recognize this pattern when it occurs (and if you keep the physical units straight), the differential equation will probably fall out in your lap.

2) *The differential equation.* The differential equation is an *instantaneous* statement, which must be valid at *any time.* This is the central part of the mathematical problem. If you have seen the key words indicating derivatives, you want to find the relation between y', y, and t. Try to focus on the overall relations in *words* first, such as "rate = input − outgo." Write *this* down, and then make sure you fill in all the pieces listed.

3) *Units.* Once you have identified which terms go into the differential equation, make sure each term has the same physical units (e.g., gal/min or furlongs/fortnight), for real life (and textbook) exercises do not usually happen that way by themselves. Attention to the physical units can often help you out in completing the differential equation itself (as you shall see in Example 5 below).

4) *Given conditions.* These are the bits of information about what happens to the system *at a specific time.* They are held *out* of the differential equation. They are used to evaluate all the constants hanging around after the differential equation has been solved. These are the constants of proportionality or whatever in the original differential equation plus the constants of integration which occur in its solution.

These given conditions should be written *with* the differential equation in order to give completely and succinctly the mathematical statement of the problem. (In Example 2 and 3 they were delayed to illustrate their necessity.)

5) *Conceptual framework*. As noted, a bullet (•) signals the key mathematical statements in all our examples, to help you pick up the main steps that constitute a solution. In a typical problem, the key steps are completed as you successively obtain the following:

a verbal equation conceptualizing the situation in words;
a statement of any principle or physical law involved;
the differential equation;
the given conditions, initial or otherwise;
solution to the differential equation;
solution with the constants evaluated;
answer to the question of the problem.

You need to look for all of these key steps. We call this collection the "framework." The goal at each step in the problem is the completion of the next framework piece.

The fact that the computational pieces of these applied problems tend to get lengthier and more complicated as you delve more deeply into the world of differential equations (DE's) is what causes unorganized souls to lose the war with word problems. You can easily spend so many pages on a calculation that when you finally get it simplified, you have forgotten that it was only a little piece of the initial question. This is particularly true for higher order nonhomogeneous linear DE's, or for systems of DE's, or for partial DE's. So *now* is the time to learn to be organized.

You want to know at a glance *where* you are in the problem. It gets increasingly risky to do pieces in your head, but if you must, a written framework is important. If you expect others to be able to read your work, a clearly written framework is *essential*.

One good way to start a word problem is to *write down* everything you know about it. Box in, bullet, or otherwise note the key statements that contribute to the framework. Then go after the other parts of the framework list.

We have barely mentioned the solution of the differential equations in these guidelines because that is not the goal of this chapter. We are concentrating on translating the applied problem into a form whereby only routine calculations remain. Our techniques may seem overdone for simpler problems, but the aim is to help you conquer the tough ones.

This is about all that can be said in general on setting up word problems requiring first-order differential equations. Keep in mind the five guidelines—

translate
make an instantaneous statement
match physical units
state given conditions
write a clear framework

—as you work toward a solution. The only way to get more of a feeling for the process is to try working problems.

You should be able to dig into the exercises right now. Some examples are given in the following section which may or may not be helpful at the places where you get stuck. We can only show *one* possible train of thought.

3. Examples

You will get the most out of the examples if you will *try* them before reading the solutions. The aim is for *you* to be able to do word problems, not just to read one way someone else can do them. You may well create a correct solution which does not look like ours (as in Exercise 1). The check is whether you get the same *answer* to the question. If you do not, try to see what you forgot to include or where you went wrong. If you have no luck, *ask* someone.

EXAMPLE 4. An indoor thermometer, reading 60°, is placed outdoors. In 10 min it reads 70°; in another 10 min it reads 76°. Using no calculations, guess the outdoor temperature. Then calculate the proper answer, assuming Newton's Law of Cooling.

Newton's Law of Cooling—or Warming—says that when an object at any temperature T is placed in a surrounding medium at constant temperature m, then T changes at a rate which is proportional to the difference of T from the temperature of the surrounding medium. The assumption for this mathematical model is that the medium is large enough so that m is essentially not disturbed by the introduction of the warmer or colder object. Experiments have shown that this is a good approximation.

Solution: Obviously, the first order of business for the word problem is to find out what Newton's Law means; that has been provided. So we have two paragraphs from which to construct our solution.

You should zero in on the key words "changes at a rate" in the second paragraph. That sentence says that dT/dt is proportional to $T - m$, giving
● $dT/dt = k(T - m)$. Three specific conditions are provided:

$$● T(0) = 60, \ T(10) = 70, \ T(20) = 76,$$

using minutes for t and degrees for T. The solution to the differential equation is ● $T = A \ e^{kt} + m$, and the three given conditions will be enough to evaluate the three constants A, k, m. (See Exercise 2.)

Now what was the question the problem asked? Simply, ● what is m? As we have noted, sufficient information is given. We can also note that, at least for this problem, you need not bother finding A or k (unless you need them on the way to m). You should note, however, that k must be negative in order for T to approach a constant m as t increases indefinitely, so it would be smart to confirm that this is indeed the case.

EXAMPLE 5. A man eats a diet of 2500 cal/day; 1200 of them go to basal metabolism (i.e., get used up automatically). He spends approximately 16 cal/kg/day times his body weight (in kilograms) in weight-proportional exercise. Assume that the storage of calories as fat is 100% efficient and that 1 kg fat contains 10,000 cal. Find how his weight varies with time.

This particular application is probably less familiar to you than the earlier examples. Therefore, it is an excellent one to try by yourself. Cover up the rest of this example and try working it as programmed learning. Whenever you get stuck, move your covering paper down till you are unstuck and see if you can go on from there.

Solution: None of our super key "derivative" words appear, but we can focus on the final question, which tells us that ● we want to get weight (call it w) as a function of time. If we consider w as a *continuous* function of t, we can seek a differential equation involving dw/dt.

Time comes into the problem only as "per day," so you can focus on one day and try for conceptual statements such as

each day, change in weight = input − outgo;
input will be net weight intake, above and beyond basal metabolism;
outgo will be loss due to weight-proportional exercise (WPE).

Since we are aiming for a derivative, the above can be combined in a better conceptual statement:

● change in weight/day = net intake/day − WPE/day.

This has fine form for a framework statement; we can start filling in the pieces.

daily net intake = 2500 cal/day eaten − 1200 cal/day used in basal
= 1300 cal/day.   metabolism

daily net outgo = 16 (cal/kg)/day × w kg in WPE
= $16w$ cal/day.

change in weight/day = $\dfrac{\Delta w}{\Delta t}$ kg/day

= $\dfrac{dw}{dt}$ kg/day, in the limit as $\Delta t \to 0$
(which is what we need for an *instantaneous* statement about a continuously changing function $w(t)$).

As you may have noticed, some of these quantities are given in terms of *energy* (calories) and others in terms of *weight* (kilograms). What are you going to do about the fact that the units on the left of the framework statement (kg/day) do not match, those arising on the right (cal/day)? That is where the last sentence of information comes in, giving cal/kg. We can use

$$\text{kg/day} = \frac{\text{net cal/day}}{10{,}000 \text{ cal/kg}}.$$

So, filling in all the pieces gives

$$\bullet \quad \frac{dw}{dt} = \frac{(2500 - 1200) - 16w}{10{,}000} \qquad (1)$$

which in physical units checks out as follows:

$$\frac{\text{kg}}{\text{day}} = \frac{\text{cal/day} - ((\text{cal/kg})/\text{day})(\text{kg})}{\text{cal/kg}}.$$

How many constants are you going to have in your solution? One, from integration. So how many given conditions would you need to give a numerical answer for the man's weight on a given day? One, e.g., that at
$\bullet$ $t = 0$, $w = w_0$, giving his weight at the beginning.

The problem of Example 5 is now completely set up for the routine calculations to take over, but we shall follow through on the solution to this problem in order to pursue some questions of interpretation and their consequences. The differential equation (1) can easily be solved by separation of variables—go ahead:

$$\frac{dw}{1300 - 16w} = \frac{dt}{10{,}000}$$

$$-\frac{1}{16}\ln|1300 - 16w| = \frac{t}{10{,}000} + C.$$

(Physically, we must have intake $\geq$ outgo, so we can drop absolute value sign.)[1]

$$1300 - 16w = e^{-16((t/10{,}000)+c)}$$

$$= Q\,e^{-16t/10{,}000}$$

$$= (1300 - 16w_0)\,e^{-16t/10{,}000}.$$

(The given condition is an initial one, with $t = 0$, which makes the constant especially easy to evaluate.)

Solving for w,

$$\bullet \quad w = \frac{1300}{16} - \left(\frac{1300 - 16w_0}{16}\right)e^{-16t/10{,}000} \text{ kg.} \qquad (2)$$

Thus we have answered the question posed by the problem, but consider one likely additional question: "Does the man reach an equilibrium weight?"

---

[1] Alternatively, the reader may keep the absolute value sign and show that $|1300 - 16w| = |1300 - 16w_0|\exp(-16t/10{,}000)$. From this, one concludes that $1300 - 16w$ has the sign of $1300 - 16w_0$ since the exponential factor is positive.

This question can be answered from (2) by noting that as $t \to \infty$, the right-hand term of this expression for w goes to 0, so $w \to 1300/16$ kg. However, we can also answer this last question directly from the differential equation (1). At an equilibrium, w does not change, so $dw/dt = 0$. This gives very directly that

$$w_{\text{equil}} = \frac{1300}{16} \text{ kg}.$$

So, if the equilibrium were all we needed to know, we would not have had to solve the differential equation! We would have been finished one line after (1).

EXAMPLE 6. What rate of interest payable annually is equivalent to 6% continuously compounded?

This problem can be solved *very* quickly using the differential equation idea, as shown below. However, some people do not find this line of thought very natural in the context of bank interest. The problem can also be attacked directly and more traditionally as an extension of simple interest; this method is outlined in Exercise 3.

Solution: • Let $S(t)$ be saving at time t. $S(t)$ includes the interest continuously compounded. At • $t = 0$, let $S = S_0$, which simply labels the initial amount of money, the principal.

Now then, what is the question? Behind the scenes, the problem is asking how much money will have been gained in one year. If at • $t = 1$ we let $S = S_1$, then the

$$\bullet \text{ equivalent annual rate} = \frac{\text{moncy carned}}{\text{original amount}} = \frac{S_1 - S_0}{S_0}.$$

"Rate of interest" means dS/dt, for the instantaneous change per unit time in savings S is due only to the calculation of the interest at that instant. Hence, because of the continuous compounding, • $dS/dt = 0.06S$ at any instant. [Another way of stating this differential equation is as follows: dS/dt = rate of change for total amount in savings, so the rate per dollar in the account $= 0.06 = (dS/dt)/S$.] The differential equation has a general solution • $S = A e^{0.06t}$. (You can verify this in Exercise 1, letting $y = S$ and $k = 0.06$.)

We need a condition to evaluate A, which we provided by setting $S = S_0$ at $t = 0$. Plugging in this initial condition gives • $S = S_0 e^{0.06t}$. Recall that we are looking for the annual rate $(S_1 - S_0)/S_0$. First find $S_1 = S_0 e^{0.06} = 1.0618S_0$. Then

$$\frac{S_1 - S_0}{S_0} = \frac{(1.0618 - 1)S_0}{S_0} = 0.0618$$

or • 6.18% annual rate. Notice that, for this particular question, we do not need a value for S_0, one of the constants. Mathematically, that is to say that the answer requested is *independent* of S_0.

EXAMPLE 7. Human skeletal fragments showing ancient Neanderthal characteristics are found in a Palestinian cave and are brought to a laboratory for carbon dating. Analysis shows that the proportion of C^{14} to C^{12} is only 6.24% of the value in living tissue. How long ago did this person live?

   (Carbon dating: The carbon in living matter contains a minute proportion of the radioactive isotope C^{14}. This radiocarbon arises from cosmic ray bombardment in the upper atmosphere and enters living systems by exchange processes, reaching an equilibrium concentration in these organisms. This means that in living matter, the amount of C^{14} is in constant ratio to the amount of the stable isotope C^{12}. After the death of an organism, exchange stops, and the radiocarbon decreases at the rate of one part in 8000 per yr.)

   Solution: Carbon dating enables calculation of the moment when an organism *died*; therefore, our question actually means "how long ago did this person die?" If we let • t = year *after death*, and • $y(t)$ = ratio C^{14}/C^{12} (say in $mg\ C^{14}/mg\ C^{12}$), then the last sentence of the carbon dating paragraph yields our differential equation (identified by the key word "rate"):

$$\bullet\ \frac{dy}{dt} = -\frac{1}{8000}\text{yr}\quad (\textit{decreases})$$

in $(mg\ C^{14}/mg\ C^{12})$/yr. (Another way of stating this property of radioactive disintegration is that "the rate of disintegration of a radioactive substance is proportional at any instant to the amount of the substance present." The constant of proportionality for C^{14} was given by the "one part in 8000/yr.")

   Our solution will have but one constant, from integration, so one given condition will suffice for evaluation. This can be provided by noting that at the time of death of the organism, when • $t = 0$, then $y = y_0$, the proportion of C^{14} in living matter.

   The general solution to the differential equation is

$$\bullet\ y = k\,e^{-t/8000}.$$

Two steps remain; to evaluate k and to answer the question.

   The initial condition tells us that $k = y_0$, so we have

$$\bullet\ y = y_0\,e^{-t/8000}.$$

The question asks us to • find t when $y = 0.0624y_0$:

$$0.0624y_0 = y_0\,e^{-t/8000}$$

$$t = -8000\ln 0.0624 \approx \bullet\ 22{,}400\ \text{yr ago}\quad \text{(that is, the}$$
number of years before the
analysis that death occurred).

Note: Recently, the practice of carbon dating has been questioned—dates between 2500 and 10,000 years ago have been in discrepancy with other dating methods. In 1966 Minze Stuiver of the Yale Laboratory and Hans E. Suess of the University of California at San Diego reported establishment of the nature of errors in carbon dating during this period. Evidently, cosmic ray activity decreased at the time, with the peak discrepancies occurring about 6000 years ago. The researchers' conclusions were the result of carbon dating of Bristlecone pine wood, which also provided accurate tree-ring dating. They suggested an apparently successful formula for correcting the carbon-dating calculation between 2300 and 6000 years ago[2]:

$$\text{true time} = C^{14} \text{ yr} \times 1.4 - 900.$$

EXAMPLE 8. A right circular cylinder of radius 10 ft and height 20 ft is filled with water. A small circular hole in the bottom is of 1-in diameter. How long will it take for the tank to empty?

We need a physical assumption about the velocity with which the water leaves the hole. Even if you feel far removed from physics, consider the following. It certainly is reasonable to assume this velocity will depend on $h(t)$, the height of the water remaining in the tank at time t. After all, the water will flow faster when the tank is full than when it is nearly empty (the greater depth of water exerts more pressure to push water out of the hole). Furthermore, if one assumes no energy loss, then the potential energy lost at the top when a small amount of water has left the tank must equal the kinetic energy of an equal amount of water leaving the bottom of the tank through the hole. That is,

$$mgh = \tfrac{1}{2}mv^2,$$

at any instant, so

$$\bullet \; v = \sqrt{2gh},$$

where g is the acceleration due to gravity, which is exactly the relation cited in physics as *Torricelli's Law*. Physically, this model may be an oversimplification for the situation in question, but at the very least we can agree that the dependence on height seems reasonable. Further physical argument might possibly produce a better formula for velocity, but it would not otherwise change the *mathematical* analysis, which proceeds as follows.

---

[2] An interesting, readable, and detailed account of dating procedures is contained in Louis Brennan's *American Dawn, A New Model of American Prehistory*; (New York: Macmillan, 1970, ch. 3). Stuiver and Suess originally reported their work in the professional journal *Radiocarbon*; the results are summarized in *American Antiquity*, July, 1966. For our purposes, we merely take note of the reassurance that for dating more than 10,000 years or less than 2500, the carbon dating model of this Example is found to be quite accurate (that is, within 200 years).

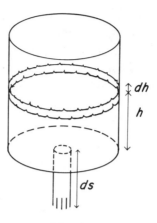

Figure 1.1

Solution: The situation says to look at *volumes.* The volume of water in the tank decreases by the volume which escapes through the hole. Letting A be the horizontal area of the tank and B be the horizontal area of the hole,

$$-A\,dh = +B\,ds$$

(volume in   (volume leaving
tank decreases)   tank increases)

in any interval dt. Now what is our question? • When (i.e., at what t) does $h = 0$? Therefore, we want to find $h(t)$. So far we have but one relation, which can be rewritten $dh = -(B/A)\,ds$. We can calculate A and B, but what can we do with ds?

The other information we have is the *velocity* creating ds, so we can make the following substitution:

$$ds = \left(\frac{ds}{dt}\right)dt = v\,dt$$

which gives

$$\bullet\ dh = -\frac{B}{A}v\,dt.$$

We can calculate the following:

$$A = \pi(10)^2\ \text{ft}^2$$

$$B = \pi(1/24)^2\ \text{ft}^2$$

$$v = \sqrt{2gh} = \sqrt{2(32)h} = 8h^{1/2}\ \text{ft/s}.$$

Substitution and separation of variables gives

$$\bullet\ h^{-1/2}\,dh = \frac{-8(1/24)^2}{(10)^2}\,dt$$

so

$$2h^{1/2} = \frac{-8(1/24)^2}{(10)^2} t + K.$$

When $\bullet\, t = 0$, $h = 20$, so $K = 2\sqrt{20}$. We want t when $h = 0$, so

$$t = \frac{(100)}{8(1/24)^2}(2\sqrt{20}) \approx \bullet\ 18\ hours\ (\text{or } 64,800\ \text{s}).$$

Exercises

(The most challenging exercises are denoted by a dagger.)

1. Fill in the following details of Example 3:
   a) Solve $dy/dt = ky$ (by separation of variables) to get $y = A\,e^{kt}$.
   b) Show that $y(24) = 400$ implies that $k = (\ln 4)/24$.
   c) Show that $y(12) = 200$.
   d) Show that you get the same answer of population $= 200$ after 12 h if you let the unit of time be one *day*.

2. Fill in the following details of Example 4:
   a) Solve $dT/dt = k(T - m)$ to get $T = A\,e^{kt} + m$.
   b) Using the three given conditions, find m.

3. Try another method to handle Example 6. Again, let $S(t)$ be savings as a function of time t, *measured in years*.
   a) Then *simple* interest, payable *annually*, at rate r would mean $S_1 = S_0 + rS_0$. Show that after n years $S_n = (1 + r)^n S_0$.
   b) Interest compounded *quarterly* at an annual rate r would mean after one quarter that $S(1/4) = S_0 + (r/4)S_0$. Show that after n quarters, $S(n/4) = (1 + r/4)^n S_0$.
   c) Interest compounded *daily* at an annual rate r would mean after one day that $S(1/365) = S_0 + (r/365)S_0$. Show that after n days, $S(n/365) = (1 + r/365)^n S_0$.
   d) Interest *continuously compounded* at a rate r is computed by $\lim_{m \to \infty} S(n/m)$. Show that this implies, at an instant when $t = n/m$, that $S(n/m) = (e^r)^{n/m} S_0$, i.e., that $S(t) = S_0\,e^{rt}$.

4. What is the half-life of C^{14}? (See Example 7. Half-life is the time required for half the amount of the radioactive isotope to disintegrate.)

5. At Cro Magnon, France, human skeletal remains were discovered in 1868 in a cave where a railway was being dug. Philip van Doren Stern, in a book entitled *Prehistoric Europe, from Stone Age Man to the Early Greeks* (New York: W. W. Norton, 1969), asserts that the best estimates of the age of these remains range from 30,000 to 20,000 B.C. What range of laboratory C^{14} to C^{12} ratios would be represented by that range of dates? (See Example 7.)

6. By Newton's law, the rate of cooling of some body in air is proportional to the difference between the temperature of the body and the temperature of the air. If the temperature of the air is 20°C and boiling water cools in 20 min to 60°C, how long will it take for the water to drop in temperature to 30°C?

7. A fussy coffee brewer wants his water at 185°F, but he almost always forgets and lets it boil. Having broken his thermometer, he asks you to calculate how long he should wait for it to cool from 212° to 185°. Can you solve his problem? If you answer "yes," do so. If "no," explain why.

8. Water at temperature 100°C cools in 3 min to 90°C, in a chamber at 60°C. Temperature changes most rapidly when the temperature difference between the water and the room is the greatest. Experiments show the rate of change is linearly proportional to this difference.
   a) Find the water temperature after 6 min.
   b) When is the temperature 75°C? 61°C?

†9. One ounce of water at 90°C is set afloat in a plastic cup in a photographer's chemical solution of exactly 100 oz at 10°C. This is an effort to warm the solution without diluting it.
   a) Express the temperature of the solution as a function of time.
   b) Reconcile your model with the temperature model used in Example 4.

10. A spherical raindrop evaporates at a rate proportional to its surface area. Find a formula for its volume V as a function of time.

11. A 100-gal tank is filled with water and 20 lb of salt. Fresh water is pumped in at a rate of 2 gal/min. The mixture is continuously stirred, and overflows to keep the tank at the 100-gal level. How much has the concentration of salt been diluted after one hour?

12. Water pollution can be diminished by treatment of raw sewage before it reaches the water supply. A common method is to use an activated sludge aeration tank containing a concentration c (which varies with time) of pollutant. Raw sewage containing a greater concentration c_1 of the pollutant enters the tank, bacteria digest some of the sewage, and the resultant cleaner mixture is dumped into a water body. The concentration of pollutant in the discharge must not exceed a certain safe level, say $0.30c_1$, so the problem is to find the time when that level is reached. In practice, at that t the raw sewage can be diverted to another tank, while this one is aerated to reduce c to c_0, the reasonable minimal level. Assume that the tank receives input at a rate of r_1 gal/min and the effluent leaves at a rate of r_2 gal/min. At $t = 0$ the tank holds V_0 gal of sewage containing z_0 lb of pollutant. Set up the problem for a mathematician to solve in order to find the time at which this tank should be bypassed.

13. If a savings account, with interest continuously compounded, doubles in 16 yr, what is the interest rate?

†14. A college education fund is begun with $\$P$ invested to grow at a rate r continuously compounded. In addition, new capital is added every year on the anniversary of the opening of the account, at a rate of $\$A$/yr. Find the accumulated amount after t yr.

15. A tank is filled with 10 gal of brine in which 5 lb of salt is dissolved. Brine having 2 lb of salt per gallon enters the tank at 3 gal/min, and the well-stirred mixture leaves at the same rate.
   a) What is the concentration of salt in the water leaving the tank after 8 min?
   b) How much salt is in the tank after a long time?

16. Neutrons in an atomic pile increase at a rate proportional to the number of neutrons present at any instant (due to nuclear fission). If N_0 neutrons are initially present and N_1 and N_2 neutrons are present at times T_1 and T_2, respectively, show that

$$\left(\frac{N_2}{N_0}\right)^{T_1} = \left(\frac{N_1}{N_0}\right)^{T_2}.$$

17. Water containing 2 oz of pollutant/gal flows through a treatment tank at a rate of 500 gal/min. In the tank, the treatment removes 2% of the pollutant per minute, and the water is thoroughly stirred. The tank holds 10,000 gal of water. On the day the treatment plant opens, the tank is filled with pure water. Find the function which gives the concentration of pollutant in the outflow.

†18. During what time t will the water flow out of an opening 0.5 cm² at the bottom of a conic funnel 10 cm high, with the vertex angle $\theta = 60°$?

19. At time $t = 0$, two tanks each contain 100 gallons of brine, the concentration of which then is one half pound of salt per gallon. Pure water is piped into the first tank at 2 gal/min, and the mixture, kept uniform by stirring, is piped into the second tank at 2 gal/min. The mixture in the second tank, again kept uniform by stirring, is piped away at 1 gal/min. How much salt is in the water leaving the second tank at any time $t > 0$?

20. Modeling glucose concentration in the body after glucose infusion: Infusion is the process of admitting a substance into the veins at a steady rate [this is what happens during intravenous feeding from a hanging bottle by a hospital bed]. As glucose is admitted, there is a drop in the concentration of free glucose (brought about mainly by its combination with phosphorous); the concentration will decrease at a rate proportional to the amount of glucose. Denote by G the concentration of glucose, by A the amount of glucose admitted (in mg/min), and by B the volume of liquid in the body (in the blood vessels). Find whether and how the glucose concentration reaches an equilibrium level.

†21. A criticism of the model of Exercise 20 is that it assumes a constant volume of liquid in the body. However, since the human body contains about 8 pt of blood, infusion of a pint of glucose solution would change this volume significantly. How would you change this model to account for variable volume? I.e., how would you change the differential equation? Will this affect your answer about an equilibrium level? How? What are the limitations of *this* model? (Aside from the fact you may have a differential equation which is hideous to solve or analyze, what criticisms or limitations do you see physically to the variable volume idea?) What sort of questions might you ask of a doctor or a biologist in order to work further on this problem?

22. A chemical A in a solution breaks down to form chemical B at a rate proportional to the concentration of uncoverted A. Half of A is converted in 20 min. Express the concentration y of B as a function of time and plot it.

23. A limnology class is presented with a laboratory exercise concerning continuous culture of algae in a chemostat. The apparatus consists of a culture vessel (with a constant level overflow tube to keep the volume at 8 liters) into which a fresh culture medium is continuously fed by a constant metered gas flow. A page of instructions is handed to the students. It contains physical and numerical data for the experiment, and concludes with the following paragraph:

"With the pumping of a fresh culture medium into the culture chamber, it is possible to calculate the theoretical percentage concentration of the medium created in the culture chamber after any given number of hours. The following mathematical relationships are used for the calculations:

where

$$C_T = C_0 + (C_i - C_0)(1 - e^{-(T-T_0)(R/V)}$$
C_T = outflow concentration at an arbitrary moment
C_0 = concentration at $T = T_0$
C_i = concentration of inflow
R = flow rate (ml/h)
V = volume of chamber (ml)
T = time at arbitrary moment
T_0 = starting time."

Show that you can rather easily justify this somewhat horrendous "out-of-the-magic-hat" formula.

†24. A snowfall begins sometime in the forenoon, and snows steadily on into the afternoon. At noon a man begins to clear the sidewalk on a certain street, shoveling at a constant rate (in cubic feet per hour) and at a constant width. He shovels two blocks by 2 p.m. and one block more by 4 p.m. At what time did the snow begin to fall? (You may assume he does not go back to clear the snow that has fallen behind him.)

Solutions

1. a) $dy/dt = ky$, so

$$\frac{dy}{y} = kdt \qquad \text{(separating variables)}$$
$$\ln y = kt + C \quad \text{(integrating both sides)}$$
$$y = e^{kt+C} \qquad \text{(expressing exponentially)}$$
$$= e^C e^{kt}$$
$$= A e^{kt} \qquad \text{(renaming single constant).}$$

b) $y(24) = 100 e^{24k} = 400$
$$e^{24k} = 4 \qquad \text{(isolating factor with } e\text{)}$$
$$24k = \ln 4 \quad \text{(taking ln both sides)}$$
$$\bullet\, k = \frac{\ln 4}{24}.$$

c) $y(12) = 100 e^{12\left(\frac{\ln 4}{24}\right)}$
$$= 100(e^{\ln 4})^{1/2} = 100(4)^{1/2} = 200.$$

d) $y = A^* e^{k^* t}$

    $y(0) = 100 \Rightarrow A^* = 100$

    $y(1) = 400 \Rightarrow 400 = 100\,e^{k^*} \Rightarrow e^{k^*} = 4 \Rightarrow k^* = \ln 4$

    $y(\tfrac{1}{2}) = 100\,e^{(1/2)\ln 4} = 100(e^{\ln 4})^{1/2} = 100(4)^{1/2} = 200.$

2. a) From $dT/dt = kT - km$, you can see that you cannot separate T from m, but $[dT/(T - m)] + K\,dt$ works very nicely.

  b) $m = 85°$.

3. a) $S(1) = S_0 + rS_0 = (1 + r)S_0$

    $S(2) = S_1 + rS_1 = (1 + r)S_1 = (1 + r)^2 S_0$

    $\vdots$

    $S(n) = S_{n-1} + rS_{n-1} = (1 + r)S_{n-1} = (1 + r)^n S_0.$

  d) $S\left(\dfrac{n}{m}\right) = \left(1 + \dfrac{r}{m}\right)^n S_0$

$$\lim_{m \to \infty} S\left(\frac{n}{m}\right) = \lim_{m \to \infty} \left(\left(1 + \frac{r}{m}\right)^m\right)^{n/m} S_0$$

$$= \lim_{m \to \infty} (e^r)^{n/m} S_0,$$

    so $S(t) = S_0\,e^{rt}$.

4. $\dfrac{y_0}{2} = y_0\,e^{-t/8000}$

  $\underbrace{\ln 0.5}_{\approx 0.7} = -t/8000$

  so • $t \approx 5600$ yr.

5. $y/y_0 = e^{-t/8000}$. In 1868, $t - 31{,}868$ represents 30,000 B.C., $t = 21{,}868$ represents 20,000 B.C.

$$e^{-31,868/8000} < \text{lab.}\ \frac{y}{y_0} < e^{-21,868/8000}$$

$$e^{-3.9835} < \text{lab.}\ \frac{y}{y_0} < e^{-2.7335}$$

$$\bullet \text{ approx. } \underbrace{0.019}_{1.9\%} < \text{lab.}\ \frac{y}{y_0} < \text{approx. } \underbrace{0.065}_{6.5\%}.$$

6. $T = 20 + 80(\tfrac{1}{2})^{t/20}$. • When $T = 30°$, $t = 60$ min. (Assuming boiling $T = 100°$ at $t = 0$, $T = 60°$ at $t = 20$.)

7. First you will have to ask your friend the room temperature R. Then $dT/dt = -k(T - R)$ which has solution • $T - R = Ce^{-kt}$. At $t = 0$, $T = 212$, so $T - R = 212\,e^{-kt}$. At $t = ?$, $T = 185$, but you are still stuck because you do not know k. You will have to ask him for one more bit of information. Unless he can recall something like "the day I was on the phone for half an hour, the stupid water cooled down to 95° ...," you are probably best off fetching him a new thermometer.

8. a) After 6 min, $T = 82.5°C$.
   b) at $75°$, $T = 10.2$ min. $\left.\begin{array}{l}\\ \\ \\\end{array}\right\}$ $T = 60 + 40(\frac{3}{4})^{t/3}$.
      at $61°$, $T = 38.5$ minutes.

9. Obviously, this is *not* a situation in which the surrounding medium remains at a constant temperature, so we need a modification of Newton's Law of Cooling as seen in Example 4. If you are not an expert in physics, all is *not* lost—think a bit about what might make *sense*. You should realize that the volume of the objects is somehow involved, for the small one (the ounce of hot water) will obviously (intuitively) cool off more than the large one (the 100-oz solution) will warm up. You can make a proposal based on this, and one way to check it will be to see if it reduces to the Newton model of Example 4 in the case where the surrounding medium is large enough to remain at constant temperature.

   Heat flow is what is conserved in a situation where objects of finite volume and different temperatures meet. *Heat flow* is what is experimentally found to be linearly proportional to the temperature difference.

   This means that $\bullet\ \dfrac{d}{dt}V_1 T_1 = -\dfrac{d}{dt}V_2 T_2 = k(T_2 - T_1)$.

   $\underbrace{\phantom{\dfrac{d}{dt}V_1 T_1}}$ $\underbrace{\phantom{\dfrac{d}{dt}V_2 T_2}}$

   T_1 going down; T_2 going up;
   deriv. will be   deriv. will
   negative   be positive

   If this did not occur to you, do you agree that it sounds reasonable? You should be able to find a convincing argument in any elementary physics text, under heat flow or heat transfer.

   In this problem, $V_1 = 1$ oz and $V_2 = 100$ oz; both are constant. So we have

   $$\frac{dT_1}{dt} = -100\frac{dT_2}{dt} = K(T_2 - T_1).$$

   Since T_1 and T_2 are both functions of t, this is a simple *system* of two differential equations in $T_1(t)$ and $T_2(t)$. Details of solving such a *system* are beyond the scope of this chapter (though *setting up* the system is not). The answer is $T_2 = (1090/101)(1 - e^{-kt}) + 10\,e^{-kt}$.

10. Volume depends on r^3, surface area on r^2. Therefore, surface area depends on $V^{2/3}$; $V = k_1 r^3$; $S = k_2 r^2 = k_2(\sqrt[3]{V/k_1})^2 = KV^{2/3}$. Differentiating, $dV/dt = -cV^{2/3}$ (negative to show *decrease* in V). By separation of variables,

    $$3V^{1/3} = -ct + Q$$

    $$V = \left(\frac{-ct + Q}{3}\right)^2.$$

    At $t = 0$, $V = V_0$, so $\bullet\ V = \left(-\dfrac{c}{3}t + V_0^{1/3}\right)^3$.

11. Let $S = $ amount of salt; rate $= $ input $-$ outgo.

    $$\frac{dS}{dt} = \left(0\,\frac{\text{lb}}{\text{gal}}\right)\left(2\,\frac{\text{gal}}{\text{min}}\right) - \left(\frac{S\,\text{lb}}{100\,\text{gal}}\right)\left(2\,\frac{\text{gal}}{\text{min}}\right) = -\frac{S}{50}$$

    $$S = ke^{-(1/50)t} \quad \text{at } t = 0, s = 20, \text{ so}$$

$$S = 20e^{-0.02t}$$

$$\bullet\, S(60) = 20e^{-1.2}.$$

12. $\dfrac{dz}{dt} = \text{input} - \text{outgo} = \left(r_1\,\dfrac{\text{gal}}{\text{min}}\right)\left(c_1\,\dfrac{\text{lb}}{\text{gal}}\right) - \left(r_2\,\dfrac{\text{gal}}{\text{min}}\right)\left(\dfrac{z\ \text{lb}}{V_0 + (r_1 - r_2)t\ \text{gal}}\right)$

$$= r_1 c_1 - \dfrac{r_2 z}{V_0 + (r_1 - r_2)t}.$$

At $t = 0$, $z = C_0$. Use to evaluate constant of integration. The question asks for

$$\bullet\ t \text{ when } \underbrace{\dfrac{z}{V_0 + (r_1 - r_2)t}}_{\substack{\text{concentration} \\ \text{of effluent}}} = 0.30C.$$

13. $dS/dt = rS$, $S = S_0\, e^{rt}$. When savings double, $t = 16$ implies

$$2S_0 = S_0\, e^{16r}$$

$$2 = e^{16r}$$

$$r = \dfrac{\ln 2}{16} = 0.0433 = \bullet\, 4.33\%.$$

14. Both the differential equation and its solution are going to have discontinuities every time that $t = $ an integer number of years.

$$\dfrac{dS}{dt} = \begin{cases} rS, & \text{for } t \neq n \\ rS + A, & \text{for } t = n \end{cases}$$

$$S(t) = S_0\, e^{rt} + A\, e^{rt}\left(\dfrac{1}{e^r} + \dfrac{1}{e^{2r}} + \dfrac{1}{e^{3r}} + \cdots + \dfrac{1}{e^{kr}}\right)$$

where k is the greatest integer less than t and $t \geq 1$ (for $0 \leq t < 1$, $S = S_0\, e^{rt}$). (See Fig. 1.2).

15. $S = $ amount of salt,

$$dS/dt = \text{input} - \text{outflow}$$

$$= \left(2\,\dfrac{\text{lb}}{\text{gal}}\right)\left(3\,\dfrac{\text{gal}}{\text{min}}\right) - \left(\dfrac{S}{10}\,\dfrac{\text{lb}}{\text{gal}}\right)\left(3\,\dfrac{\text{gal}}{\text{min}}\right)$$

$$= 6 - \tfrac{3}{10}S$$

$S = 20(1 - Ce^{-(3/10)t})$; $S(0) = 5$, so $C = \tfrac{3}{4}$. $\bullet\, S = 20(1 - (3/4)\,e^{-(3/10)t})$

a) After 8 min, $concentration = S(8)/10 = 2(1 - (3/4)\,e^{-2.4})\text{lb/gal}$.

b) Long term, as $t \to \infty$, $S \to 20$ lb in tank.

16. Let $N(t) = $ number of neutrons present. Then $dN/dt = kN$, so $N(t) = N_0\, e^{kt}$. Then

$$\left(\dfrac{N_2}{N_0}\right)^{T_1} = \left(\dfrac{N(T_2)}{N_0}\right)^{T_1} = (e^{kT_2})^{T_1} = e^{kT_1 T_2} = (e^{kT_1})^{T_2} = \left(\dfrac{N_1}{N_0}\right)^{T_2}.$$

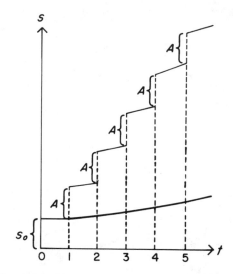

graph of $S(t)$ for this account, with $A \neq 0$.

(each vertical jump = A)

regular compound interest account if $A = 0$.

(no annual increase in principal)

Figure 1.2. The lower curve shows regular compound interest account if $A = 0$ (no annual increase in principal). The stepped curves above show the graph of $S(t)$ for this account, with $A \neq 0$. Each vertical jump = A.

17. Let $P(t)$ = amount of pollutant in tank,

$$\frac{dP}{dt} = \text{input} - \text{outgo}$$

$$= \underbrace{\left(2\frac{oz}{gal}\right)\left(500\frac{gal}{min}\right)}_{\text{inflow}} - \underbrace{\left(\frac{P \ oz}{10,000 \ gal}\right)\left(500\frac{gal}{min}\right)}_{\text{outflow}} - \underbrace{0.02P\frac{oz}{min}}_{\text{treatment}}.$$

So $dP/dt = 1000 - 0.07P$ and $P = (100,000/7)(1 - c e^{-0.07t})$. At $t = 0$, $P = 0$, so $C = 1$.

18. As in Example 8,

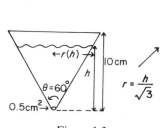

Figure 1.3

area of surface
of water

area of hole

$$-\pi\left(\frac{h}{\sqrt{3}}\right)^2 dh = 0.5\sqrt{2gh}\,dt$$

$$h^{3/2}\,dh = -\frac{1.5}{\pi}\sqrt{2g}\,dt$$

$$\int_{10}^{0} h^{3/2}\,dh = -\frac{1.5}{\pi}\sqrt{2g}\int_{0}^{T} dt$$

$$\tfrac{2}{5}h^{5/2}]_{10}^{0} = -\frac{1.5}{\pi}\sqrt{2g}\,t]_{0}^{T}$$

$$-\tfrac{2}{5}(10)^{5/2} = -\frac{1.5}{\pi}\sqrt{2g}\,T$$

$$T = \frac{2\pi}{7.5\sqrt{2g}}(10)^{5/2}\text{ s.}$$

19. Let $y_1(t)$ and $y_2(t)$ be the amount of salt, in pounds, in the first and second vats, respectively, at time t.

First tank: $\dfrac{dy_1}{dt}$ = input − outgo

$$= \left(0 \, \frac{lb}{gal}\right)\left(2 \, \frac{gal}{min}\right) - \left(\frac{y_1 \; lb}{100 \; gal}\right)\left(2 \, \frac{gal}{min}\right)$$

$$= -\frac{y_1}{50}$$

$$y_1 = C e^{-t/50}; \; y_1(0) = 50; \; \bullet \; y_1 = 50 \, e^{-t/50}.$$

Second tank: $\dfrac{dy_2}{dt}$ = input − outgo

$$= \left(\frac{y_1 \; lb}{100 \; gal}\right)\left(2 \, \frac{gal}{min}\right) - \left(\frac{y_2 \; lb}{(100+t) \; gal}\right)\left(1 \, \frac{gal}{min}\right)$$

$$= e^{-t/50} - \frac{y_2}{100 + t}.$$

Solving as linear equation with integrating factor,

$$y_2 = k - 50(150 + t)\,e^{-t/50}.$$

$y_2(0) = 50$, so $k = 12{,}500$. $\bullet$ The *concentration* of salt in water leaving second tank:

$$\frac{y_2}{100 + t} = \frac{12{,}500}{100 + t} - 50\left(\frac{150 + t}{100 + t}\right)e^{-t/50}.$$

20. $G(t)$ = concentration of glucose,
       A = rate at which glucose is admitted, in mg/min,
       rate = input − outgo,

$$\frac{dG}{dt} = \frac{A}{V} - KG \quad \text{(decrease proportional to } G)$$

$$\underset{\frac{mg/cm^3}{min}}{\uparrow} \quad \underset{\frac{mg/min}{cm^3}}{\uparrow} \quad \left(\frac{1}{min}\right)\frac{mg}{cm^3}.$$

It is not necessary to *solve* this differential equation to answer the question of the problem. An equilibrium level will be a *constant* concentration G such that G no longer changes, i.e., $dG/dt = 0$.

$$\frac{dG}{dt} = \frac{A}{V} - kG = 0$$

where $G = A/KV$ is equilibrium concentration. Still without solving, you can see from the differential equation that

$$G > \frac{A}{KV} \text{ gives negative } \frac{dG}{dt} \text{ (decreasing } G)$$

$$G < \frac{A}{KV} \text{ gives positive } \frac{dG}{dt} \text{ (increasing } G).$$

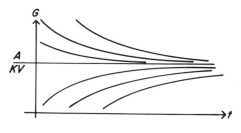

Figure 1.4

The solutions for $G(t)$ must look like those in Fig. 1.4 (with different values of the constant arising from integration corresponding to different solution curves). These solution curves are further confirmed by noting that if

$$\frac{dG}{dt} = \frac{A}{V} - KG,$$

then

$$\frac{d^2G}{dt^2} = -K\frac{dG}{dt} = -K\left(\frac{A}{V} - KG\right),$$

positive if $G > A/KV$ (then G is concave up), negative if $G < A/KV$ (then G is concave down) (Actual solution: $G = (A/KV)(1 - e^{-kt})$.)

21. Possible variable volume model: V is no longer constant, S is the volume of solution per minute being infused;

$$\bullet \quad V = V_0 + St.$$
$$\text{cm}^3 \quad \text{cm}^3 \quad \left(\frac{\text{cm}^2}{\text{min}}\right)\text{min}$$

There is a relation between S and A:

$$\bullet \frac{A}{S} = \left(\frac{\text{mg/min}}{\text{cm}^3/\text{min}}\right)\text{glucose solution} = \frac{\text{mg}}{\text{cm}^3} = C,$$

is the constant concentration of glucose in solution being infused. So $V = V_0 + (1/c)At$, and the differential equation of Exercise 20 becomes

$$\bullet \frac{dG}{dt} = \frac{A}{V_0 + \dfrac{A}{C}t} - kG.$$

There now is *no* equilibrium level of G because no *constant* G produces $dG/dt = 0$.

This model appears messy either to solve *or* to sketch solutions as we did in Exercise 20. As with many differential equations, a numerical approach to a solution seems the only sensible one, but at this point you might easily question the value of the whole variable volume consideration, since it has not produced any immediate physical insights.

Thinking about variable volume further, you would realize that B cannot physically increase this way for very long—the body does not have unlimited capacity for increasing volume of liquid in the veins. Other mechanisms such as urination will work to keep the volume in line, so you would want to ask the medical and biological experts about the realistic limits of variation in volume; how long a glucose solution may be steadily infused; whether an equilibrium concentration is actually observed, etc. One reference on this topic which provides some actual data is Defares and Sneddon, *The Mathematics of Medicine and Biology* (Chicago: 1961).[3]

22. Let y be the concentration of B and A_0 be the initial concentration of A:

$$\frac{dy}{dt} = c(A_0 - y) \qquad\qquad y(0) = 0; \; y(20) = \frac{A_0}{2}$$

$$-\ln|A_0 - y| = ct + K$$

$\underbrace{}$
always
positive

$$A_0 - y = Q e^{-ct} \qquad\qquad Q = A_0; \; e^{-20c} = \tfrac{1}{2}$$

$$\text{so } c = \frac{\ln 2}{20}$$

$$\bullet \; y = A_0(1 - (\tfrac{1}{2})^{t/20})$$

$$y' = c(A_0 - y) \Big| \;\; \text{always}$$
$$\quad\uparrow\quad\uparrow\quad \; \text{positive}$$
$$\text{pos. pos.}$$

$$y'' = -cy' = -c^2(A_0 - y)$$

$$\underbrace{}_{\text{neg.}} \quad \underbrace{}_{\text{pos.}}$$

always negative;
y concave down

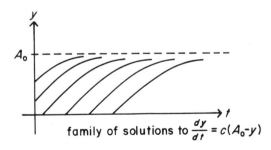

family of solutions to $\dfrac{dy}{dt} = c(A_0\text{-}y)$

Figure 1.5

The family of solutions to $y' = c(A_0 - y)$ is shown in Fig. 1.5.

[3] They refer to S. G. Jokipii *et al.*, *J. Clin. Invest.*, vol. 34, 1954, pp. 331, 452, 458.

VISTA

23. C_T = concentration in the outflow,

    = concentration of the medium in the vessel at any moment T.

Therefore, VC_T = *amount* of the medium in the vessel at T.

Rate of change = input − outflow

$$\frac{d}{dt}(VC_T) = \left(R\frac{ml}{h}\right)\left(C_i\frac{mg}{ml}\right) - \left(R\frac{ml}{h}\right)\left(\frac{VC_T\,mg}{V\,ml}\right).$$

V, R, and C_i are all constant, so

$$V\frac{d}{dt}C_T = RC_i - RC_T$$

or

$$\bullet\ \frac{d}{dt}C_T = \frac{R}{V}(C_i - C_T).$$

Solution by separation of variables (the use of $T - T_0$ in the "formula" is a clue that definite integrals may lead us most quickly to the result):

$$\int_{C_0}^{C_T} \frac{dC_T}{C_i\left(1 - \frac{1}{C_i}C_T\right)} = \int_{T_0}^{T} \frac{R}{V}\,dt$$

$$\left.-\ln\left|1 - \frac{C_T}{C_i}\right|\right]_{C_0}^{C_T} = \left.\frac{R}{V}t\right]_{T_0}^{T}$$

$$\ln\left|1 - \frac{C_T}{C_i}\right| - \ln\left|1 - \frac{C_0}{C_i}\right| = -\frac{R}{V}(T - T_0)$$

$$\frac{\left|1 - \frac{C_T}{C_i}\right|}{\left|1 - \frac{C_0}{C_i}\right|} = \exp\left[-\frac{R}{V}(T - T_0)\right].$$

Because of the absolute value signs, the resulting formula is only valid if $C_T > C_i$ and $C_0 < C_i$ or if $C_T < C_i$ and $C_0 < C_i$. This does not seem to be *required* of the chemostat, but perhaps in practice it is the case.

$$C_i - C_T = (C_i - C_0)\exp\left[-\frac{R}{V}(T - T_0)\right]$$

$$C_T = C_i - (C_i - C_0)\exp\left[-\frac{R}{V}(T - T_0)\right]$$

which can indeed be shown to equal

$$C_0 + (C_i - C_0)\left(1 - \exp\left[-\frac{R}{V}(T - T_0)\right]\right).$$

24. *Hint:* Look at the *velocity* at which the shoveler *travels*, which is inversely proportional to the volume of snow that has fallen on any spot at any instant. Answer: $\sqrt{5} - 1$ h before noon.

Notes for the Instructor

Objectives. Applied problems requiring differential equations seem to be harder for many students to translate into mathematical terms than problems met heretofore. Extra attention is needed to the following facts.

(1) The differential equation is an *instantaneous* statement, which must be valid *at any time*.
(2) Not all the numbers given in a problem go into the differential equation. Some must be held out to provide the conditions necessary to evaluate the constants of integration, or other parameters, such as those of proportionality.
(3) Matching physical dimensions for each term of the differential equation is essential and not usually automatic.
(4) An organized and clearly written framework is helpful to the student and necessary for those who read his work.

Prerequisites. Elementary calculus, specifically

(1) familiarity with differentiation and basic integration,
(2) some experience with simple differential equations and their solutions.

Time. Part of one class hour should suffice for the introduction of a few varied examples. A later recitation, *after* students have worked on the exercises, can discuss those problems where they have encountered difficulties.

Remark. This module has been developed within our third semester course in calculus and differential equations which attends to the *applicability* of the mathematics. We have written especially for students who may be majoring in biological or social sciences and who may not feel agile with mathematics or physics. The module was written primarily for independent use by the students. It has focused on *setting up* the mathematical models, *not* on the subsequent solutions; some more difficult exercises (denoted by dagger) are included for this purpose.

CHAPTER 2
Qualitative Solution Sketching for First-Order Differential Equations

Beverly Henderson West*

1. Introduction

Qualitative solution sketching, as explained in this module, can yield very useful information about the solutions $y = f(x)$ to a given differential equation $y' = g(x, y)$. In particular, it usually allows you to examine the limiting or long-range behavior for y as $x \to \infty$ without actually coming up with an explicit expression for the solution. Frequently, an explicit solution is unnecessary or technically difficult, or it might not exist at all in terms of elementary functions (polynomial, trigonometric, logarithmic, exponential). In these cases the qualitative approach may be a lifesaver—or at least a worksaver.

Recall that the differential equation $y' = g(x, y)$ has a whole *family* of solutions of the form $y = f(x)$. Different solutions in the family result from different values of the constant of integration, e.g., $y' = 1$ has a family of solutions $y = x + c$ (see Fig. 2.1). For a particular problem, the value of c is determined by some given condition $f(x_0) = y_0$; this determines which member of the family of solutions is *the* solution. However, we shall not be looking for explicit solutions, much less particular solutions, so we shall be sketching the *family* of solutions to each differential equation.

In elementary calculus you learned how derivatives could be used in graphing a function. Now we shall capitalize on those skills and show how to do it even when you do not have the explicit equation $y = f(x)$ to help you (Sections 2–4). Furthermore, we shall see that much additional information on equilibria and stability of solutions can be extracted from these sketches (Section 5).

* Department of Mathematics, Cornell University Ithaca, NY 14853.

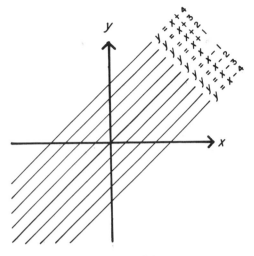

Figure 2.1

2. Direction Field

If a first-order differential equation can be put in the form $y' = g(x, y)$, then we can determine the *slope* of the solution $y = f(x)$ through any point (x, y). Graphically, we can draw a short line of the proper slope through each of many points (x, y) in the plane. This is called a *direction map* or *direction field*.

EXAMPLE 1. Consider

$$\frac{dy}{dx} = -xy. \tag{1}$$

You might start off with a few simple calculations, as tabulated here.

	x	y	$y' = \dfrac{dy}{dx} = -xy$
if $x = 0$ or if $y = 0$, $y' = 0$, so the direction lines along both axes are all horizontal	0	anything	0
	anything	0	0
if $x = 1$, $y' = -1$, so	1	1	-1
	1	2	-2
if $x = 1/2$, $y' = -y/2$, so	1/2	1	$-1/2$
	1/2	2	-1

With very little further precise calculation, you can continue to fill in graphically the direction field by noting facts as the following.

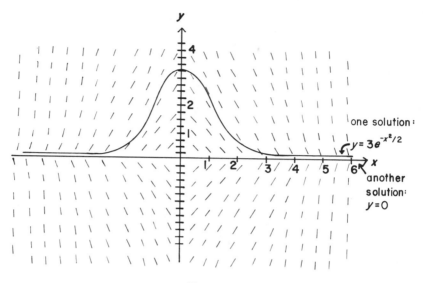

Figure 2.2

(1) Symmetry exists about the origin and about both axes, so considering the first quadrant in detail gives you all the information for the other three.
(2) For fixed x, the slopes get steeper as y increases.
(3) For fixed y, the slopes get steeper as x increases.

Thus you can arrive rather quickly at a direction field (Fig. 2.2).

A direction field such as this gives a visual indication of the family of all possible solutions to the differential equation. Any one solution must be tangent to these direction lines for each point through which it passes. Usually, the solutions can be put in the form $y = f(x)$, with each member of the family having different values of the constant which results from integration. For instance, the general solution to (1) is, in fact, $y = ae^{-x^2/2}$, and one of these solutions, with $a = 3$, has been drawn in on the direction field so that you may see how it fits all the direction lines of the map. The actual solution is easily obtained by separation of variables:

$$y' = -xy$$

$$\int \frac{dy}{y} = \int -x\,dx$$

$$\ln y = -\frac{x^2}{2} + c$$

$$y = e^{-x^2/2+c} = e^c e^{-x^2/2} = ae^{-x^2/2}.$$

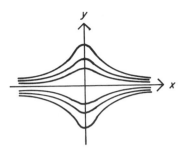

Figure 2.3

3. Relevance of Uniqueness Theorems

Whenever a uniqueness theorem applies to a first-order differential equation, there can be only *one* solution through any point (x, y), so *no two solutions can ever cross*. This is terribly helpful information when you are trying to sketch a family of solutions. For instance, in Example 1, this means the *only* way to draw in a family of solutions on that direction field is as illustrated in Figure 2.3.

You will now want to know "when does a uniqueness theorem apply?" The closest we can come to an easy answer is "usually" for first-order ordinary differential equations. To be mathematically precise is rather complicated. Any elementary text on differential equations will focus on this equation and will state some theorem such as the following.

Theorem. *For a first-order differential equation* $y' = g(x, y)$ *with initial condition* $y(x_0) = y_0$, *a sufficient (though not necessary) condition that a unique solution* $y = f(x)$ *exist is that* g *and* $\partial g/\partial y$ *be real, finite, single-valued, and continuous over a rectangular region of the plane containing the point* (x_0, y_0).

Seldom do elementary texts *prove* such theorems. A notable exception is Martin Braun's *Differential Equations and their Applications* (New York: Springer–Verlag, 1978, 2nd ed.).[1] In Chapter 1.10 Braun gives an excellent discussion, with proofs, of questions of uniqueness and existence. This treatment is easily accessible to anyone with a simple elementary calculus background.

In any case, you can see that the criterion for uniqueness of solutions to $y' = g(x, y)$ with a given condition is roughly that g and $\partial g/\partial y$ are "nice." We shall now look at some of the many cases where this is so.

[1] See also W. E. Boyce and R. C. Di Prima, *Elementary Differential Equations and Boundary Value Problems* (New York: Wiley, 1977, 3rd ed.)

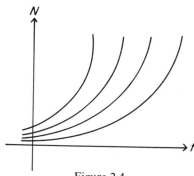

Figure 2.4

4. Sketching of Solutions

From Sections 2 and 3 we have derived two guidelines for sketching solutions of first-order differential equations $y' = g(x, y)$:

(1) solutions must be consistent with the direction field;
(2) for most of these (e.g., with g and $\partial g/\partial y$ continuous), uniqueness tells us that no two solutions will ever cross.

Keep these guidelines in mind as you consider the following example.

EXAMPLE 2. Consider the simplest population growth assumption, that the per capita rate of growth is constant. This means that if $N(t)$ = population as a function of time,

$$\frac{1}{N}\frac{dN}{dt} = r = \text{const.} \tag{2}$$

To get a quick qualitative picture of possible solutions, rewrite as

$$\frac{dN}{dt} = rN$$

and consider the graphical behavior of $N(t)$.

For this physical problem, we need only $N > 0$, $t > 0$. You can see that slope $= dN/dt = rN = 0$ when $N = 0$, and that slope increases as N increases. Presto, you have shown that the solutions must look *approximately* like Figure 2.4.

This rough sketch is quite satisfactory for showing clearly that for *all* solutions, N rises ever faster, with no end in sight. You can easily confirm the sketch by solving (2) to get $N = N_0 e^{rt}$.

We note two additional facts which will give a more accurate graphical picture and clarify the role of the constants r and N_0: the effect of r is to increase the slope as r increases, at a given N (Fig. 2.5). In this chapter we want to consider the constants in the differential equation as fixed and to sketch only the family of solutions arising from different constants of in-

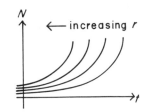

Figure 2.5

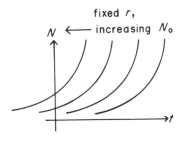

Figure 2.6

tegration. For a *given fixed r*, look at a given N. Since no t is explicit in dN/dt, the *same* slopes occur at that N for all t. Hence all solution curves are horizontal translates of one another (Fig. 2.6).

EXAMPLE 3. Consider qualitatively what happens when the basic assumption of (2) in the above example is modified to show the decrease in per capita rate of growth due to *crowding*. The simple Verhulst–Pearl expression of this modification is

$$\frac{1}{N}\frac{dN}{dt} = r\left(1 - \frac{N}{k}\right). \tag{3}$$

Here the per capita rate $(1/N)/(dN/dt)$ decreases linearly, diminished by a crowding term directly proportional to N, the size of the population. The positive constant r still represents the per capita rate of growth when no crowding occurs (when $N = 0$).

To look at the solutions graphically, we rewrite

$$\frac{dN}{dt} = rN\left(1 - \frac{N}{k}\right) = \text{slope.}$$

We can see that slope is zero (horizontal) for $N = 0$ and for $N = k$. We can also show that the slope is positive for $0 < N < k$ and negative for $N > k$, but that does not seem to be enough information to tell us how to graph solutions that do not cross (for uniqueness). See Figure 2.7.

Our other graphing trick from calculus is the determination of *concavity*, which depends on the *second* derivative. This we can get:

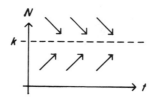

Figure 2.7

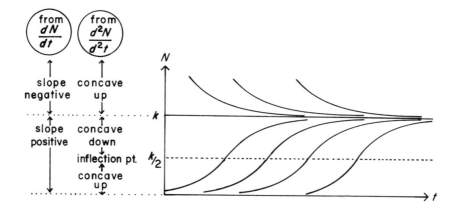

Figure 2.8

$$\frac{d^2N}{dt^2} = \frac{d}{dt}\frac{dN}{dt} = r\frac{dN}{dt}\left(1 - \frac{N}{k}\right) + rN\left(-\frac{1}{k}\frac{dN}{dt}\right)$$

$$= r\frac{dN}{dt}\left(1 - \frac{N}{k} - \frac{N}{k}\right) = r\frac{dN}{dt}\left(1 - \frac{2N}{k}\right).$$

We want to know where the second derivative is 0, positive, and negative. Considering dN/dt and $(1 - (2N/k))$ as the two factors (multiplied by positive r), we can show that $d^2N/dt^2 = 0$ when $N = 0$, k, or $k/2$, that it is positive (N concave upward) for $N < k/2$ and for $N > k$, and that it is negative (N concave downward) for $k/2 < N < k$.

All this information fits together as illustrated in Figure 2.8.

No t is explicit in dN/dt, so again (for fixed r and k), the *same* slopes occur at a given N for all t. Hence all solution curves are horizontal translates of one another.

Because (3) cannot be solved by simple integration after separation of variables, we are already into the area where our sketching technique becomes valuable. (Equation (3) *can* be solved by partial fraction integration, or by making the substitution $M = 1/N$ which gives a linear first-order differential equation for M. Exercise 22 also illustrates a surprisingly good

Figure 2.9

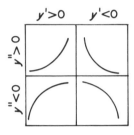

Figure 2.10

fit of this model to actual population data of the United States from 1790 to 1910.)

In general, equations of the form of (3) arise frequently and are called *logistic*. This means that any equation of the form $dx/dt = x(a - bx)$ is called logistic and is known to have solutions of the form illustrated in Figure 2.9.

These examples lead us to the formulation of a general approach. We went to combine information from the first and second derivatives to graph solutions which exhibit approximately the proper slope and concavity. (This process has been likened to the construction of a composite sketch of a crime suspect from a Dragnet Idento-Kit. You can consider the drawing below as a solution-sketching idento-kit.) The four basic types of curves are shown in Figure 2.10. The problem is simply to determine which pieces go where in the coordinate plane. The following example illustrates a good procedure.

EXAMPLE 4. Sketch a number of solutions to the following differential equation. Your sketches should exhibit approximately the right slope and concavity,

$$y' = (y + 3)(y - 2). \tag{4}$$

1) *Look at y' (slope):* $y' = (y + 3)(y - 2)$ is a product of two factors, so we need to determine the positivity and negativity of the factors. You can keep this straight on a number line (for y), as shown.

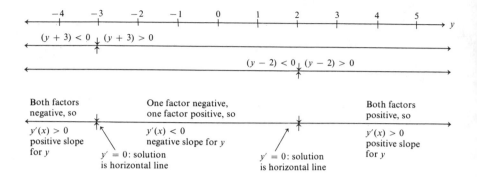

2) *Look at* y'' *(concavity):* Using implicit differentiation and the product rule on (4), we get

$$y''(x) = y'(y - 2) + (y + 3)y'$$
$$= (2y + 1)y'$$
$$= (2y + 1)(y + 3)(y - 2).$$

Use the number line again to deal with this product of three factors.

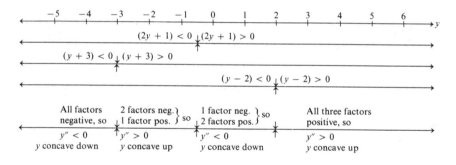

3) *Put it all together:* Sketch in the horizontal solutions $y = -3$ and $y = 2$. Use the information from 1) and 2) to block off the coordinate plane wherever changes occur in y' or y''—in this case, at $y = -3$, $y = -1/2$, $y = 2$. Then put in the right curve pieces. Figure 2.11 results.

Here are some notes on the above solution sketches for Example 4.

a) Where the ⌢ piece meets the ⌣ piece, an inflection point occurs.

b) The solution curves must approach the equilibria asymptotically because you cannot have two solutions passing through the same point (i.e., no two solution curves can intersect).

c) Again, no x is explicit in y', so horizontal translates of solution curves are also solution curves.

d) Using the differential equation (4), we *could* calculate the slope exactly

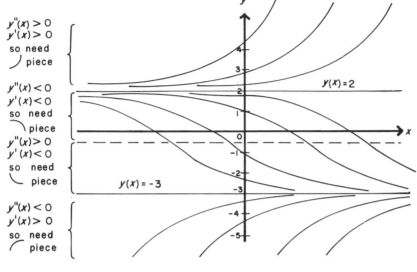

$y''(x) > 0$
$y'(x) > 0$
so need
⟋ piece

$y''(x) < 0$
$y'(x) < 0$
so need
⟍ piece

$y''(x) > 0$
$y'(x) < 0$
so need
⟍ piece

$y''(x) < 0$
$y'(x) > 0$
so need
⟋ piece

$y(x) = 2$

$y(x) = -3$

Figure 2.11

at any point if we so desire, but we are able to get this beautiful picture even without a point-by-point plot of the direction field.

You may streamline the details of this last example in your own work, but headings 1), 2), and 3) list the key steps to obtaining good solution sketches. In the next section we shall examine these sketches to see what they tell us.

5. Discussion of Equilibria and Stability

An *equilibrium* solution to a differential equation is one that *does not change* or *is constant.* That means an equilibrium is a solution for which $y' = g(x, y) \equiv 0$ or $y = f(x) \equiv c$, for all real numbers x. In Example 4, the solutions $y = -3$ and $y = 2$ are equilibrium solutions.

Equilibria may be classified as stable or unstable. The next illustration may help in distinguishing the two. Consider two cones, one sitting on its base, and the other balancing on its point (Fig. 2.12). Both cones are in equilibrium: their positions are not changing. If you jiggle the left cone a bit with your finger, it will quickly return to its original position. This is a *stable* equilibrium. If you jiggle the right-hand cone, however, it will promptly fall down, demonstrating that the equilibrium was *unstable.*

Equilibrium solutions to differential equations are classified as stable or unstable depending on whether, graphically, nearby solutions converge to the equilibrium or diverge from the equilibrium, respectively, *as $x \to \infty$.*

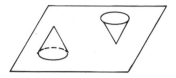

Figure 2.12

The graph of Example 4 demonstrates a stable equilibrium at $y = -3$ and an unstable equilibrium at $y = 2$. What does this mean physically? A possible interpretation of this unstable equilibrium is given by the following. Suppose $y(x)$ represents millions of bacteria on a plate at time x. Suppose you have a plate with 2,000,000 bacteria at time x_0. Then you are on the $y(x) = 2$ equilibrium solution. If someone sneezes on the plate at time x_0, suddenly giving you 2,000,126 bacteria, you now go on to one of the $y' > 0$, $y'' > 0$ solutions, and the number of bacteria increases drastically. Similarly, if you had removed 617 of the bacteria at time x_0, your population of 1,999,383 would now be represented by one of the $y' < 0$, $y'' < 0$ curves, and the number of bacteria will decrease as time passes. So, when we jiggle or *perturb* this equilibrium solution $y = 2$, the conditions change drastically. The new situation never returns to $y = 2$.

On the other hand, if a system is perturbed near a stable equilibrium, such as $N = k$ of Example 3, it tends to come right back to the equilibrium. A population of $N = k$ would tend to remain at $N = k$, and one with N close to k will move toward the level $N = k$. This question of stability is crucial in many applied problems (see Exercises 5–8, 17, 18).

Now it is time for *you* to try some problems. Again, the point of this chapter is to teach *you* how to do it, not to show that *some* people can. Try some of the Exercises 1–19. Talk with someone else if you get stuck. *When* you feel you have conquered some of those, try the following.

6. A More Difficult Example

EXAMPLE 5. Sketch solutions for

$$y' = 2x - y. \tag{5}$$

Try this problem before reading further. *Then* the following explanations are more apt to clarify and less apt to confuse or complicate your thinking.

1) *Look at y':*

$$y' = 0, \quad \text{for} \quad y = 2x$$

$$y' > 0, \quad \text{for} \quad y < 2x$$

$$y' < 0, \quad \text{for} \quad y > 2x$$

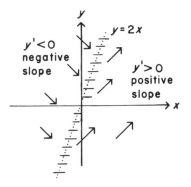

Figure 2.13

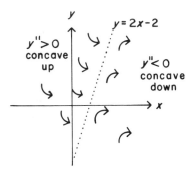

Figure 2.14

which gives us, so far, Figure 2.13. (Note: Where $y' = 0$, y is not a constant. Therefore, there is no equilibrium, no horizontal solution.)

2) *Look at y'':*

$$y'' = 2 - y' = 2 - 2x + y$$
$$y'' = 0, \text{ for } y = 2x - 2$$
$$y'' > 0, \text{ for } y > 2x - 2$$
$$y'' < 0, \text{ for } y < 2x - 2$$

giving us, in addition to the information in 1), Figure 2.14.

3) *Put it together:*

y	y'	y''	Conclusion
Greater than (to the left of) $y = 2x$	$-$	$+$	⌇
Between $y = 2x$ and $y = 2x - 2$	$+$	$+$	⌣
Less than (to the right of) $y = 2x - 2$	$+$	$-$	⌢

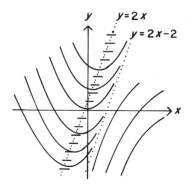

Figure 2.15

The tabular results are graphed in Figure 2.15. Remember, it all has to fit together. Further questions arise before you can be sure you have finished.

a) The solutions on the left of the last diagram will fit without crossing and destroying uniqueness *if* $y = 2x - 2$ is an *asymptote*. Is it? Recall that an asymptote is a line which is approached, but never reached, as $x \to \infty$. If $y = 2x - 2$, then $y' = 2$, and the original equation (5) is also satisfied. So, $y = 2x - 2$ is a solution, with constant slope 2, and the other solutions can never cross it.

b) What exactly happens to all those positive-slope, concave-downward solutions on the right of $y = 2x - 2$? How quickly do they bend over? Do they approach a limit as $x \to \infty$?

At this point we would be smart to go back to the direction field. We know where $y' = 0$ (along $y = 2x$) and where $y' = 2$ (along $y = 2x - 2$). From there, it is easy to consider

a fixed y since $y' = 2x - y$, as $x \uparrow$, $y' \uparrow$ (slopes get steeper as you move to the right along a fixed y);

a fixed x since $y' = 2x - y$, as $y \uparrow$, $y' \downarrow$ (slopes diminish as you move higher along a vertical line).

Alternatively, you might note that along any line $y = 2x + k$, the slope $y' = k$.

So, putting all this together also, we can improve the sketch in the final result, note that $y = 2x - 2$ is an asymptote for the solutions on the right as well. The sketch of solutions to (5)

$$y' = 2x - y$$

looks like Figure 2.16 (negative C's give the solutions below the asymptote).

The line $y = 2x - 2$, which is an asymptote, can be considered an equilibrium solution in a generalized sense. Solutions near this line tend toward it as $x \to \infty$. Actual solution of this differential equation (5) yields

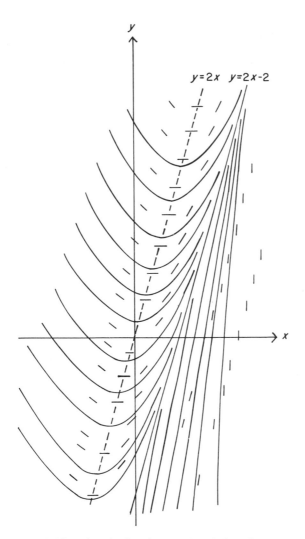

Figure 2.16. Negative c's give the solutions below the asymptote.

$$y = Ce^{-x} + 2x - 2.$$

Sure enough, as $x \to \infty$, $Ce^{-x} \to 0$. You can note that when x is quite negative, the solution is mostly Ce^{-x} (i.e., y is dominantly exponential); when x is quite positive, the solution is nearly $2x - 2$ (a straight line). The "interesting" part is where these two run into each other.

Example 5 has shown that such an innocent-looking differential equation as $y' = 2x - y$ can have a complex picture of the family of solutions. Nevertheless, that picture can be obtained simply by following through on the y', y'', put-together technique until all the pieces do fit together and everything is nailed down.

Exercises

(The most challenging exercises are denoted by a dagger.)

For Exercises 1–4, sketch solution to the given differential equations.

1. $\dfrac{dy}{dx} = -\dfrac{y}{x}.$

2. $\dfrac{dy}{dx} = y.$

3. $x + y' - 1 = -yy'.$

4. $y' = \dfrac{2y}{x}.$

5. Nutrients flow into a cell at a constant rate of R molecules per unit time and leave it at a rate proportional to the concentration, with constant of proportionality K. Let N be the concentration at time t. Then the mathematical description of the rate of change of nutrients in the above process is

$$\frac{dN}{dt} = R - KN;$$

that is, the rate of change of N is equal to the rate at which nutrients are entering the cell minus the rate at which they are leaving. Will the concentration of nutrients reach an equilibrium? If so, what is it and is it stable? Explain, using a graph of the solutions to this equation.

6. Suppose that the average new professor at a university begins checking books out of the library at the rate of one per day. Suppose further that the library recalls, in an average *week*, 1/10 of the books checked out. How many books does the average professor than have checked out at any one time after he has been around several years?

7. Suppose that an island is colonized by immigration from the mainland. Suppose further that there are S species on the mainland and, at time t, $N(t)$ on the island. The rate that new species immigrate to the island and colonize is proportional to the number of species on the mainland which are not already established on the island $(S - N(t))$, with constant of proportionality I. Moreover, on the island species become extinct at a rate proportional to the number of species on the island, with constant of proportionality E. Show that the number of species on the island will reach an equilibrium number approximately $= [I/(I + E)]S$. Sketch the curve of N as a function of t.

8. Suppose a hard rock fan club starts out at time $t = 0$ with N_0 fanatical members. The club would no doubt grow at a rate proportional to the membership, except there are at most M people who are at all interested in such music. Hence as the membership approaches M the rate decreases because new recruits become harder to find. So in actuality, the rate of increase is proportional to the product of the number of members and the number of remaining interested people. Give the differential equation involving membership, $N(t)$. How many memberships can the organizers expect to sell per year, assuming the constants remain steady for a few years?

Sketch a number of solutions to the following equations. Your sketches should exhibit the correct slope and concavity and indicate points of inflection.

9. $y' = y^2 - 1$.

10. $y' = y(y - 2)(y - 4)$.

11. $y' = (e^{-x} - 1)y$.

12. $y' = (y - 1)(3 + y)$. Include the solution where $y(0) = 1.5$.

13. $dy/dx = x(2 - y)/(x + 1)$, $x > 0$. Find all equilibrium points, and classify them as stable or unstable. Sketch the solutions which cross the y axis at 0, 1, 2, and 3. Make sure your curves have the correct slope at $x = 0$. However, y'' is very complicated to find, so you might try to do without looking at it.

14. Sketch a solution to the equation $y' + y^4 = 16$ when $y(0) = 0$. Find the equilibrium values, if any, and tell whether they are stable or unstable.

15. Sketch a solution to the equation $y' + y^3 = 8$ when $y(0) = 0$. Find equilibrium values, if any, and tell whether they are stable or unstable.

16. Without solving the differential equation $y' = x + y$, sketch a solution between $x = 0$ and $x = 1$ satisfying $y(0) = 0$. Make sure your curve has the correct slope at the origin and the correct concavity.

17. Water flows into a conical tank at a rate of k_1 units of volume per unit time. Water evaporates from the tank at a rate proportional to $V^{2/3}$, where V is the volume of water in the tank. Let the constant of proportionality be k_2. Find the differential equation satisfied by V. Without solving it, sketch some solutions. Is there an equilibrium? Is it stable?

18. A population of bugs on a plate tend to live in a circular colony. If N is the number of bugs and r_1 is the per capita growth rate, then $dN/dt = r_1 N$ is the Malthusian growth rule. However, those bugs on the perimeter suffer from cold, and they die at a rate proportional to their number, which means that they die at a rate proportional to $N^{1/2}$. Let this constant of proportionality be r_2. Find the differential equation satisfied by N. Without solving it, sketch some solutions. Is there an equilibrium? If so, is it stable?

19. Let $dy/dx = y((1/x) - 1)$, $x \geq 1$. Find all equilibrium points and classify them as stable or unstable. Sketch solutions with $y(1) = 1$ and $y(1) = -1$. Indicate points of inflection. Your curves should have the correct slope at $x = 1$ and the correct concavity. For a general solution $y(x)$, what happens as $x \to \infty$?

†20. Consider the differential equation $y' = e^x - y$. Without solving this equation, sketch solutions passing through the following points: $(0, 1)$, $(1, e)$, $(-1, 1/e)$, $(0, 0)$, $(1, 0)$, $(-1, 0)$. In addition to these curves and coordinate axes, sketch $y = e^x$. Your curves should exhibit the correct slope at the given points and the correct concavity. You should write down your reasoning. Do you see any sort of stable solution?

†21. Sketch the curve $y(x) = \int_0^x e^{-t^2} dt$, paying careful attention to slope, concavity, and initial value $y(0)$. Hint: Use the Fundamental Theorem of Calculus to notice that $y' = e^{-x^2}$.

Sidelight:

22. Solve the differential equation $dN/dt = rN(1 - (N/k))$ from Example 3. (Separate variables and use partial fractions, or, if you have had some differential equations, substitute $M = 1/N$ and solve the resulting linear differential equation in M.) Confirm that one way of writing the answer is

$$N = \frac{k}{1 + \dfrac{e^{-rt}}{c}}.$$

Discussion: The formula of Exercise 22 was used successfully by R. L. Pearl and L. J. Reed (*Proceedings of the National Academy of Sciences,* 1920, p. 275) to demonstrate a rather good fit with the population data of the United States gathered in the decennial census from 1790 to 1910. Using 1790, 1850, and 1910 as the points by which to evaluate the constants, they obtained

$$N = \frac{197{,}273{,}000}{1 + e^{-0.0313395t}} \tag{6}$$

and then calculated a predicted $N(t)$ for each of the decades between, to compare with the census figures. The results are given in the table, with four more decades added by the Dartmouth College Writing Group in 1967.

Year	Population from Decennial Census	Population from Formula (6)	Error	% Error
1790	3,929,000	3,929,000	0	0.0
1800	5,308,000	5,336,000	28,000	0.5
1810	7,240,000	7,228,000	− 12,000	− 0.2
1820	9,638,000	9,757,000	119,000	1.2
1830	12,866,000	13,109,000	243,000	1.9
1840	17,069,000	17,506,000	437,000	2.6
1850	23,192,000	23,192,000	0	0.0
1860	31,443,000	30,412,000	− 1,031,000	− 3.3
1870	38,558,000	39,372,000	814,000	2.1
1880	50,156,000	50,177,000	21,000	0.0
1890	62,948,000	62,769,000	− 179,000	− 0.3
1900	75,995,000	76,870,000	875,000	1.2
1910	91,972,000	91,972,000	0	0.0
1920	105,711,000	107,559,000	1,848,000	1.7
1930	122,775,000	123,124,000	349,000	0.3
1940	131,669,000	136,653,000	4,984,000	3.8
1950	150,697,000	149,053,000	− 1,644,000	− 1.1
1960	179,300,000[1]			
1970	204,000,000[1]			
1980	226,500,000[1]			

[1] Rounded to the nearest hundred thousand.

Solutions

1. $\dfrac{dy}{dx} = -\dfrac{y}{x}$.

2. $\dfrac{dy}{dx} = y$.

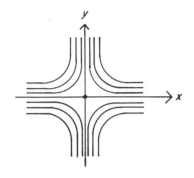

Figure 2.17

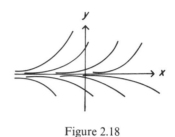

Figure 2.18

3. $\dfrac{dy}{dx} = \dfrac{-(x-1)}{y+1} = \dfrac{1-x}{1+y}$.

4. $\dfrac{dy}{dx} = \dfrac{2y}{x}$.

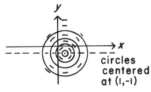

Figure 2.19

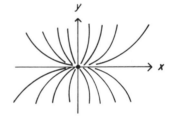

Figure 2.20

Note: Solutions 1 and 4 show "holes" at the origin where a unique solution does not exist; solution 3 has such a hole at $(1, -1)$. In all these cases, at the designated point, $g(x, y)$ and $\partial g/\partial y$ are "not nice" (g is undefined). See the discussion following the uniqueness theorem near the beginning of the chapter.

5. $dN/dt = R - KN$.

6. 70 books.

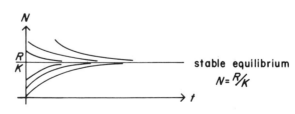

Figure 2.21

7. $dN/dt = I(S - N) - EN = IS - N(I + E) = 0$, when $N = IS/(I + E)$. Figure 2.22.

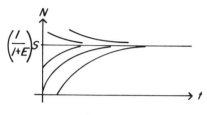

Figure 2.22

8. $dN/dt = kN(M - N)$, equilibria graphically at $N = 0$ and $N = M$. The latter is the stable one for memberships.

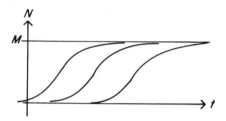

Figure 2.23

9. $y' = y^2 - 1 = 0$, for $y = \pm 1$
$\qquad\qquad > 0$, for $|y| > 1$
$\qquad\qquad < 0$, for $|y| < 1$.

$y'' = 2yy'$
$\quad = 2y(y + 1)(y - 1)$
$\quad = 0$, for $y = 0, y = \pm 1$
$\quad > 0$, for $y > 1$ and $-1 < y < 0$
$\quad < 0$, for $y < -1$ and $0 < y < 1$.

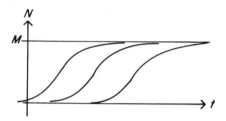

Figure 2.24

10. $y' = y(y - 2)(y - 4)$
    $y' = 0$, for $y = 0, 2, 4$
    $\quad > 0$, for $y > 4$ and $0 < y < 2$
    $\quad < 0$, for $2 < y < 4$ and $y < 0$.

    $y'' = (3y^2 - 12y + 8)y' = 0$, for $y' = 0$ or $y = 2 \pm (2\sqrt{3}/3)$.

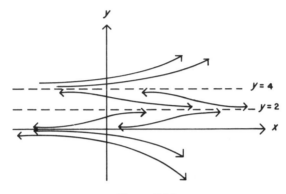

Figure 2.25

11. $y' = (e^{-x} - 1)y = 0$, when $y = 0$ or $x = 0$ ($y = 0$ is a solution)
    $\quad > 0$, for 2nd and 4th quadrants
    $\quad < 0$, for 1st and 3rd quadrants.

    $y'' = (e^{-2x} - 3e^{-x} + 1)y$
    $\quad = 0$, when $y = 0$ or when $e^{-2x} - 3e^{-x} + 1 = 0$.

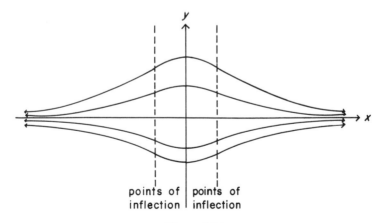

points of | points of
inflection | inflection

Figure 2.26

This last equation has two solutions for x, one positive and one negative:

$$3e^{-x} = e^{-2x} + 1;$$

multiply by e^x:

$$3 = e^x + e^{-x}.$$

The graph of the sum of e^x and e^{-x} crosses $y = 3$ twice and is symmetric about the y axis.

12. $y' = (y - 1)(3 + y) = 0$, when $y = 1$ or -3
$y'' = [(3 + y) + (y - 1)]y' = (2y + 2)y'$
$= 0$, when $y = -1$.

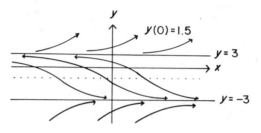

Figure 2.27

13. $y' = x(2 - y)/(x + 1)$, $x > 0$. $y' = 0$, at $x = 0$ or at $y = 2$.

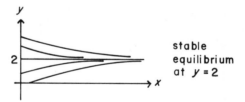

Figure 2.28

14. $y' + y^4 = 16$.
$y' = 16 - y^4 = (2 + y)(2 - y)(4 + y^2) = 0$, at $y = 2, y = -2$

always pos.

$y'' = -4y^3 y' = 0$, at $y = 0, 2, -2$.

	y'	y''
$y > 2$	$-$	$+$
$0 < y < 2$	$+$	$-$
$-2 < y < 0$	$+$	$+$
$y < -2$	$-$	$-$

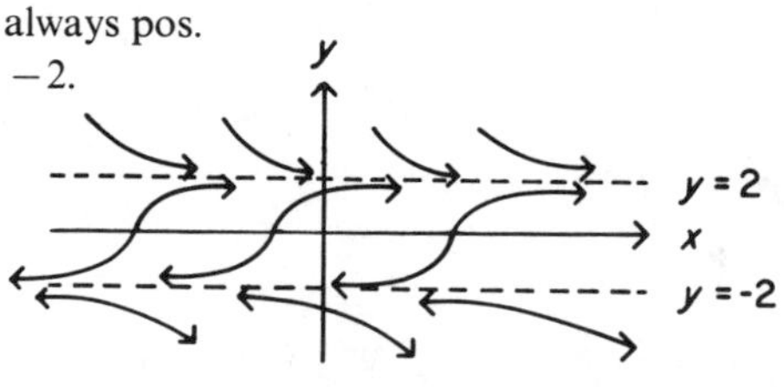

Figure 2.29

15. $y' = 8 - y^3 = (2 - y)\underbrace{(4 + 2y + y^2)}_{\text{always pos.}} = 0$, when $y = 2$

$y'' = -3y^2 y'$.

	y'	y''
$y > 2$	$-$	$+$
$y < 2$	$+$	$-$

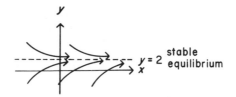

Figure 2.30

16. $y' = x + y$
   $y'' = 1 + y' = 1 + x + y$
   $y(0) = 0$

so $y'(0) = 0$ (horizontal) and $y''(0) = 1$ (concave up).

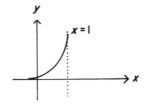

Figure 2.31

17. $\dfrac{dV}{dt} = k_1 - k_2 V^{2/3}$.

$V'' = -K_2(\tfrac{2}{3})\dfrac{V'}{(V)^{1/3}}$

(so of opposite sign from V')

$V' = 0$, at $V^{2/3} = \dfrac{K_1}{K_2}$ or $V = \left(\dfrac{K_1}{K_2}\right)^{3/2}$.

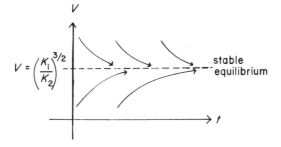

Figure 2.32

18. $\dfrac{dN}{dt} = r_1 N - r_2 N^{1/2} = N^{1/2}(r_1 N^{1/2} - r_2)$

$$= 0, \text{ at } N = 0, N = \left(\dfrac{r_2}{r_1}\right)^2$$

$$N'' = r_1 N' - \dfrac{r_2}{2N^{1/2}} N' = N'\left(r_1 - \dfrac{r_2}{2N^{1/2}}\right)$$

$$= 0, \text{ at } N' = 0, N = \left(\dfrac{r_2}{2r_1}\right)^2.$$

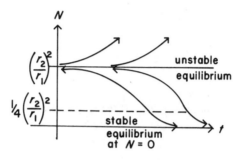

Figure 2.33

19. $y' = y((1/x) - 1)$, $x \geq 1$; horizontal slope at $x = 1$, inflection points at $x = 2$ as $x \to \infty$, $y(x) \to 0$.

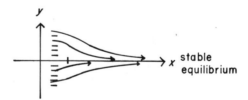

Figure 2.34

20. $y' = e^x - y$.

$$y' = e^x - y = 0, \text{ when } y = e^x$$
$$> 0, \text{ for } y < e^x$$
$$< 0, \text{ for } y > e^x$$

$$y'' = e^x - y'$$
$$= e^x - (e^x - y)$$
$$= y = 0, \text{ at } y = 0$$
$$> 0, \text{ for } y > 0$$
$$< 0, \text{ for } y < 0.$$

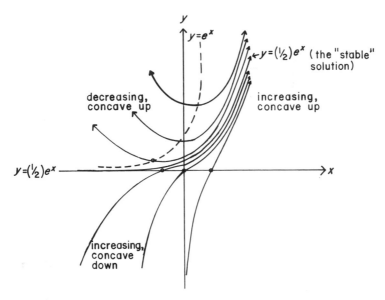

Figure 2.35

21. $y(x) = \int_0^x e^{-t^2}\, dt;\; y' = e^{-x^2};\; y(0) = 0.$

$y' = e^{-x^2} \neq 0,$ for all x
$\phantom{y' = e^{-x^2}} \to 0,$ as $x \to \pm\infty$ (so there *are* horizontal asymptotes)

$y''(x) = e^{-x^2}(-2x) = 0,$ for $x = 0$
$\phantom{y''(x) = e^{-x^2}(-2x)} > 0,$ for $x < 0$
$\phantom{y''(x) = e^{-x^2}(-2x)} < 0,$ for $x > 0$

$y'(0) = 1;\; y''(0) = 0.$

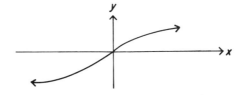

Figure 2.36

22. $\displaystyle \int\left(\frac{1}{N - N^2/k}\right) dN = \int r\, dt$

$\displaystyle = \int\left(\frac{1/k}{1 - (N/k)} + \frac{1}{N}\right) dN = rt + a$

$\displaystyle = -\ln\left|1 - \frac{N}{k}\right| + \ln N = rt + a,$

so

$$\frac{N}{1 - \dfrac{N}{k}} = (\text{const.})\, e^{rt} \quad \text{or} \quad \frac{N}{K - N} = C e^{rt} \Rightarrow N = \frac{kCe^{rt}}{1 + Ce^{rt}} = \frac{K}{1 + \dfrac{e^{-rt}}{c}}$$

$$\left(\text{or } M = \frac{1}{N}; \frac{dM}{dt} = -\frac{1}{N^2}\frac{dN}{dt} \Rightarrow \frac{dM}{dt} + rM = \frac{r}{k} \right.$$

$$\left. \Rightarrow M = \frac{1}{k} + C e^{-rt}, \text{etc.} \right).$$

Notes for the Instructor

Objectives. Qualitative solution sketching is used primarily in the writing of mathematical models in fields such as the biological or social sciences. The module details the technique of solution sketching without actually deriving an explicit expression for the solution, and considers the long-term behavior of the solution as $x \to \infty$ (this provides all one needs to know for many applied problems). Equilibria and the stability of long-range solutions are also discussed.

Prerequisites. Elementary calculus—specifically curve sketching (slope and concavity), the recognition of a general solution for ordinary first-order differential equations, and the simple integration of separable equations.

Time. Less than one class period can be used to introduce students to the material; the more difficult problems can be discussed when the class reconvenes.

Remarks. Solutions to all the exercises are provided, but the following comments may be helpful. Exercises 1–19 are not difficult—they come from straight homework assignments and exams. Exercises 1–4 do have points at which the uniqueness theorems do not hold. Half the word problems involve finding the differential equation, the others state it explicitly. Exercises 20 and 21 are trickier. Exercise 22 is a sidelight following up Example 3: the student can solve the differential equation (by partial fractions or by a substitution leading to a linear differential equation) if he has been taught these topics; if not, the answer is explicitly given and he might be asked to confirm that it satisfies the differential equation. An additional table shows a nice set of data that fit the model.

Students with access to a graphics terminal and a DE solving program might experiment with qualitative solution sketching. We would welcome reports on this and any other additions, corrections, or comments on this chapter.

CHAPTER 3

Difference and Differential Equation Population Growth Models

James C. Frauenthal*

1. Introduction

Ordinarily, the derivative is defined by the following limit:

$$\frac{dy}{dx} = \lim_{h \to 0} \frac{y(x + h) - y(x)}{h}. \tag{1a}$$

Similarly, when a computer is used to "solve" a differential equation numerically, derivatives are ordinarily replaced by finite difference approximations such as

$$\frac{dy}{dx} \simeq \frac{y(x + h) - y(x)}{h}. \tag{1b}$$

These two operations are really just inverses of one another. At times, the conversion of a difference equation into the analogous differential equation is convenient because the calculus can be employed, so the finite interval of the independent variable is made to vanish. At other times, this limit is "undone" so that numerical methods can be used on the difference equation analog of a differential equation. Unfortunately, these inverse operations have a profound effect upon the nature of the solutions found. This frequently neglected point is the main topic of this chapter.

The implications of dealing with a difference equation, as opposed to dealing with the analogous differential equation, will be illustrated by studying a well-known equation which can be interpreted as describing the

* Department of Applied Mathematics and Statistics, State University of New York, Stony Brook, NY 11794.

growth of a single animal species within a limited environment. In its difference equation form it will be written

$$\frac{N(t + \Delta t) - N(t)}{\Delta t} = r_0 \left[1 - \frac{N(t)}{K} \right] N(t), \qquad \begin{cases} t \geq 0 \\ N(0) = N_0 \end{cases} \qquad (2)$$

where $N(t)$ is the population size at time t, r_0 and K are positive constants which parameterize how the population grows, and N_0 is a positive constant which tells the initial population size. (For the sake of completeness, the assumptions which lead to (2) are listed and the equation is derived with considerable detail in the Appendix.) If we allow $\Delta t \to 0$ in (2), the differential equation analogous to (2) results,

$$\frac{dN(t)}{dt} = r_0 \left[1 - \frac{N(t)}{K} \right] N(t), \qquad \begin{cases} t \geq 0 \\ N(0) = N_0. \end{cases} \qquad (3)$$

This is a nonlinear differential equation which is often called the logistic equation.

Note that although a particular pair of equations is used for illustration, the points expounded in this chapter are quite general and carry over to a lesser or greater degree to all difference and differential equations.

2. Solution of the Difference Equation

We begin by considering how we might solve the difference equation (2). Since this equation is nonlinear, it cannot be solved directly. However, with the aid of a pocket calculator or, better yet, a computer, we can determine the sequence of numbers $N(n\Delta t)$, $n = 0, 1, 2, \cdots$, for any positive choice of Δt with r_0, K, and N_0 given. We would do so as follows. First, by setting $t = 0$ in (2) we find

$$N(\Delta t) = N_0 + \{r_0[1 - N_0/K]N_0\}\Delta t. \qquad (4a)$$

Next, we set $t = \Delta t$ in (2) to find

$$N(2\Delta t) = N(\Delta t) + \{r_0[1 - N(\Delta t)/K]N(\Delta t)\}\Delta t. \qquad (4b)$$

By successively setting $t = n\Delta t$, with $n = 0, 1, 2, \cdots$, we could "step along," finding each new value of $N(n\Delta t)$ in terms of the previous one.

Note that this is not a solution in the usual sense but rather just a realization of the population size at successive instants of time. Although r_0, K, and N_0 are presumably given, the choice of Δt is quite arbitrary, and no rational method appears to exist for selecting Δt. By choosing Δt very small, we get more detail—but at the expense of using more calculator or computer time. An equally serious shortcoming of this method is that if we wish to find answers for various choices of r_0, K, and N_0, the entire solution must be recalculated for each such choice. It might therefore appear desirable to seek a method which provides a "closed-form" solution. Since no general

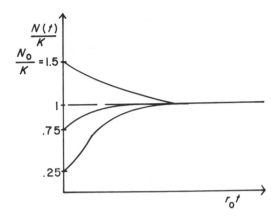

Figure 3.1. Solution to Logistic Differential Equation for Three Values of Initial Population Size

analytic technique is known for finding closed-form solutions to nonlinear difference equations, we might decide to look at the analogous differential equation (3).

3. Solution of the Differential Equation

Although the logistic equation (3) is also nonlinear, its special form permits us to find an analytic solution. The method we employ takes advantage of the fact that the dependent (N) and independent (t) variables can be moved algebraically to opposite sides of the equation. Equations which can be separated in this way can be solved by direct integration from the initial conditions to an arbitrary point in time. Thus

$$\int_{N_0}^{N(t)} \frac{dN}{N(1 - N/K)} = r_0 \int_0^t dt. \tag{5}$$

The integral on the left-hand side is easily evaluated by expanding the integrand by partial fractions (see Exercise 1). The result, following a bit of algebra, is

$$\frac{N(t)}{K} = \frac{e^{r_0 t}}{\dfrac{K}{N_0} + e^{r_0 t} - 1}, \qquad t \geq 0. \tag{6}$$

It is apparent from this form of the solution that the naturally occurring variables are $N(t)/K$ and $r_0 t$. This means that our solution only needs to be evaluated for different choices of N_0. This is clear from the plot of the solution in Figure 3.1, which shows three possible choices of N_0. (Note that r_0 and K are always positive on physical grounds—see the Appendix if this is not obvious.)

The most important thing to notice is that $N(t)/K \to 1$ as $r_0 t \to \infty$, regardless of the initial population size N_0. This is, of course, equivalent to saying that $N(t) \rightarrow K$ as $t \rightarrow \infty$ since $r_0 > 0$. This can be easily confirmed by examining the solution directly. Note further (see Exercise 2) that the solution never crosses the line $N/K = 1$; if $N_0/K > 1$, the solution approaches $N/K = 1$ from above, if $N_0/K < 1$, the solution approaches $N/K = 1$ from below. (K is called the "carrying capacity" of the environment by biologists.)

Clearly, converting the difference equation to a differential equation has provided a means for finding an analytic solution which is easily reinterpreted if the value of r_0, K, or N_0 is changed. In fact, it serves to illustrate why calculus is so important. However, the process of letting $\Delta t \rightarrow 0$ has had a profound effect on the nature of the solution. The truth of this could be illustrated by comparing results for various numerical solutions to the difference equation (4) with the solution to the differential equation (6). However, a more informative procedure is available.

4. Linearized Stability Analysis

The method which we will employ is called linearized stability analysis. It is a very general method for finding out the nature of the solution of both nonlinear difference and nonlinear differential equations. In order to understand the method, it is first necessary to understand the meaning of the word equilibrium.

Very simply, a system is in equilibrium if it is in a state of balance. A more useful, functional definition of equilibrium for our differential equation (3) is that $dN/dt = 0$. The analogous definition for the difference equation (2) is that $N(t + \Delta t) = N(t)$. Note that for both of our equations, this means that

$$ r_0 \left(1 - \frac{N}{K} \right) N = 0. \tag{7} $$

The solutions to this algebraic equation are $N = 0$ and $N = K$ and are called the equilibrium points.

Linearized stability analysis will provide us with a mathematical procedure for determining whether the equilibrium points of the equations are stable or unstable if disturbed a small amount. This is an extremely important thing to know because in nature systems are found at (or near) stable equilibrium points but not at (or near) unstable equilibrium points. This idea can be clarified by means of a simple physical example.

Consider a pendulum made of a long, straight, rigid rod, hinged at one end and with a mass on the other. Two equilibrium points exist for this system, one with the mass directly above the pivot, the other with the mass directly below the pivot. The configuration on the left is called *unstable*. The slightest disturbance of the mass away from the equilibrium position causes

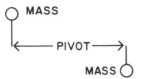

a force to develop which drives the system still further from equilibrium. Note that one would not ordinarily expect to find the system in this position. The configuration on the right is *stable*, though the nature of the stability depends upon how much friction there is in the pivot. If there is no friction and the mass is disturbed a little, it will swing back and forth through the equilibrium point, returning each time to the point of maximum displacement. This situation is called *neutrally stable and oscillatory*. If a tiny bit of friction exists and the mass is disturbed a little, it will swing back and forth, but on each swing it will go a bit less far until it finally comes to rest at the equilibrium point. This situation is called *(asymptotically) stable and oscillatory*. Finally, if quite a lot of friction is present and the mass is disturbed a little, it will slowly return to the equilibrium point, and it will do so without ever passing through equilibrium. This situation is called *(asymptotically) stable and nonoscillatory*. Note that the stable positions of the pendulum are the ones we would ordinarily expect to encounter.

Let us next investigate the nature of the equilibrium at $N = K$ for the differential and the difference equation version of our mathematical model. We will not bother to look at the equilibrium point at $N = 0$, as our model is really only valid for fairly large populations. (This is discussed in the Appendix.) Note that since we already know the analytic solution (6) to the differential equation, we can anticipate the answer. A glance at the plot of the solution confirms that the population always goes asymptotically to $N/K = 1$ and does so without passing through the equilibrium point. We therefore expect to find $N = K$ to be a stable nonoscillatory equilibrium point.

The analytic procedure for performing a linearized stability analysis of an equilibrium point proceeds as follows. First, define a new dependent variable whose origin (zero point) occurs at $N = K$. For example, we might let $X(t)$ be the new variable and define $N(t) = K + X(t)$. Although this is perfectly valid (see Exercise 6), the arguments which we will make shortly are slightly neater if we instead use the new dependent variable $x(t) = X(t)/K$, hence $N(t) = K[1 + x(t)]$. Substituting this expression into the differential equation (3), we get

$$\frac{dx}{dt} = -r_0 x - r_0 x^2, \qquad \begin{cases} t \geq 0 \\ x_0 \text{ given} \end{cases} \tag{8}$$

where Kx_0 is the size of the initial disturbance away from equilibrium. Note that (8) is really identical to (3), expect that it is written in terms of the new variable $x(t)$. We next linearize. To do so, we demand that $|x|$ be very small, that is, $|x| \ll 1$, but if this is so, $|x^2| \ll |x|$. We therefore choose to ignore the x^2 term (and all higher power terms if there are any) relative to the linear term in x on the right-hand side of (8), thus

$$\frac{dx}{dt} \simeq -r_0 x, \qquad \begin{cases} t \geq 0 \\ x_0 \text{ given} \\ |x| \ll 1 \end{cases} \tag{9}$$

where we must also demand that $|x_0| \ll 1$ for consistency. This is a linear differential equation which is easily solved by separating variables and integrating:

$$\int_{x_0}^{x(t)} \frac{dx}{x} \simeq -r_0 \int_0^t dt. \tag{10}$$

The solution is easily seen to be

$$x(t) \simeq x_0 e^{-r_0 t}. \tag{11}$$

Notice that regardless of the sign of x_0, $x(t) \to 0$ as $t \to \infty$ since $r_0 > 0$. In addition, the sign of $x(t)$ for $t \geq 0$ does not change. This allows us to conclude that the equilibrium of the differential equation (3) at $N = K$ is stable and nonoscillatory. This is what we anticipated earlier.

We next repeat the linearized stability analysis, this time for the equilibrium at $N = K$ of the difference equation (2). The method is the exact analog of the one employed for the differential equation. Thus let $N(t) = K[1 + x(t)]$ in (2), which leads to

$$\frac{x(t + \Delta t) - x(t)}{\Delta t} = -r_0 x(t) - r_0 x^2(t), \qquad \begin{cases} t \geq 0 \\ x_0 \text{ given} \end{cases} \tag{12}$$

and again demand that $|x| \ll 1$. We can then linearize to get

$$x(t + \Delta t) \simeq (1 - r_0 \Delta t) x(t), \qquad \begin{cases} t \geq 0 \\ x_0 \text{ given} \\ |x| \ll 1. \end{cases} \tag{13}$$

Since this is a linear difference equation, it can be solved simply. This, of course, is the reason why we like linearized stability analysis. Perhaps the easiest way to develop the solution is by induction. First, set $t = 0$ in (13) to get

$$x(\Delta t) \simeq (1 - r_0 \Delta t) x_0. \tag{14a}$$

Then set $t = \Delta t$ in (13) to get

$$x(2\Delta t) \simeq (1 - r_0 \Delta t) x(\Delta t) \tag{14b}$$

and use (14a) to eliminate $x(\Delta t)$ from (14b)

$$x(2\Delta t) \simeq (1 - r_0 \Delta t)^2 x_0.$$

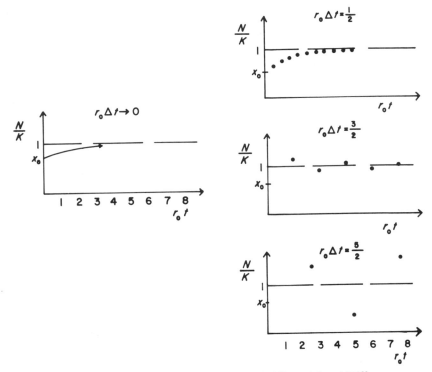

Figure 3.2. Selected Solutions to Linearized Form of Differential and Difference Equations Near $N = K$

Repeating this procedure inductively leads to the general solution

$$x(n\Delta t) \simeq (1 - r_0\Delta t)^n x_0, \qquad n = 0, 1, 2, \cdots. \tag{15}$$

Interpreting this result requires some care. Since x_0 is just a small constant. the nature of the solution is provided by the behavior of the term multiplying x_0 as n takes on successive positive integer values. Clearly, if $|1 - r_0\Delta t| < 1$, then $x(n\Delta t) \to 0$ as $n \to \infty$. This means that the equilibrium is stable if $0 < r_0\Delta t < 2$. Similarly, the equilibrium is neutrally stable if $r_0\Delta t = 2$ and unstable if $r_0\Delta t > 2$. (Note that we do not allow $r_0\Delta t \leq 0$ since both r_0 and Δt must be positive.) Notice also that if the quantity $(1 - r_0\Delta t) < 0$, then successive values of $x(n\Delta t)$ have opposite signs. This means that the solution is oscillatory if $r_0\Delta t > 1$ and nonoscillatory otherwise.

The results determined above are now illustrated using a number line.

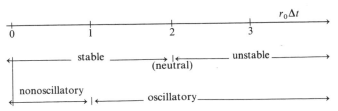

From this we can read off the following results. If $0 < r_0\Delta t < 1$ the equilibrium is stable and nonoscillatory; if $1 < r_0\Delta t < 2$, the equilibrium is stable and oscillatory; and if $r_0\Delta t > 2$, the equilibrium is unstable and oscillatory. Note also that if $r_0\Delta t = 2$ the equilibrium is neutrally stable and oscillatory. (If $r_0\Delta t = 1$, a pathological situation occurs as can be seen by looking at the linearized difference equation (13), and an alternative method of solution is required.)

In Figure 3.2, the results of the stability analysis are summarized. All of the illustrations are drawn for a small negative initial disturbance x from the equilibrium at $N = K$. Note that the results are all plotted in $(N/K, r_0 t)$ coordinates and follow directly by transforming the solutions (11) and (15) using the definition $N(t) = K[1 + x(t)]$. The differential equation has a stable nonoscillatory equilibrium, while the difference equation has, respectively, from top to bottom, a stable nonoscillatory, a stable oscillatory, and an unstable oscillatory equilibrium. This last illustration requires a little further observation. Recall that the ability to linearize the difference (or the differential) equation was predicated upon having $|x| \ll 1$. As the linearized solution diverges from the equilibrium point, $|x|$ is no longer small. Thus our solution ceases to be accurate. In other words, when a linearized solution is divergent, it really only informs us of how the instability starts to occur. It does not help us predict how the instability progresses or where the system is going.

The most important thing to realize in assessing each of the solutions in Figure 3.2 is that *all* correspond to identical choices of the governing parameters of the model, r_0, K, and N_0. The only distinction is in how we have treated time.

5. Interpretation of Results

Another look at the exact solution to the nonlinear difference equation (2) will help to clarify the reason for the unstable behavior found when Δt is too large. Insight is most easily gained by replacing (2) by the algorithm

$$N_{\text{new}} = N_{\text{old}} + \left\{ r_0 \left[1 - \frac{N_{\text{old}}}{K} \right] N_{\text{old}} \right\} \Delta t \qquad (16)$$

and interpreting this as the equation for a straight line in $(N, \Delta t)$ coordinates. The algorithm tells us that the value of N_{new} is found by starting from the value of N_{old} and projecting a distance Δt along a line with slope given by the quantity in braces $\{\ \}$. The critical point to notice is that the slope is entirely determined by N_{old} (and the constants r_0 and K) and is independent of Δt. Hence the larger the value of Δt, the further we project before reaching N_{new}. Only then do we rename N_{new} as N_{old}, reevaluate the slope, and repeat the process.

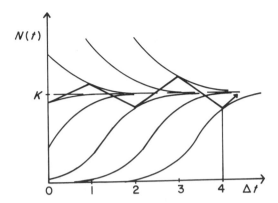

Figure 3.3. Exact Graphical Solution to Nonlinear Difference Equation for Value of Δt which Leads to Instability

In Figure 3.3 this procedure is illustrated for a choice of Δt which is so large that it causes instability. Note that the curved lines superimposed on the figure have as their tangent at any value of N the quantity in braces in (16). Thus the solution is generated by starting at some known value of N and progressing for $\Delta t = 1$ along the tangent to the curved line at N. Only then is the slope adjusted to its new value which is followed for another Δt. The reader is encouraged to repeat this construction for a smaller value of Δt so as to confirm that stable solutions can occur if Δt is small enough.

A physical analog of the above solution procedure might help clarify the reason for the instability. Imagine that you are driving a car but are blindfolded. The person sitting next to you is watching the road and telling you how to adjust your direction. Your speed is the analog of the constant r_0, and staying on the road is analogous to approaching the equilibrium at $N = K$. If you are going slowly (small r_0) and your assistant can give you instructions very frequently (small Δt), you can stay on the road, but if either your speed (r_0) or the time interval between instructions (Δt) becomes big, you will progressively do worse and worse, eventually going off the road to one side or the other.

6. Concluding Remarks

It is hoped that the reader will carry two things away from this exercise: first, the very useful technique of linearized stability analysis for anticipating the behavior of nonlinear difference and differential equations, and second, a healthy respect for the implications of how time (the independent variable) is treated in a mathematical model.

To help put the findings of this chapter in perspective, one final remark is in order. It is simple to check a numerical solution to determine if oscillations have occurred as a consequence of taking the increment in the independent variable too large. Merely recalculate using a smaller step size and compare results. If these are effectively identical, then the numerical scheme is almost certainly not at fault.

The dynamics of populations that genuinely live in discrete time, such as some species of insects and fishes, is a fascinating subject. Extremely complicated and unusual results can be found in the mathematics as well as in nature. The interested reader is advised to consult the paper by May [3] or the book by Frauenthal [1].

7. Appendix: The Basic Model

Let us consider an animal population. Our goal is to construct a simple mathematical model for determining the future size of the population. In order to do so we must make the following assumptions.

(i) The population is closed to migration, both in and out, and the initial size of the population is known.
(ii) Each animal in the population has the same chance of dying or of reproducing as any other animal.

The first assumption is equivalent to saying that the only way to leave the population is to die, and the only way to enter is either to be born or else to be present when the model begins. Note that although the first assumption is mathematically simple, it might be difficult to require in the field. The second assumption can be made to appear a bit more reasonable by stating it as follows. We attribute to each animal the average reproductive and survival traits for all animals in the population. This, of course, does away with all individual variation. Implicit in the second assumption is a somewhat deeper consequence: both age and sex have been eliminated from the model.

We must next define our variables. It seems clear that the independent variable should be time t. We will choose as the dependent variable $N(t)$, the number of animals in the population at time t. Finally, we let b and d represent the number of births and death, respectively, per animal in the population per unit of time. Thus, since the population numbers $N(t)$ at time t, there will be $bN(t)\Delta t$ births during the next small unit of time Δt.

In order to formulate the model, we assume that we know the population size at time t, $N(t)$, and then ask what our assumptions demand for the size $N(t + \Delta t)$ a short time Δt later. Clearly, from assumption (i) the population size at time $t + \Delta t$ must equal the size at time t plus the number of animals born during the interval Δt less the number that died. Using assumption (ii) to describe the births and deaths leads to the symbolic form of the model

$$N(t + \Delta t) = N(t) + bN(t)\Delta t - dN(t)\Delta t. \tag{A1}$$

Already an inconsistency has arisen. Animals occur in integer units. Yet there is no reason to expect in general that $N(t + \Delta t)$ will be an integer, even if $N(t)$ was one. This problem need not cause us serious alarm. Note that the larger the size of the population, the smaller the percentage difference between successive integers. Thus, if we demand that the total population size be large, without seriously altering the results, we may round our answer to the nearest integer. This innocent ploy provides a means for interpreting the population size when it assumes any positive real value. Since there cannot meaningfully be negative numbers of animals, $N \leq 0$ implies extinction of the population.

Next, rearrange (A1) into the form

$$\frac{N(t + \Delta t) - N(t)}{\Delta t} = r N(t) \tag{A2}$$

where $r \equiv b - d$ is called the intrinsic rate of population growth. Note that one indirect consequence of assumption (ii) is that only the difference between the birth and death rates enters the equation.

As a consequence of assumption (i) we assume the size of the population at time $t = 0$ is known to be $N(0) = N_0$. Our goal is to find solutions to (A2) for $t \geq 0$, given that N_0 is known. Note that we might refer to our formulation as being "exact" in the sense that it is consistent with our assumptions concerning reality. That is, up to this point no mathematical approximations have been introduced.

So far, little has been said about r, the intrinsic rate of growth. Presumably r will vary with time or with population size, or both. The analytic description of this variation is our next task. Intuition suggests that we make the following assumptions.

(iii) The individual birth and death rates b and d do not depend explicitly on time.

(iv) As the population size increases, b tends to decrease and d tends to increase.

Assumption (iii) really just admits that since we do not know any better, we choose to ignore the time variation of b and d. The second assumption results from the effects of malnutrition and crowding which occur in any limited environment.

Assumption (iii) implies that $r = r(N)$ only and (iv) suggests that $dr/dN < 0$. However, the assumptions are not sufficient to define a functional form for $r(N)$. As an approximation to the true form of $r(N)$, we choose the simplest (i.e., linear) expression consistent with the assumptions.

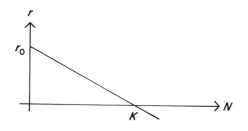

$$r = r_0 + r_1 N = r_0 \left(1 - \frac{N}{K}\right). \tag{A3}$$

The second form of the expression, written in terms of r_0 and K, has been introduced as it will be convenient for future interpretation. Note that we require that r_0 and K be positive constants, thus (A3) appears as illustrated here.

Note that when the population is smaller than K, r is positive and when it is larger than K, r is negative. In addition, the decrease occurs linearly. By replacing r in (A2) by the approximate expression in (A3) (written in terms of r_0 and K), we arrive at the equation

$$\frac{N(t + \Delta t) - N(t)}{\Delta t} = r_0 \left[1 - \frac{N(t)}{K}\right] N(t), \quad \begin{cases} t \geq 0 \\ N(0) = N_0. \end{cases} \tag{A4}$$

This is a nonlinear difference equation if Δt is finite, and becomes the well-known nonlinear differential equation

$$\frac{dN}{dt} = r_0 \left[1 - \frac{N(t)}{K}\right] N(t), \quad \begin{cases} t \geq 0 \\ N(0) = N_0 \end{cases} \tag{A5}$$

called the logistic equation if $\Delta t \to 0$. Note that (A4) and (A5) are reproduced in the body of the chapter where they are called, respectively, (2) and (3).

Exercises

1. Solve the logistic differential equation:

$$\frac{dN}{dt} = r_0 \left(1 - \frac{N}{K}\right) N, \quad \begin{cases} t \geq 0 \\ N(0) = N_0. \end{cases}$$

Use separation of variables and integration by partial fractions to get the result

$$\frac{N(t)}{K} = \frac{e^{r_0 t}}{\dfrac{K}{N_0} + e^{r_0 t} - 1}.$$

2. Show that the solution to the logistic differential equation can never cross the line $N/K = 1$. *Hint:* Look directly at the differential equation for proof.

3. In Figure 1, only the solution associated with a small value of N_0/K exhibits a point of inflection. Find the population size at the point of inflection.

4. Classify the stability of the equilibrium point $N = 0$ for the difference equation

$$\frac{N(t + \Delta t) - N(t)}{\Delta t} = r_0 \left[1 - \frac{N(t)}{K}\right] N(t).$$

Be careful that the prescribed initial disturbance is physically meaningful.

5. Repeat Exercise 4 for the logistic differential equation

$$\frac{dN}{dt} = r_0 \left[1 - \frac{N}{K}\right] N.$$

6. Repeat the linearized stability analysis of the logistic differential equation at $N = K$, using the new dependent variable $X(t) = N(t) - K$. Be sure to state all assumptions carefully.

7. Consider the differential equation

$$\frac{1}{K} \frac{dN}{dt} = 1 - e^{-r_0(1 - (N/K))}, \qquad \begin{cases} t \geq 0 \\ r_0, K > 0 \end{cases}.$$

a) Find the equilibrium point(s).
b) Determine the nature of the stability at the equilibrium point(s). (Note: In order to linearize, you will have to use a Taylor series expansion.)

Solutions

1. $\dfrac{dN}{dt} = r_0 \left(1 - \dfrac{N}{K}\right) N, \qquad \begin{cases} t \geq 0 \\ N(0) = N_0 \end{cases}$

$$\Rightarrow \int_{N_0}^{N(t)} \frac{dN}{N(1 - (N/K))} = r_0 \int_0^t dt = r_0 t.$$

Partial fractions:

$$\frac{1}{N(1 - (N/K))} = \frac{A}{N} + \frac{B}{(1 - (N/K))} = \frac{A - AN/K + BN}{N(1 - (N/K))}$$

$$\Rightarrow A = 1, \quad B = \frac{A}{K} = \frac{1}{K},$$

thus

$$r_0 t = \int_{N_0}^{N(t)} \frac{dN}{N} + \int_{N_0}^{N(t)} \frac{dN/K}{(1 - (N/K))}$$

$$= \ln N \big|_{N_0}^{N(t)} - \ln (1 - (N/K)) \big|_{N_0}^{N(t)}$$

$$= \ln \frac{N(t)[1 - (N_0/K)]}{N_0[1 - (N(t)/K)]}$$

$$\Rightarrow \frac{N(t)[1 - (N_0/K)]}{N_0[1 - (N(t)/K)]} = e^{r_0 t}$$

$$N(t)\{[1 - (N_0/K)] + (N_0/K)e^{r_0 t}\} = N_0 e^{r_0 t}$$

$$\frac{N(t)}{K} = \frac{e^{r_0 t}}{(K/N_0) + e^{r_0 t} - 1}.$$

2. $dN/dt = r_0(1 - (N/K))N$, thus at $N = K$, $dN/dt = 0 \Rightarrow N(t)$ cannot cross $N = K$.

3. $\dfrac{dN}{dt} = r_0(1 - (N/K))N = r_0 N - r_0 N^2/K$

$$\frac{d^2 N}{dt^2} = r_0 \left\{\frac{dN}{dt} - \frac{2N}{K}\frac{dN}{dt}\right\}$$

$$= r_0 \left\{1 - \frac{2N}{K}\right\}\frac{dN}{dt}.$$

Point of inflection occurs where $d^2 N/dt^2 = 0$ and $dN/dt \neq 0 \Rightarrow 1 - (2N/K) = 0$. Thus $N = K/2$ or $N/K = 1/2$.

4. Stability analysis: Let $N(t) = Kx(t)$

$$\Rightarrow K\frac{x(t + \Delta t) - x(t)}{\Delta t} = r_0 \left[1 - \frac{Kx(t)}{K}\right]Kx(t)$$

$$\frac{x(t + \Delta t) - x(t)}{\Delta t} = r_0 x(t) - r_0 x^2(t).$$

Assume $|x| \ll 1 \Rightarrow |x^2| \ll |x|$. Linearize to get

$$x(t + \Delta t) \simeq (1 + r_0 \Delta t)x(t).$$

Initial disturbance: $x(0) = x_0 > 0$.
Solution to linear difference equation:

$$x(n\Delta t) = (1 + r_0 \Delta t)^n x_0$$

but $r_0 \Delta t > 0 \Rightarrow (1 + r_0 \Delta t) > 1$, $\therefore$ equilibrium is unstable and nonoscillatory.

5. Stability analysis: Let $N(t) = Kx(t)$

$$\Rightarrow K\frac{dx}{dt} = r_0 \left(1 - \frac{Kx}{K}\right)Kx$$

$$\frac{dx}{dt} = r_0 x - r_0 x^2.$$

Assume $|x| \ll 1 \Rightarrow |x^2| \ll |x|$. Linearize to get

$$\frac{dx}{dt} \simeq r_0 x.$$

Initial disturbance: $x(0) = x_0 > 0$.
Solution to linear differential equation:

$$x(t) = x_0 e^{r_0 t}$$

but $r_0 > 0$, $\therefore$ equilibrium is unstable and nonoscillatory.

6. Stability analysis: Let $N(t) = K + X(t)$

$$\frac{dX}{dt} = -r_0 X - r_0 \frac{X^2}{K}.$$

Assume $|X| \ll K \Rightarrow |X^2/K| \ll |X| \ll K$. Linearize to get

$$\frac{dX}{dt} \simeq -r_0 X.$$

Initial disturbance: $X(0) = X_0$.
Solution to linearized differential equation:

$$X(t) = X_0 e^{-r_0 t}$$

since $r_0 > 0$, $X(t) \rightarrow 0$ as $t \rightarrow \infty$, $\therefore$ equilibrium is stable and nonoscillatory.

7. $\dfrac{1}{K}\dfrac{dN}{dt} = 1 - e^{-r_0(1-(N/K))}.$

a) Equilibrium: $dN/dt = 0$

$$\Rightarrow 1 - e^{-r_0(1-(N/K))} = 0 \Rightarrow N = K.$$

b) Stability: Let $N(t) = K[1 + x(t)]$

$$\Rightarrow \frac{dx}{dt} = 1 - e^{+r_0 x}.$$

Series expansion: $e^{r_0 x} = 1 + r_0 x + 1/2 r_0 x^2 + \cdots$, thus

$$\frac{dx}{dt} = -r_0 x - 1/2 r_0 x^2 - \cdots.$$

Assume $|x| \ll 1 \Rightarrow |x^n| \ll |x|$, $n = 2, 3, \cdots$. Linearize to get

$$\frac{dx}{dt} \simeq -r_0 x, \quad x(0) = x_0.$$

Solution:

$$x(t) \simeq x_0 e^{-r_0 t}$$

since $r_0 > 0$, $x(t) \rightarrow 0$ as $t \rightarrow 0$, $\therefore$ equilibrium is stable and nonoscillatory.

References

[1] J. C. Frauenthal, *Introduction to Population Modeling.* UMAP Monograph Series, Birkhauser-Boston, 1979, pp. 59–73.
[2] R. M. May, *Stability and Complexity in Model Ecosystems.* Princeton, NJ: Princeton Univ. Press, 1973, pp. 26–30.
[3] ——, "Biological populations with nonoverlapping generations: Stable points, stable cycles and chaos," *Science, 186,* 645–647, Nov. 15, 1974.
[4] J. Maynard Smith, *Mathematical Ideas in Biology.* New York: Cambridge Univ. Press, 1971, pp. 20–25, 40–44.
[5] E. O. Wilson and W. H. Bossert, *A Primer of Population Biology.* Stamford, CT: Sinauer Associates, 1971, pp. 14–19, 102–111.

Notes for the Instructor

Objectives. The mathematical consequences of taking the limit $\Delta t \to 0$ to convert a difference equation into a differential equation are studied. In the process, the method of linearized stability analysis is developed in detail. This chapter could be employed equally well in an introductory course in differential equations, numerical analysis, or mathematical population biology.

Prerequisites. One year of calculus. A cursory knowledge of population biology and Newtonian mechanics would be helpful but could be supplemented by the instructor.

Time. The material in this chapter could be covered in a one hour class meeting. It would be desirable to have students read the chapter in advance.

PART II

GROWTH AND DECAY MODELS: FIRST-ORDER DIFFERENTIAL EQUATIONS

CHAPTER 4

The Van Meegeren Art Forgeries

Martin Braun*

After the liberation of Belgium in World War II, the Dutch Field Security began its hunt for Nazi collaborators. They discovered, in the records of a firm which had sold numerous works of art to the Germans, the name of a banker who had acted as an intermediary in the sale to Goering of the painting '*Woman Taken in Adultery*' by the famed 17th-century Dutch painter Jan Vermeer. The banker in turn revealed that he was acting on behalf of a third-rate Dutch painter H. A. Van Meegeren, and on May 29, 1945, Van Meegeren was arrested on the charge of collaborating with the enemy. On July 12, 1945, Van Meegeren startled the world by announcing from his prison cell that he had never sold *Woman Taken in Adultery* to Goering. Moreover, he stated that this painting and the very famous and beautiful *Disciples at Emmaus*, as well as four other presumed Vermeers and two de Hooghs (another 17th-century Dutch painter) were his own works. Many people, thought that Van Meegeren was lying to save himself from the charge of treason. To prove his point, Van Meegeren began, while in prison, to forge the Vermeer painting *Jesus Amongst the Doctors* to demonstrate to the skeptics just how good a forger of Vermeer he really was. The work was nearly completed when Van Meegeren learned that a charge of forgery had been substituted for that of collaboration. He therefore refused to finish and age the painting in the hope that investigators would not uncover his secret of aging his forgeries. To settle the question, an international panel of distinguished chemists, physicists, and art historians was appointed to investigate the matter. The panel took x-rays of the paintings to determine whether other paintings were underneath them. In addition, they analyzed the pigments

---

* Department of Mathematics, Queens College, Flushing, NY 11367.

(coloring materials) used in the paint and examined the paintings for certain signs of age.

Van Meegeren was well aware of these methods. To avoid detection, he scraped the paint from old paintings that were not worth much just to get the canvas, and he tried to use pigments that Vermeer would have used. Van Meegeren also knew that old paint was extremely hard and impossible to dissolve. Therefore, he cleverly mixed a chemical (phenoformaldehyde) into his paint, and this hardened into Bakelite when the finished painting was heated in an oven.

However, Van Meegeren was careless with several of his forgeries, and the panel of experts found traces of the modern pigment cobalt blue. In addition, they also detected the phenoformaldehyde (which was first discovered at the close of the 19th century) in several of the paintings. On the basis of this evidence Van Meegeren was convicted on October 12, 1947, and sentenced to one year in prison. While in prison he suffered a heart attack and died on December 30, 1947.

Despite the evidence gathered by the panel of experts, many people still refused to believe that the famed *Disciples at Emmaus* was forged by Van Meegeren. Their contention was based on the fact that the other alleged forgeries and Van Meegeren's nearly completed *Jesus Amongst the Doctors* were of a very inferior quality. Surely, they said, the creator of the beautiful *Disciples at Emmaus* could not produce such inferior pictures. Indeed, the *Disciples at Emmaus* was certified as an authentic Vermeer by the noted art historian A. Bredius and was bought by the Rembrandt Society for $170,000. The answer of the panel to these skeptics was that because Van Meegeren was keenly disappointed by his lack of status in the art world, he worked on the *Disciples at Emmaus* with the fierce determination of proving that he was better than a third-rate painter. After producing such a masterpiece his determination was gone. Moreover, after seeing how easy it was to dispose of the *Disciples at Emmaus* he devoted less effort to his subsequent forgeries. This explanation failed to satisfy the skeptics. They demanded a thoroughly scientific and conclusive proof that the *Disciples at Emmaus* was indeed a forgery. This was done in 1967 by scientists at Carnegie-Mellon University, and we would now like to describe their work.

The key to the dating of paintings and other materials such as rocks and fossils lies in the phenomenon of radioactivity discovered at the turn of the century. The physicist Rutherford and his colleagues showed that the atoms of certain "radioactive" elements are unstable and that within a given time period a fixed proportion of the atoms spontaneously disintegrates to form atoms of a new element. Because radioactivity is a property of the atom, Rutherford showed that the radioactivity of a substance is directly proportional to the number of atoms of the substance present. Thus, if $N(t)$ denotes the number of atoms present at time t, then dN/dt, the number of atoms that disintegrate per unit time, is proportional to N; that is,

$$\frac{dN}{dt} = -\lambda N. \tag{1}$$

The constant λ, which is positive, is known as the decay constant of the substance. The larger λ is, of course, the faster the substance decays. One measure of the rate of disintegration of a substance is its *half-life* which is defined as the time required for half of a given quantity of radioactive atoms to decay. To compute the half-life of a substance in terms of λ, assume that at time t_0, $N(t_0) = N_0$. Then the solution of the initial value problem

$$dN/dt = -\lambda N, \; N(t_0) = N_0$$

is

$$N(t) = N_0 \, e^{-\lambda \int_{t_0}^{t} ds} = N_0 \, e^{-\lambda(t-t_0)}$$

or $N/N_0 = e^{-\lambda(t-t_0)}$. Taking logarithms of both sides we obtain that

$$-\lambda(t - t_0) = \ln \frac{N}{N_0}. \tag{2}$$

Now, if $N/N_0 = 1/2$, then $-\lambda(t - t_0) = \ln 1/2$, so that

$$(t - t_0) = \frac{\ln 2}{\lambda} = \frac{0.6931}{\lambda}. \tag{3}$$

Thus the half-life of a substance is $\ln 2$ divided by the decay constant λ. The dimension of λ, which we suppress for simplicity of writing, is reciprocal time. If t is measured in years, then λ has the dimension of reciprocal years, and if t is measured in minutes, then λ has the dimension of reciprocal minutes. The half-lives of many substances have been determined and recorded. For example, the half-life of carbon-14 is 5568 years, and the half-life of uranium-238 is 4.5 billion years.

Now the basis of "radioactive dating" is essentially the following. From (2) we can solve for $t - t_0 = (1/\lambda) \ln N_0/N$. If t_0 is the time the substance was initially formed or manufactured, then the age of the substance is $(1/\lambda) \ln N_0/N$. The decay constant λ is known or can be computed in most instances. Moreover, we can usually evaluate N quite easily. Thus, if we knew N_0, we could determine the age of the substance, but this is the real difficulty, since we usually do not know N_0. In some instances though, we can either determine N_0 indirectly, or else determine certain suitable ranges for N_0, and such is the case for the forgeries of Van Meegeren.

We begin with the following well-known facts of elementary chemistry. Almost all rocks in the earth's crust contain a small quantity of uranium. The uranium in the rock decays to another radioactive element, and that one decays to another, and another, and so forth, in a series of elements that results in lead (see Figure 4.1), which is not radioactive. The uranium (whose

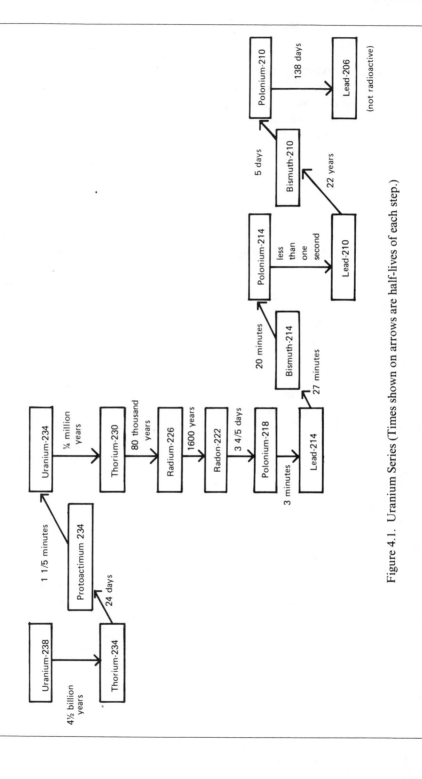

Figure 4.1. Uranium Series (Times shown on arrows are half-lives of each step.)

half-life is over 4 billion years) keeps feeding the elements following it in the series, so that as fast as they decay, they are replaced by the elements before them.

Now, all paintings contain a small amount of the radioactive element lead-210 (Pb^{210}) and an even smaller amount of radium-226 (Ra^{226}). These elements both occur in white lead (lead oxide), which is a pigment that artists have used for over 2000 years. For the analysis which follows, it is important to note that white lead is made from lead metal which, in turn, is extracted from a rock called lead ore in a process called smelting. In this process, the lead-210 in the ore goes along with the lead metal. However, 90–95% of the radium and its descendants are removed with other waste products in a material called slag. Thus most of the "supply" of lead-210 is cut off, and it begins to decay very rapidly, with a half-life of 22 yr. This process continues until the lead-210 in the white lead is once more in radioactive equilibrium with the small amount of radium present, i.e., the disintegration of lead-210 is exactly balanced by the disintegration of radium.

Let us now use this information to compute the amount of lead-210 present in a sample in terms of the amount originally present at the time of manufacture. Let $y(t)$ be the amount of lead-210 per gram of white lead at time t, y_0 the amount of lead-210 per gram of white lead present at the time of manufacture t_0, and $r(t)$ the number of disintegrations of radium-226 per-minute per gram of white lead at time t. If λ is the decay constant for lead-210, then

$$\frac{dy}{dt} = -\lambda y + r(t) \qquad y(t_0) = y_0. \tag{4}$$

Since we are only interested in a time period of at most 300 yr, we may assume that the radium-226, whose half-life is 1600 yr, remains constant, so that $r(t)$ is a constant r. Multiplying both sides of the differential equation by the integrating factor $\mu(t) = e^{\lambda t}$, we obtain $d(e^{\lambda t}y)/dt = re^{\lambda t}$. Hence

$$e^{\lambda t}y(t) - e^{\lambda t_0}y_0 = (r/\lambda)(e^{\lambda t} - e^{\lambda t_0}),$$

or

$$y(t) = \frac{r}{\lambda}\left[1 - e^{-\lambda(t-t_0)}\right] + y_0 e^{-\lambda(t-t_0)}. \tag{5}$$

Now $y(t)$ and r can be easily measured. Thus, if we knew y_0 we could use (5) to compute $(t - t_0)$, and consequently, we could determine the age of the painting. As we pointed out, though, we cannot measure y_0 directly. One possible way out of this difficulty is to use the fact that the original quantity of lead-210 was in radioactive equilibrium with the larger amount of radium-226 in the ore from which the metal was extracted. Let us, therefore, take samples of different ores and compute the rate of disintegration of radium-226. This was done for a variety of ores and the results are given in Table 1. These

Table 1. Ore and Ore Concentrate Samples

Description and Source	Disintegrations/min of Ra^{226}
Ore concentrate (Oklahoma–Kansas)	4.5
Crushed raw ore (S. E. Missouri)	2.4
Ore concentrate (S. E. Missouri)	0.7
Ore concentrate (Idaho)	2.2
Ore concentrate (Idaho)	0.18
Ore concentrate (Washington)	140
Ore concentrate (British Columbia)	1.9
Ore concentrate (British Columbia)	0.4
Ore concentrate (Bolivia)	1.6
Ore concentrate (Australia)	1.1

All disintegration rates are per minute per gram of white lead.

numbers vary from 0.18 to 140. Consequently, the number of disintegrations of the lead-210 per minute per gram of white lead at the time of manufacture will vary from 0.18 to 140. This implies that y_0 will also vary over a large interval, since the number of disintegrations of lead-210 is proportional to the amount present. Thus we cannot use (5) to obtain an accurate—or even a crude—estimate of the age of a painting. However, we can still use (5) to distinguish between a 17th-century painting and a modern forgery. The basis for this statement is the simple observation that if the painting is very old compared to the 22-year half-life of lead, then the amount of radioactivity from the lead-210 in a sample of paint will be nearly equal to the amount of radioactivity from the radium in the sample. On the other hand, if the painting is modern (20 years old or so) then the amount of radioactivity from the lead-210 will be much greater than the amount of radioactivity from the radium.

We can make this argument precise in the following manner. Let us assume that the painting in question is either very new or about 300 years old. Set $t - t_0 = 300$ in (5). Then after some simple algebra, we see that

$$\lambda y_0 = \lambda y(t) e^{300\lambda} - r(e^{300\lambda} - 1). \tag{6}$$

If our painting is indeed a modern forgery, then λy_0 should be absurdly large. To determine what is an absurdly high disintegration rate we observe (see Exercise 1) that if the lead-210 in a sample of white lead decays originally (at the time of manufacture) at the rate of 100 disintegrations (dis)/min per gram of white lead, then the ore from which it was extracted had a uranium content of 0.014%. This is a fairly high concentration of uranium since the average amount of uranium in rocks of the earth's crust is about 2.7 parts per million (ppm). On the other hand, some very rare ores exist in the western hemisphere whose uranium content is 2–3%. To be on the safe side, we will say that a disintegration rate of lead-210 is certainly absurd if it exceeds 30,000 dis/min per gram of white lead.

Table 2. Paintings of Questioned Authorship

Description	Po^{210} Disintegration	Ra^{226} Disintegration
Disciples at Emmaus	8.5	0.8
Washing of Feet	12.6	0.26
Woman Reading Music	10.3	0.3
Woman Playing Mandolin	8.2	0.17
Lace Maker	1.5	1.4
Laughing Girl	5.2	6

All disintegration rates are per minute per gram of white lead.

To evaluate λy_0, which is the number of disintegrations of the lead-210 per minute per gram of white lead at the time of manufacture, we must evaluate the present disintegration rate $\lambda y(t)$ of the lead-210, the disintegration rate r of the radium-226, and $e^{300\lambda}$. Since the disintegration rate of polonium-210 (Po^{210}) equals that of lead-210 after several years, and since it is easier to measure the disintegration rate of polonium-210, we substitute these values for those of lead-210. To compute $e^{300\lambda}$, observe from (3) that $\lambda = \ln 2/22$. Hence

$$e^{300\lambda} = e^{(300/22)\ln 2} = 2^{150/11}.$$

The disintegration rates of polonium-210 and radium-226 were measured for the *Disciples at Emmaus* and various other alleged forgeries and are given in Table 2.

If we now evaluate λy_0 from (6) for the white lead in the painting *Disciples at Emmaus*, we obtain that

$$\lambda y_0 = 2^{150/11}(8.5) + 0.8(2^{150/11} - 1) \text{ dis/min/g of Pb}$$

$$= 98,050$$

which is unacceptably large. Hence this painting must be a modern forgery. By a similar analysis (see Exercises 2–4) the paintings *Washing of Feet*, *Woman Reading Music*, and *Woman Playing Mandolin* were indisputably shown to be faked Vermeers. On the other hand, the paintings *Lace Maker* and *Laughing Girl* cannot be recently forged Vermeers, as claimed by some experts, since for these two paintings, the polonium-210 is very nearly in radioactive equilibrium with the radium-226, and no such equilibrium has been observed in any samples from 19th- or 20th-century paintings.

Exercises

1. In this exercise we show how to compute the concentration of uranium in an ore from the disintegration of the lead-210 in the ore.
   a) The half-life of uranium-238 is 4.51×10^9 yr. Since this half-life is so large,

we may assume that the amount of uranium in the ore is constant over a period of 200–300 yr. Let $N(t)$ denote the number of atoms of U^{238}/g of ordinary lead in the ore at time t. Since the lead-210 is in radioactive equilibrium with the uranium-238 in the ore, we know that $dN/dt = -\lambda N = -100$ dis/min/g of Pb at time t_0. Show that there are 3.42×10^{17} atoms of uranium-238/g of ordinary lead in the ore at time t_0. *Hint:* 1 yr = 525,600 min.

b) Using the fact that one mole of uranium-238 weighs 238 g and that a mole contains 6.02×10^{23} atoms, show that the concentration of uranium in the ore is approximately 0.014%.

For the paintings in Exercises 2, 3, and 4 use the data in Table 2 to compute the disintegrations per minute of the original amount of white lead per gram of ordinary lead, and conclude that each of these paintings is a forged Vermeer.

2. *Washing of Feet.*

3. *Woman Reading Music.*

4. *Woman Playing Mandolin.*

5. The following problem describes a very accurate derivation of the age of uranium.
   a) Let $N_{238}(t)$ and $N_{235}(t)$ denote the number of atoms of U^{238} and U^{235} at time t in a given sample of uranium, and let $t = 0$ be the time this sample was created. By the radioactive decay law,

   $$\frac{d}{dt} N_{238}(t) = \frac{-\ln 2}{(4.5)10^9} N_{238}(t)$$

   $$\frac{d}{dt} N_{235}(t) = \frac{-\ln 2}{0.707(10)^9} N_{235}(t).$$

   Solve these equations for $N_{238}(t)$ and $N_{235}(t)$ in terms of their original numbers $N_{238}(0)$ and $N_{235}(0)$.
   b) In 1946 the ratio of U^{238}/U^{235} in any sample was 137.8. Assuming that equal amounts of U^{238} and U^{235} appeared in any sample at the time of its creation, show that the age of uranium is 5.96×10^9 yr. This figure is universally accepted as the age of uranium.

6. In a samarskite sample discovered recently, there was 3 g of Thorium (Th^{232}). Thorium decays to lead-208 (Pb^{208}) through the reaction $Th^{232} \rightarrow Pb^{208} + 6(4He^4)$. It was determined that 0.0376 g of lead-208 was produced by the disintegration of the original Thorium in the sample. Given that the half-life of Thorium is 13.9 billion years, derive the age of this samarskite sample. (Hint: 0.0376 g of Pb^{208} is the product of the decay of $232/208 \times 0.0376$ g of Thorium.

One of the most accurate ways of dating archaeological finds is the method of carbon-14 (C^{14}) dating discovered by Willard Libby around 1949. The basis of this method is delightfully simple. The atmosphere of the earth is continuously bombarded by cosmic rays. These cosmic rays produce neutrons in the earth's atmosphere, and these neutrons combine with nitrogen to produce C^{14}, which is usually called radiocarbon since it decays radioactively. This radiocarbon is incorporated in carbon dioxide and thus moves through the atmosphere to be absorbed by plants. Animals, in turn, build radiocarbons into their tissues by eating the plants. In living tissue, the rate of ingestion of C^{14} exactly balances the rate of disintegration of C^{14}. When an organism dies, though, it ceases to ingest carbon-14, and thus its C^{14} concentration begins to decrease through disintegration of the C^{14} present. Now, it is a fundamental assumption of physics that the rate of bombardment of the earth's atmosphere by cosmic rays has always been constant. This implies that the original rate of disintegration of the C^{14} in a sample such as charcoal is the same as the rate measured today.[1] This assumption enables us to determine the age of a sample of charcoal. Let $N(t)$ denote the amount of carbon-14 present in a sample at time t, and N_0 the amount present at time $t = 0$ when the sample was formed. If λ denotes the decay constant of C^{14} (the half-life of carbon-14 is 5568 yr) then $dN/(t)/dt = -\lambda N(t)$, $N(0) = N_0$. Consequently, $N(t) = N_0 e^{-\lambda t}$. Now the present rate $R(t)$ of disintegration of the C^{14} in the sample is given by $R(t) = \lambda N(t) = \lambda N_0 e^{-\lambda t}$ and the original rate of disintegration is $R(0) = \lambda N_0$. Thus $R(t)/R(0) = e^{-\lambda t}$ so that $t = (1/\lambda)\ln(R(0)/R(t))$. Hence if we measure $R(t)$, the present rate of disintegration of the C^{14} in the charcoal and observe that $R(0)$ must equal the rate of disintegration of the C^{14} in a comparable amount of living wood, then we can compute the age t of the charcoal. The following two problems are real-life illustrations of this method.

7. Charcoal from the occupation level of the famous Lascaux Cave in France gave an average count in 1950 of 0.97 dis/min/g. Living wood gave 6.68 disintegrations. Estimate the date of occupation and hence the probable date of the remarkable paintings in the Lascaux Cave.

8. In the 1950 excavation at Nippur, a city of Babylonia, charcoal from a roof beam gave a count of 4.09 dis/min/g. Living wood gave 6.68 disintegrations. Assuming that this charcoal was formed during the time of Hammurabi's reign, find an estimate for the likely time of Hammurabi's succession.

References

[1]   P. Coremans, *Van Meegeren's Faked Vermeers and De Hooghs.*   Amsterdam: Meulenhoff, 1949.
[2]   B. Keisch, R. L. Feller, A. S. Levine, and P. R. Edwards, "Dating and authenticating works of art by measurement of natural alpha emitters, *Science, 155*, 1967, pp. 1238–1241.
[3]   ——, "Dating works of art through their natural radioactivity: Improvements and applications," *Science, 160*, 1968, pp. 413–415.
[4]   ——, *The Mysterious Box: Nuclear Science and Art*, a *World of the Atom* series booklet.

---

[1] Since the mid-1950's, the testing of nuclear weapons has significantly increased the amount of radioactive carbon in our atmosphere. Ironically, this unfortunate state of affairs provides us with yet another powerful method of detecting art forgeries. To wit, many artists' materials, such as linseed oil, canvas, paper, and so on, come from plants and animals, and so will contain the same concentration of carbon-14 as the atmosphere at the time the plant or animal dies. Therefore, linseed oil (which is derived from the flax plant), for example, that was produced during the last few years will have a much greater concentration of carbon-14 in it than linseed oil produced before 1950.

Notes for the Instructor

Objectives. The module shows how the radioactive decay of certain substances in white lead was used to prove that the famed painting *Disciples at Emmaus* bought by the Rembrandt Society for $170,000 was a forged Vermeer.

Prerequisites. First-order linear nonhomogeneous differential equations.

Time. The module can be covered in one or two lectures.

CHAPTER 5
Single Species Population Models

Martin Braun*

In this module we will study first-order differential equations which govern the growth of various species. At first glance it would seem impossible to model the growth of a species by a differential equation since the population of any species always changes by integer amounts. Hence the population of any species can never be a differentiable function of time. However, if a given population is very large and it is suddenly increased by one, then the change is very small compared to the given population. Thus we make the approximation that large populations change continuously and even differentiably with time.

Let $p(t)$ denote the population of a given species at time t and let $r(t, p)$ denote the difference between its birth rate and its death rate. If this population is isolated, that is, no net immigration or emigration occurs, then dp/dt, the rate of change of the population, equals $rp(t)$. In the most simplistic model we assume that r is constant, that is, it does not change with either time or population. Then, we can write down the following differential equation governing population growth:

$$\frac{dp(t)}{dt} = ap(t), \qquad a = \text{const.}$$

This is a linear equation and is known as the Malthusian law of population growth. If the population of the given species is p_0 at time t_0, then $p(t)$ satisfies the initial value problem $dp(t)/dt = ap(t), p(t_0) = p_0$. The solution of this initial value problem is $p(t) = p_0 e^{a(t-t_0)}$. Hence any species satisfying the Malthusian law of population growth grows exponentially with time.

We have just formulated a very simple model for population growth—so

* Department of Mathematics, Queens College, Flushing, NY 11367.

simple, in fact, that we have been able to solve it completely in a few lines. It is important, therefore, to see if this model, with its simplicity, has any relationship at all with reality. Let $p(t)$ denote the human population of the earth at time t. It was estimated that the earth's human population in 1961 was 3,060,000,000 and that during the past decade the population was increasing at a rate of 2%/yr. Thus $t_0 = 1961$, $p_0 = (3.06)10^9$, and $a = 0.02$, so that

$$p(t) = (3.06) \, 10^9 \, e^{0.02(t-1961)}.$$

We can certainly check this formula out for past populations.

Result. It reflects with surprising accuracy the population estimate for the period 1700–1961. The population of the earth has been doubling about every 35 years, and our equation predicts a doubling of the earth's population every 34.6 years. To prove this, observe that the human population of the earth doubles in a time $T = t - t_0$ where $e^{0.02T} = 2$. Taking logarithms of both sides of this equation gives $0.02T = \ln 2$ so that $T = 50 \ln 2 \simeq 34.6$. However, let us look into the distant future. Our equation predicts that the earth's population will be 200,000 billion in the year 2510, 1,800,000 billion in the year 2635, and 3,600,000 billion in the year 2670. These are astronomical numbers whose significance is difficult to gauge. The total surface of this planet is approximately 1,860,000 billion square feet. Eighty percent of this surface is covered by water. Assuming that we are willing to live on boats as well as land, it is easy to see that by the year 2510 there will be only 9.3 square feet per person; by 2635 each person will have only one square foot on which to stand; and by 2670 we will be standing two deep on each other's shoulders.

It would seem therefore, that this model is unreasonable and should be thrown out. However, we cannot ignore the fact that it offers exceptional agreement in the past. Moreover, we have additional evidence that populations do grow exponentially. Consider the Microtus Arvallis Pall, a small rodent which reproduces very rapidly. We take the unit of time to be a month and assume that the population is increasing at the rate of 40%/mo. If two rodents are present initially at time $t = 0$, then $p(t)$, the number of rodents at time t, satisfies the initial value problem $dp(t)/dt = 0.4p(t)$, $p(0) = 2$. Consequently,

$$p(t) = 2 \, e^{0.4t}. \tag{1}$$

Table 1 compares the observed population with the population calculated from (1). As one can see, there is excellent agreement.

Remark. In the case of the Microtus Arvallis Pall, p observed is very accurate since the pregnancy period is three weeks, and the time required for the census taking is considerably less. If the pregnancy period were very short then p observed could not be accurate since many of the pregnant rodents would have given birth before the census was completed.

Table 1. The Growth of
Microtus Arvallis Pall

Months	0	2	6	10
P observed	2	5	20	109
P calculated	2	4.5	22	109.1

The way out of our dilemma is to observe that linear models for population growth are satisfactory *as long as* the population is not too large. When the population gets extremely large though, these models cannot be very accurate, since they do not reflect the fact that individual members are now competing with each other for the limited living space, natural resources, and food available. Thus we must add a competition term to our linear differential equation. A suitable choice of a competition term is $-bp^2$, where b is a constant, since the statistical average of the number of encounters of two members per unit time is proportional to p^2. We consider, therefore, the modified equation

$$\frac{dp}{dt} = ap - bp^2.$$

This equation is known as the logistic law of population growth, and the numbers a, b are called the vital coefficients of the population. It was first introduced in 1837 by the Dutch mathematical-biologist Verhulst. Now, the constant b, in general, will be very small compared to a, so that if p is not too large then the term $-bp^2$ will be negligible compared to ap and the population will grow exponentially. However, when p is very large, the term $-bp^2$ is no longer negligible and thus serves to slow down the rapid rate of increase of the population. Needless to say, the more industrialized a nation is, the more living space it has, and the more food it has, the smaller the coefficient b is.

Let us now use the logistic equation to predict the future growth of an isolated population. If p_0 is the population at time t_0, then $p(t)$, the population at time t, satisfies the initial value problem

$$\frac{dp}{dt} = ap - bp^2 \qquad p(t_0) = p_0.$$

This is a separable differential equation, and its solution is

$$\int_{p_0}^{p} \frac{dr}{ar - br^2} = \int_{t_0}^{t} ds = t - t_0.$$

To integrate the function $1/(ar - br^2)$ we resort to partial fractions. Let

$$\frac{1}{ar - br^2} \equiv \frac{1}{r(a - br)} = \frac{A}{r} + \frac{B}{a - br}.$$

To find A and B, observe that

$$\frac{A}{r} + \frac{B}{a - br} = \frac{A(a - br) + Br}{r(a - br)} = \frac{Aa + (B - bA)r}{r(a - br)}.$$

Therefore, $Aa + (B - bA)r = 1$. Since this equation is true for all values of r, we see that $Aa = 1$ and $B - bA = 0$. Consequently, $A = 1/a$, $B = b/a$, and

$$\int_{p_0}^{p} \frac{dr}{r(a - br)} = \frac{1}{a}\int_{p_0}^{p}\left(\frac{1}{r} + \frac{b}{a - br}\right)dr$$

$$= \frac{1}{a}\left[\ln\frac{p}{p_0} + \ln\left|\frac{a - bp_0}{a - bp}\right|\right] = \frac{1}{a}\ln\frac{p}{p_0}\left|\frac{a - bp_0}{a - bp}\right|.$$

Thus

$$a(t - t_0) = \ln\frac{p}{p_0}\left|\frac{a - bp_0}{a - bp}\right|. \tag{2}$$

It is a simple matter to show (see Exercise 1) that $(a - bp_0)/(a - bp(t))$ is always positive. Hence

$$a(t - t_0) = \ln\frac{p}{p_0}\frac{a - bp_0}{a - bp}.$$

Taking exponentials of both sides of this equation gives

$$e^{a(t-t_0)} = \frac{p}{p_0}\frac{a - bp_0}{a - bp} \quad \text{or} \quad p_0(a - bp)e^{a(t-t_0)} = (a - bp_0)p.$$

Bringing all terms involving p to the left-hand side of this equation, we see that

$$[a - bp_0 + bp_0 e^{a(t-t_0)}]p(t) = ap_0 e^{a(t-t_0)}.$$

Consequently,

$$p(t) = \frac{ap_0 e^{a(t-t_0)}}{a - bp_0 + bp_0 e^{a(t-t_0)}} = \frac{ap_0}{bp_0 + (a - bp_0)e^{-a(t-t_0)}}. \tag{3}$$

Let us now examine (3) to see what kind of population it predicts. Observe that as $t \to \infty$,

$$p(t) \to \frac{ap_0}{bp_0} = \frac{a}{b}.$$

Thus, *regardless of its initial value, the population always approaches the limiting value a/b*, which is called the carrying capacity of the microcosm. Next, observe that $p(t)$ is a monotonically increasing function of time if $0 < p_0 < a/b$. Moreover, since

$$\frac{d^2p}{dt^2} = a\frac{dp}{dt} - 2bp\frac{dp}{dt} = (a - 2bp)p(a - bp),$$

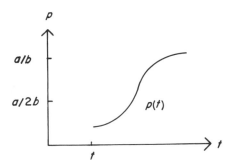

Figure 5.1. Graph of $p(t)$

we see that dp/dt is increasing if $p(t) < a/2b$, and that dp/dt is decreasing if $p(t) > a/2b$. Hence, if $p_0 < a/2b$, the graph of $p(t)$ must have the form given in Figure 5.1. Such a curve is called a logistic, or sigmoidal curve. From its shape we conclude that the time period before the population reaches half its limiting value is a period of accelerated growth. After this point, the rate of growth decreases and in time reaches zero. This is a period of diminishing growth.

These predictions are borne out by an experiment on the protozoa *Paramecium caudatum* performed by the mathematical-biologist G. F. Gause [1]. Five individuals of *Paramecium* were placed in a small test tube containing 0.5 cm^3 of a nutritive medium, and for six days the number of individuals in every tube was counted daily. The Paramecium were found to increase at a rate of 230.9%/day when their numbers were low. The number of individuals increased rapidly at first, and then more slowly, until towards the fourth day it attained a maximum level of 375, saturating the test tube. From this data we conclude that if the *Paramecium caudatum* grow according to the logistic law $dp/dt = ap - bp^2$, then $a = 2.309$ and $b = 2.309/375$. Consequently, the logistic law predicts that

$$
\begin{aligned}
p(t) &= \frac{(2.309)5}{\dfrac{(2.309)5}{375} + \left(2.309 - \dfrac{(2.309)5}{375}\right)e^{-2.309t}} \\
&= \frac{375}{1 + 74e^{-2.309t}}.
\end{aligned} \tag{4}
$$

(We have taken the initial time t_0 to be 0.) Figure 5.2 compares the graph of $p(t)$ predicted by (4) with the actual measurements, which are denoted by ∘. As can be seen, the agreement is remarkably good.

In order to apply our results to predict the future human population of the earth, we must estimate the vital coefficients a and b in the logistic equation governing its growth. Some ecologists have estimated that the natural value of a is 0.029. We also know that the human population was increasing at the rate of 2%/yr when the population was $(3.06)10^9$. Since $(1/p)(dp/dt) = a - bp$, we see that

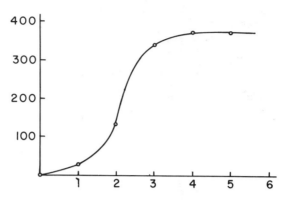

Figure 5.2. Growth of Paramecium

$$0.02 = a - b(3.06)10^9.$$

Consequently, $b = 2.941 \times 10^{-12}$. Thus, according to the logistic law of population growth, the human population of the earth will tend to the limiting value of

$$\frac{a}{b} = \frac{0.029}{2.941 \times 10^{-12}} = 9.86 \text{ billion people.}$$

Note that according to this prediction, we were still on the accelerated growth portion of the logistic curve in 1961, since we had not yet attained half the limiting population predicted for us.

As another verification of the logistic law of population growth, we consider the equation

$$p(t) = \frac{197{,}273{,}000}{1 + e^{-0.03134(t-1913.25)}} \tag{5}$$

which was introduced by Pearl and Reed as a model of the population of the United States [2]. This model was derived in the following manner. First, using the census figures for the years 1790, 1850, and 1910, Pearl and Reed found from (3) that $a = 0.03134$ and $b = (1.5887)10^{-10}$ (see Exercise 2a). Then, to simplify (3), they calculated that the population of the United States achieved half its limiting population of $a/b = 197{,}273{,}000$ in April 1913 (see Exercise 2b). Consequently (see Exercise 2c), we can rewrite (3) in the simpler form (5).

Table 2 compares Pearl and Reed's predictions with the observed values of the population of the United States. These results are remarkable, especially since we have not taken into account the large waves of immigration into the United States and the fact that the United States was involved in five wars during this period.

In 1845 Verhulst prophesied a maximum population for Belgium of 6,600,000, and a maximum population for France of 40,000,000. The

Table 2. Population of U.S. from 1790 to 1950

Year	Actual	Predicted	Error	Percent
1790	3,929,000	3,929,000	0	0.0
1800	5,308,000	5,336,000	28,000	0.5
1810	7,240,000	7,228,000	− 12,000	−0.2
1820	9,638,000	9,757,000	119,000	1.2
1830	12,866,000	13,109,000	243,000	1.9
1840	17,069,000	17,506,000	437,000	2.6
1850	23,192,000	23,192,000	0	0.0
1860	31,443,000	30,412,000	− 1,031,000	− 3.3
1870	38,558,000	39,372,000	814,000	2.1
1880	50,156,000	50,177,000	21,000	0.0
1890	62,948,000	62,769,000	− 179,000	−0.3
1900	75,995,000	76,870,000	875,000	1.2
1910	91,972,000	91,972,000	0	0.0
1920	105,711,000	107,559,000	1,848,000	1.7
1930	122,775,000	123,124,000	349,000	0.3
1940	131,669,000	136,653,000	4,984,000	3.8
1950	150,697,000	149,053,000	−1,644,000	− 1.1

(The last four entries were added by the Dartmouth College Writing Group)

population of Belgium in 1930 was already 8,092,000. This large discrepancy would seem to indicate that the logistic law of population growth is very inaccurate, at least as far as the population of Belgium is concerned. However, this discrepancy can be explained by the astonishing rise of industry in Belgium and by the acquisition of the Congo which secured for the country sufficient additional wealth to support the extra population. Thus, after the acquisition of the Congo, and the astonishing rise of industry in Belgium, Verhulst should have lowered the vital coefficient b.

On the other hand, the population of France in 1930 was in remarkable agreement with Verhulst's forecast. Indeed, we can now answer the following tantalizing paradox: why was the population of France increasing extremely slowly in 1930 while the French population of Canada was increasing very rapidly? After all, they are the same people! The answer to this paradox, of course, is that the population of France in 1930 was very near its limiting value and thus was far into the period of diminishing growth, while the population of Canada in 1930 was still in the period of accelerated growth.

Remarks.

1. Clearly, technological developments, pollution considerations, and sociological trends have significant influence on the vital coefficients a and b. Therefore, they must be reevaluated every few years.

2. To derive more accurate models for population growth, we should not

consider the population as made up of one homogeneous group of individuals. Rather, we should subdivide it into different age groups. We should also subdivide the population into males and females, since the reproduction rate in a population usually depends more on the number of females than on the number of males.

3. Perhaps the severest criticism leveled at the logistic law of population growth is that some populations have been observed to fluctuate periodically between two values, and any type of fluctuation is ruled out in a logistic curve. However, some of these fluctuations can be explained by the fact that when certain populations reach a sufficiently high density, they become susceptible to epidemics. The epidemic brings the population down to a lower value where it again begins to increase, until when it is large enough, the epidemic strikes again. In Exercise 7 we derive a model to describe this phenomenon, and we apply this model in Exercise 8 to explain the sudden appearance and disappearance of hordes of small rodents.

Exercises

1. Prove that $(a - bp_0)/(a - bp(t))$ is positive for $t_0 < t < \infty$. *Hint:* Use (2) to show that $p(t)$ can never equal a/b if $p_0 \neq a/b$.

2. a) Choose three times t_0, t_1, and t_2 with $t_1 - t_0 = t_2 - t_1$. Show that (3) determines a and b uniquely in terms of $t_0, p(t_0), t_1, p(t_1), t_2$, and $p(t_2)$.
   b) Show that the period of accelerated growth for the United States ended in April 1913 (according to Pearl and Reed's model).
   c) Let a population $p(t)$ grow according to the logistic law (3), and let $\bar{t}$ be the time at which half the limiting population is achieved. Show that

   $$p(t) = \frac{a/b}{1 + e^{-a(t-\bar{t})}}.$$

3. In 1879 and 1881 a number of yearling bass were seined in New Jersey, taken across the continent in tanks by train, and planted in San Francisco Bay. A total of only 435 Striped Bass survived the rigors of these two trips. Yet, in 1899, the commercial net catch alone was 1,234,000 lb. Since the growth of this population was so fast, it is reasonable to assume that it obeyed the Malthusian law $dp/dt = ap$. Assuming that the average weight of a bass fish is 3 lb and that in 1899 every tenth bass fish was caught, find a lower bound for a.

4. A population grows according to the logistic law, with a limiting population of 5×10^8 individuals. When the population is low, it doubles every 40 min. What will the population be after 2 hr with each of the following initial values?
   a) 10^8.
   b) 10^9?

5. A family of salmon fish living off the Alaskan Coast obeys the Malthusian law of population growth $dp(t)/dt = 0.003p(t)$, where t is measured in minutes. At time $t = 0$ a group of sharks establishes residence in these waters and begins attacking

the salmon. The rate at which salmon are killed by the sharks is $0.001p^2(t)$, where $p(t)$ is the population of salmon at time t. Moreover, since an undesirable element has moved into their neighborhood, 0.002 salmon/min leave the Alaskan waters.

a) Modify the Malthusian law of population growth to take these two factors into account.

b) Assume that at time $t = 0$ there are one million salmon. Find the population $p(t)$. What happens as $t \to \infty$?

6. The population of New York City would satisfy the logistic law

$$\frac{dp}{dt} = \frac{1}{25}p - \frac{1}{(25)10^6}p^2$$

where t is measured in years, if we neglected the high emigration and homicide rates.

a) Modify this equation to take into account the facts that 6,000 people/yr move from the city, and 4,000 people/yr are murdered.

b) Assume that the population of New York City was 8,000,000 in 1970. Find the population for all future time. What happens as $t \to \infty$?

7. We can model a population which becomes susceptible to epidemics in the following manner. Assume that our population is originally governed by the logistic law

$$\frac{dp}{dt} = ap - bp^2 \qquad\qquad (i)$$

and that an epidemic strikes as soon as p reaches a certain value Q, with Q less than the limiting population a/b. At this stage the vital coefficients become $A < a$, $B < b$, and (i) is replaced by

$$\frac{dp}{dt} = Ap - Bp^2. \qquad\qquad (ii)$$

Suppose that $Q > A/B$. The population then starts decreasing. A point is reached when the population falls below a certain value $q > A/B$. At this moment the epidemic ceases, and the population again begins to grow following (i), until the incidence of a fresh epidemic. In this way periodic fluctuations of p occur between q and Q. We now indicate how to calculate the period T of these fluctuations.

a) Show that the time T_1 taken by the first part of the cycle, when p increases from q to Q is given by

$$T_1 = \frac{1}{a}\ln\frac{Q(a - bq)}{q(a - bQ)}.$$

b) Show that the time T_2 taken by the second part of the cycle, when p decreases from Q to q is given by $T_2 = (1/A)\ln[q(QB - A)/Q(qB - A)]$. Thus the time for the entire cycle is $T_1 + T_2$.

8. It has been observed that plagues appear in mice populations whenever the population becomes too large. Further, a local increase of density attracts predators in large numbers. These two factors will succeed in destroying 97–98% of a population of small rodents in two or three weeks, and the density then falls to a level at which the disease cannot spread. The population, reduced to 2% of its maximum, finds its refuges from the predators sufficient, and its food abundant. The population

therefore begins to grow again until it reaches a level favorable to another wave of disease and predation. The speed of reproduction in mice is so great that we may set $b = 0$ in (i) of Exercise 7. In the second part of the cycle, on the contrary, A is very small in comparison with B, and it may be neglected therefore in (ii).

a) Under these assumptions, show that

$$T_1 = \frac{1}{a}\ln\frac{Q}{q} \quad \text{and} \quad T_2 = \frac{Q - q}{qQB}.$$

b) Assuming that T_1 is approximately 4 yr, and Q/q is approximately 50, show that a is approximately one. This value of a, incidentally, corresponds very well with the rate of multiplication of mice in natural circumstances.

References

[1]   G. F. Gause, *The Struggle for Existence*.   New York: Hafner, 1964.
[2]   R. Pearl and L. J. Reed, *Proc. Nat. Acad. Sci.*, Vol. 6 p. 275, 1920.

Notes for the Instructor

Objectives. The aim of the module is to derive differential equations which model population growth and thus to predict future populations. The predictions of the model are compared with known data.

Prerequisite. Prior acquaintance with separable differential equations.

Time. The module may be covered in one to two lectures.

CHAPTER 6
The Spread of Technological Innovations

Martin Braun*

Economists and sociologists have long been concerned with how a techno-logical change or innovation spreads in an industry. Once an innovation is introduced by one firm, how soon do others in the industry come to adopt it, and what factors determine how rapidly they follow? In this chapter we construct a model of the spread of innovations among farmers, and then show that this same model describes the spread of innovations in such diverse industries as bituminous coal, iron, and steel, brewing, and railroads.

Assume that a new innovation is introduced into a fixed community of N farmers at time $t = 0$. Let $p(t)$ denote the number of farmers who have adopted at time t. As in the previous chapter, we make the approximation that $p(t)$ is a continuous function of time, even though it obviously changes by integer amounts. The simplest realistic assumption that we can make concerning the spread of this innovation is that a farmer adopts the innova-tion only after he has been told of it by a farmer who had already adopted it. Then the number of farmers Δp who adopt the innovation in a small time interval Δt is directly proportional to the number of farmers p who have already adopted, and the number of farmers $N - p$ who are as yet unaware. Hence $\Delta p = cp(N - p)\Delta t$ or $\Delta p/\Delta t = cp(N - p)$ for some posi-tive constant c. Letting $\Delta t \to 0$, we obtain the differential equation

$$\frac{dp}{dt} = cp(N - p). \tag{1}$$

This is the logistic equation of the previous chapter if we set $a = cN$, $b = c$. Assuming that $p(0) = 1$; i.e., one farmer has adopted the innovation at time $t = 0$, we see that $p(t)$ satisfies the initial value problem

* Department of Mathematics, Queens College, Flushing, NY 11367.

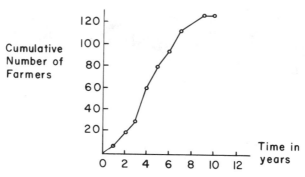

Figure 6.1. Cumulative Number of Farmers Who Adopted 2,4-D Weed Spray in Iowa

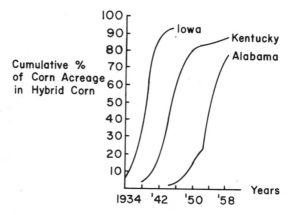

Figure 6.2. Cumulative Percentage of Corn Acreage in Hybrid Corn in Three American States

$$\frac{dp}{dt} = cp(N - p), \qquad p(0) = 1. \tag{2}$$

The solution of (2) is

$$p(t) = \frac{Ne^{cNt}}{N - 1 + e^{cNt}} \tag{3}$$

which is a logistic function.[1] Hence our model predicts that the adoption process accelerates up to that point at which half the community is aware of the innovation. After this point, the adoption process begins to decelerate until it eventually reaches zero.

Let us compare the predictions of our model with data on the spread of two innovations through American farming communities in the middle

---

[1] For an analysis of logistic functions, see Chapter 5, this volume.

1950's. Figure 6.1 represents the cumulative number of farmers in Iowa during 1944–1955 who adopted 2,4-D weed spray, and Figure 6.2 represents the cumulative percentage of corn acreage in hybrid corn in three American states during the years 1934–1958. The circles in these figures are the actual measurements, and the graphs were obtained by connecting these measurements with straight lines. As can be seen, these curves have all the properties of logistic curves and, on the whole, offer very good agreement with our model. However, there are two discrepancies. First, the actual point at which the adoption process ceases to accelerate is not always when 50% of the population has adopted the innovation. As can be seen from Figure 6.2, the adoption process for hybrid corn began to decelerate in Alabama only after nearly 60% of the farmers had adopted the innovation. Secondly, the agreement with our model is much better in the later stages of the adoption process than in the earlier stages.

The source of the second discrepancy is our assumption that a farmer only learns of an innovation through contact with another farmer. This is not entirely true. Studies have shown that mass communication media such as radio, television, newspapers, and farmers' magazines play a large role in the early stages of the adoption process. Therefore, we must add a term to the differential equation (1) to take this into account. To compute this term, we assume that the number of farmers Δp who learn of the innovation through the mass communication media in a short period of time Δt is proportional to the number of farmers who do not yet know; i.e.,

$$\Delta p = c'(N - p)\Delta t$$

for some positive constant c'. Letting $\Delta t \to 0$, we see that $c'(N - p)$ farmers per unit time learn of the innovation through the mass communication media. Thus, if $p(0) = 0$, then $p(t)$ satisfies the initial value problem

$$\frac{dp}{dt} = cp(N - p) + c'(N - p), \qquad p(0) = 0. \tag{4}$$

The solution of (4) is

$$p(t) = \frac{Nc'\left[e^{(c'+cN)t} - 1\right]}{cN + c'\,e^{(c'+cN)t}}, \tag{5}$$

and in Exercises 2 and 3 we indicate how to determine the shape of the curve (5).

The corrected curve (5) now gives remarkably good agreement with Figures 1 and 2 for suitable choices of c and c'. However (see Exercise 3c), it still does not explain why the adoption of hybrid corn in Alabama only began to decelerate after 60% of the farmers had adopted the innovation. This indicates, of course, that other factors, such as the time interval that elapses between when a farmer first learns of an innovation and when he actually adopts it, may play an important role in the adoption process and must be taken into account in any model.

We would now like to show that the differential equation $\frac{dp}{dt} = cp(N - p)$ also governs the rate at which firms in such diverse industries as bituminous coal, iron and steel, brewing, and railroads adopted several major innovations in the first part of this century. This is rather surprising, since we would expect that the number of firms adopting an innovation in one of these industries certainly depends on the profitability of the innovation and the investment required to implement it, and we have not mentioned these factors in deriving (1). However, as we shall see shortly, these two factors are incorporated in the constant c.

Let n be the total number of firms in a particular industry which have adopted an innovation at time t. Clearly, the number of firms Δp adopting the innovation in a short time interval Δt is proportional to the number of firms $n - p$ which have not yet adopted; i.e., $\Delta p = \lambda(n - p)\Delta t$. Letting $\Delta t \to 0$, we see that

$$\frac{dp}{dt} = \lambda(n - p).$$

The proportionality factor λ depends on the profitability π of installing this innovation relative to that of alternative investments, the investment s required to install this innovation as a percentage of the total assets of the firm, and the percentage of firms who have already adopted. Thus

$$\lambda = f(\pi, s, p/n).$$

Expanding f in a Taylor series and dropping terms of order three or more gives

$$\lambda = a_1 + a_2\pi + a_3 s + a_4\frac{p}{n} + a_5\pi^2 + a_6 s^2 + a_7\pi s + a_8\pi\left(\frac{p}{n}\right)$$
$$+ a_9 s\left(\frac{p}{n}\right) + a_{10}\left(\frac{p}{n}\right)^2.$$

In the late 1950's, Edwin Mansfield of Carnegie-Mellon University investigated the spread of 12 innovations in four major industries [1]. From his exhaustive studies, Mansfield concluded that $a_{10} = 0$ and

$$a_1 + a_2\pi + a_3 s + a_5\pi^2 + a_6 s^2 + a_7\pi s = 0.$$

Thus, setting

$$k = a_4 + a_8\pi + a_9 s, \tag{6}$$

we see that

$$\frac{dp}{dt} = k\frac{p}{n}(n - p).$$

(This is the equation obtained previously for the spread of innovations among farmers if we set $k/n = c$.) We assume that the innovation is first

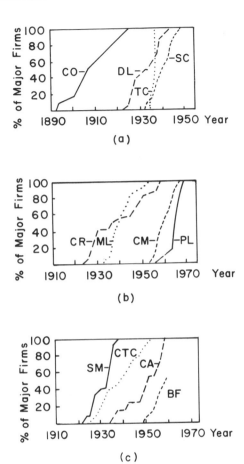

Figure 6.3. Growth in Percentage of Major Firms that Introduced 12 Innovations; Bituminous Coal, Iron and Steel, Brewing, and Railroad Industries, 1890–1958. (a) By-product coke oven (CO), diesel locomotive (DL), tin container (TC), and shuttle car (SC). (b) Car retarder (CR), trackless mobile loader (ML), continuous mining maching (CM), and pallet-loading machine (PL). (c) Continuous wide-strip mill (SM), centralized traffic control (CTC), continuous annealing (CA), and highspeed bottle filler (BF).

Table 1

Innovation	n	t_G	a_4	a_8	a_9	π	s
Diesel locomotive	25	1925	−0.59	0.530	−0.027	1.59	0.015
Centralized traffic control	24	1926	−0.59	0.530	−0.027	1.48	0.024
Car retarders	25	1924	−0.59	0.530	−0.027	1.25	0.785
Continuous wide strip mill	12	1924	−0.52	0.530	−0.027	1.87	4.908
By-product coke oven	12	1894	−0.52	0.530	−0.027	1.47	2.083
Continuous annealing	9	1936	−0.52	0.530	−0.027	1.25	0.554
Shuttle car	15	1937	−0.57	0.530	−0.027	1.74	0.013
Trackless mobile loader	15	1934	−0.57	0.530	−0.027	1.65	0.019
Continuous mining machine	17	1947	−0.57	0.530	−0.027	2.00	0.301
Tin container	22	1935	−0.29	0.530	−0.027	5.07	0.267
High speed bottle filler	16	1951	−0.29	0.530	−0.027	1.20	0.575
Pallet-loading machine	19	1948	−0.29	0.530	−0.027	1.67	0.115

adopted by one firm in the year t_0. Then $p(t)$ satisfies the initial value problem

$$\frac{dp}{dt} = \frac{k}{n}p(n - p), \qquad p(t_0) = 1 \tag{7}$$

and this implies that

$$p(t) = \frac{n}{1 + (n - 1)e^{-k(t-t_0)}}.$$

Mansfield studied how rapidly the use of 12 innovations spread from enterprise to enterprise in four major industries—bituminous coal, iron and steel, brewing, and railroads. The innovations are the shuttle car, trackless mobile loader, and continuous mining machine (in bituminous coal); the by-product coke oven, continuous wide strip mill, and continuous annealing line for tin plate (in iron and steel); the pallet-loading machine, tin container, and high-speed bottle filler (in brewing); and the diesel locomotive, centralized traffic control, and car retarders (in railroads). His results are described graphically in Figure 6.3. For all but the by-product coke oven and tin container, the percentages given are for every two years from the year of initial introduction. The length of the interval for the by-product coke oven is about six years, and for the tin container, it is six months. Notice how all these curves have the general appearance of a logistic curve.

For a more detailed comparison of the predictions of our model (7) with these observed results, we must evaluate the constants n, k and t_0 for each of the twelve innovations. Table 1 gives the value of n, t_0, a_4, a_5, a_9, π, and s for each of the 12 innovations; the constant k can then be computed from (6). As the answers to Exercises 5 and 6 will indicate, (7) predicts the rate of adoption of these 12 innovations with reasonable accuracy.

Exercises

1. Solve the initial value problem (2).

2. Let $c = 0$ in (5). Show that $p(t)$ increases monotonically from 0 to N and has no points of inflection.

3. Here is a heuristic argument to determine the behavior of the curve (5). If $c' = 0$, then we have a logistic curve, and if $c = 0$, then we have the behavior described in Exercise 2. Thus, if c is large relative to c', then we have a logistic curve, and if c is small relative to c', then we have the behavior illustrated in Exercise 2.
   a) Let $p(t)$ satisfy (4). Show that

$$\frac{d^2 p}{dt^2} = (N - p)(cp + c')(cN - 2cp - c').$$

   b) Show that $p(t)$ has a point of inflection, at which dp/dt achieves a maximum if and only if $c'/c < N$.
   c) Assume that $p(t)$ has a point of inflection at $t = t^*$. Show that $p(t^*) \leq N/2$.

4. Solve the initial value problem (7).

5. It seems reasonable to take the time span between the date when 20% of the firms had introduced the innovation and the date when 80% of the firms had introduced the innovation, as the rate of imitation.
   a) Show from our model that this time span is $4(\ln 2)/k$.
   b) For each of the 12 innovations, compute this time span from the data in Table 1 and compare with the observed value in Figure 6.3.

6. a) Show from our model that $(1/k) \ln (n - 1)$ years elapse before 50% of the firms introduce an innovation.
   b) Compute this time span for each of the 12 innovations and compare with the observed values in Figure 6.3.

Reference

[1]   E. Mansfield, "Technical Change and the Rate of Imitation," *Econometrica*, vol. 29, no. 4, Oct. 1961.

Notes for the Instructor

Objectives. The logistic model is applied to the spread of innovation in diverse industries. The predictions of the model are compared with actual data.

Prerequisites. Separable differential equations.

Time. One or two lectures are enough to cover the material.

PART III

HIGHER ORDER LINEAR MODELS

CHAPTER 7
A Model for the Detection of Diabetes

Martin Braun*

Diabetes mellitus is a disease of metabolism which is characterized by too much sugar in the blood and urine. In diabetes, the body is unable to burn off all its sugars, starches, and carbohydrates because of an insufficient supply of insulin. Diabetes is usually diagnosed by means of a glucose tolerance test (GTT). In this test the patient comes to the hospital after an overnight fast and is given a large dose of glucose (sugar in the form in which it usually appears in the bloodstream). During the next three to five hours, several measurements are made of the concentration of glucose in the patient's blood, and these measurements are used in the diagnosis of diabetes. A very serious difficulty associated with this method of diagnosis is that no universally accepted criterion exists for interpreting the results of a glucose tolerance test. Three physicians interpreting the results of a GTT may come up with three different diagnoses. In one case recently, a Rhode Island physician, after reviewing the results of a GTT, came up with a diagnosis of diabetes. A second physician declared the patient to be normal. To settle the question, the results of the GTT were sent to a specialist in Boston. After examining these results, the specialist concluded that the patient was suffering from a pituitary tumor.

In the mid-1960's Drs. Rosevear and Molnar of the Mayo Clinic and Drs. Ackerman and Gatewood of the University of Minnesota discovered a fairly reliable criterion for interpreting the results of a glucose tolerance test [1]. Their discovery arose from a very simple model they developed for the blood glucose regulatory system. Their model is based on the following simple and fairly well-known facts of elementary biology.

1. Glucose plays an important role in the metabolism of any vertebrate

* Department of Mathematics, Queens College, Flushing, NY 11367.

since it is a source of energy for all tissues and organs. For each individual there is an optimal blood glucose concentration, and any excessive deviation from this optimal concentration leads to severe pathological conditions and potentially death.

2. While blood glucose levels tend to be autoregulatory, they are also influenced and controlled by a wide variety of hormones and other metabolites. Among these are the following.

(i) *Insulin*, a hormone secreted by the β cells of the pancreas. After we eat any carbohydrates, our gastrointestinal tract sends a signal to the pancreas to secrete more insulin. In addition, the glucose in our blood directly stimulates the β cells of the pancreas to secrete insulin. It is generally believed that insulin facilitates tissue uptake of glucose by attaching itself to the impermeable membrane walls, thus allowing glucose to pass through the membranes to the center of the cells, where most of the biological and chemical activity takes place. Without sufficient insulin, the body cannot avail itself of all the energy it needs.

(ii) *Glucagon*, a hormone secreted by the α cells of the pancreas. Any excess glucose is stored in the liver in the form of glycogen. In times of need this glycogen is converted back into glucose. The hormone glucagon increases the rate of breakdown of glycogen into glucose. Evidence collected thus far clearly indicates that hypoglycemia (low blood sugar) and fasting promote the secretion of glucagon while increased blood glucose levels suppress its secretion.

(iii) *Epinephrine* (adrenalin), a hormone secreted by the adrenal medulla. Epinephrine is part of an emergency mechanism to quickly increase the concentration of glucose in the blood in times of extreme hypoglycemia. Like glucagon, epinephrine increases the rate of breakdown of glycogen into glucose. In addition, it directly inhibits glucose uptake by muscle tissue; it acts directly on the pancreas to inhibit insulin secretion; and it aids in the conversion of lactate to glucose in the liver.

(iv) *Glucocorticoids*, hormones such as cortisol which are secreted by the adrenal cortex. Glucocorticoids play an important role in the metabolism of carbohydrates.

(v) *Thyroxin*, a hormone secreted by the thyroid gland. This hormone aids the liver in forming glucose from noncarbohydrate sources such as glycerol, lactate, and amino acids.

(vi) *Growth hormone* (somatotropin), a hormone secreted by the anterior pituitary gland. Not only does growth hormone affect glucose levels in a direct manner, but it also tends to "block" insulin. It is believed that growth hormone decreases the sensitivity of muscle and adipose membrane to insulin, thereby reducing the effectiveness of insulin in promoting glucose uptake.

The aim of Ackerman *et al* was to construct a model which would accurately describe the blood glucose regulatory system during a glucose tolerance test and in which one or two parameters would yield criteria for distinguishing normal individuals from mild diabetics and prediabetics. Their model is a very simplified one, requiring only a limited number of blood samples during a GTT. It centers attention on two concentrations, that of glucose in the blood, labeled G, and that of the net hormonal concentration, labeled H. The latter is interpreted to represent the cumulative effect of all the pertinent hormones. Those hormones such as insulin which decrease blood glucose concentrations are considered to increase H, while those hormones such as cortisol which increase blood glucose concentrations are considered to decrease H. Now there are two reasons why such a simplified model can still provide an accurate description of the blood glucose regulatory system. First, studies have shown that under normal—or close to normal—conditions, the interaction of one hormone, namely insulin, with blood glucose so predominates that a simple "lumped parameter model" is quite adequate. Second, evidence indicates that normoglycemia does not necessarily depend on the normalcy of each kinetic mechanism of the blood glucose regulatory system. Rather, it depends on the overall performance of the blood glucose regulatory system, and this system is dominated by insulin-glucose interactions.

The basic model is described analytically by the equations

$$\frac{dG}{dt} = F_1(G, H) + J(t) \tag{1}$$

$$\frac{dH}{dt} = F_2(G, H). \tag{2}$$

The dependence of F_1 and F_2 on G and H signify that changes in G and H are determined by the values of both G and H. The function $J(t)$ is the external rate at which the blood glucose concentration is being increased. We assume that G and H have achieved optimal values G_0 and H_0 by the time the fasting patient has arrived at the hospital. This implies that $F_1(G_0, H_0) = 0$ and $F_2(G_0, H_0) = 0$. Since we are interested here in the deviations of G and H from their optimal values, we make the substitution

$$g = G - G_0 \qquad h = H - H_0.$$

Then

$$\frac{dg}{dt} = F_1(G_0 + g, H_0 + h) + J(t)$$

$$\frac{dh}{dt} = F_2(G_0 + g, H_0 + h).$$

Now, observe that

$$F_1(G_0 + g, H_0 + h) = F_1(G_0, H_0) + \frac{\partial F_1(G_0, H_0)}{\partial G} g + \frac{\partial F_1(G_0, H_0)}{\partial H} h + e_1$$

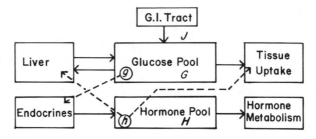

Figure 7.1. Simplified Model of Blood Glucose Regulatory System

and

$$F_2(G_0 + g, H_0 + h) = F_2(G_0, H_0) + \frac{\partial F_2(G_0, H_0)}{\partial G} g + \frac{\partial F_2(G_0, H_0)}{\partial H} h + e_2$$

where e_1 and e_2 are very small compared to g and h. Hence, assuming that G and H deviate only slightly from G_0 and H_0, and therefore neglecting the terms e_1 and e_2, we see that

$$\frac{dg}{dt} = \frac{\partial F_1(G_0, H_0)}{\partial G} g + \frac{\partial F_1(G_0, H_0)}{\partial H} h + J(t) \qquad (3)$$

$$\frac{dh}{dt} = \frac{\partial F_2(G_0, H_0)}{\partial G} g + \frac{\partial F_2(G_0, H_0)}{\partial H} h. \qquad (4)$$

No means exist of determining *a priori* the numbers $\partial F_1(G_0, H_0)/\partial G$, $\partial F_1(G_0, H_0)/\partial H$, $\partial F_2(G_0, H_0)/\partial G$, and $\partial F_2(G_0, H_0)/\partial H$. However, we can determine their signs. Referring to Figure 7.1, we see that dg/dt is negative for $g > 0$ and $h = 0$, since the blood glucose concentration will be decreasing through tissue uptake of glucose and the storing of excess glucose in the liver in the form of glycogen. Consequently, $\partial F_1(G_0, H_0)/\partial G$ must be negative. Similarly, $\partial F_1(G_0, H_0)/\partial H$ is negative since a positive value of h tends to decrease blood glucose levels by facilitating tissue uptake of glucose and by increasing the rate at which glucose is converted to glycogen. The number $\partial F_2(G_0, H_0)/\partial G$ must be positive, since a positive value of g causes the endocrine glands to secrete those hormones which tend to increase H. Finally, $\partial F_2(G_0, H_0)/\partial H$ must be negative, since the concentration of hormones in the blood decreases through hormone metabolism.

Thus we can write (3) and (4) in the form

$$\frac{dg}{dt} = -m_1 g - m_2 h + J(t) \qquad (5)$$

$$\frac{dh}{dt} = -m_3 h + m_4 g \qquad (6)$$

where m_1, m_2, m_3, and m_4 are positive constants. Equations (5) and (6) are two first-order equations for g and h. However, since we only measure the concentration of glucose in the blood, we would like to remove the variable h.

This can be accomplished as follows. Differentiating (5) with respect to t gives

$$\frac{d^2g}{dt^2} = -m_1\frac{dg}{dt} - m_2\frac{dh}{dt} + \frac{dJ}{dt}.$$

Substituting for dh/dt from (6) we obtain that

$$\frac{d^2g}{dt^2} = -m_1\frac{dg}{dt} + m_2m_3h - m_2m_4g + \frac{dJ}{dt}. \tag{7}$$

Next, observe from (5) that $m_2h = -(dg/dt) - m_1g + J(t)$. Consequently, $g(t)$ satisfies the second-order linear differential equation

$$\frac{d^2g}{dt^2} + (m_1 + m_3)\frac{dg}{dt} + (m_1m_3 + m_2m_4)g = m_3J + \frac{dJ}{dt}.$$

We rewrite this equation in the form

$$\frac{d^2g}{dt^2} + 2\alpha\frac{dg}{dt} + \omega_0^2g = S(t) \tag{8}$$

where $\alpha = (m_1 + m_3)/2$, $\omega_0^2 = m_1m_3 + m_2m_4$, and $S(t) = m_3J + (dJ/dt)$.

Notice that the right side of (8) is identically zero except for the very short time interval in which the glucose load is being ingested. Such functions can be dealt with very effectively by introducing the Dirac delta function. For our purposes here, let $t = 0$ be the time at which the glucose load has been completely ingested. Then, for $t \geq 0$, $g(t)$ satisfies the second-order linear homogeneous equation

$$\frac{d^2g}{dt^2} + 2\alpha\frac{dg}{dt} + \omega_0^2g = 0. \tag{9}$$

This equation has positive coefficients. Hence (see Exercise 5) $g(t)$ approaches zero as t approaches infinity. Thus our model certainly conforms to reality in predicting that the blood glucose concentration tends to return eventually to its optimal concentration.

The solutions $g(t)$ of (9) are of three different types, depending as to whether $\alpha^2 - \omega_0^2$ is positive, negative, or zero. We will assume that $\alpha^2 - \omega_0^2$ is negative; the other two cases are treated in a similar manner. If $\alpha^2 - \omega_0^2 < 0$, then the characteristic equation of (9) has complex roots. It is easily verified in this case (see Exercise 1) that every solution $g(t)$ of (9) is of the form

$$g(t) = A\,e^{-\alpha t}\cos(\omega t - \delta), \qquad \omega^2 = \omega_0^2 - \alpha^2. \tag{10}$$

Consequently,

$$G(t) = G_0 + A\,e^{-\alpha t}\cos(\omega t - \delta). \tag{11}$$

Now (11) contains five unknowns: G_0, A, α, ω_0, and δ. One way of determining them is as follows. The patient's blood glucose concentration before the glucose load is ingested is G_0. Hence we can determine G_0 by measuring the patient's blood glucose concentration immediately upon his arrival at the hospital. Next, if we take four additional measurements G_1, G_2, G_3, and G_4 of the patient's blood glucose concentration at times t_1, t_2, t_3, and t_4, then we can determine A, α, ω_0, and δ from the four equations

$$G_j = G_0 + A e^{-\alpha t_j} \cos(\omega t_j - \delta), \qquad j = 1, 2, 3, 4.$$

A second and better method of determining G_0, A, α, ω_0, and δ is to take n measurements $G_1, G_2, \cdots, G_n$ of the patient's blood glucose concentration at times $t_1, t_2, \cdots, t_n$. Typically, n is 6 or 7. We then find optimal values for G_0, A, α, ω_0, and δ such that the least-square error

$$E = \sum_{j=1}^{n} [G_j - G_0 - A e^{-\alpha t_j} \cos(\omega t_j - \delta)]^2$$

is minimized. The problem of minimizing E can be solved on a digital computer, and Ackerman et al. [1] provide a complete Fortran program for determining optimal values for G_0, A, α, ω_0, and δ. This method is preferable to the first method since (11) is only an approximate formula for $G(t)$. Consequently, it is possible to find values G_0, A, δ, ω_0, and δ so that (11) is satisfied exactly at four points t_1, t_2, t_3, and t_4 but yields a poor fit to the data at other times. The second method usually offers a better fit to the data on the entire time interval since it involves more measurements.

In numerous experiments, Ackerman et al. observed that a slight error in measuring G could produce a very large error in the value of α. Hence any criterion for diagnosing diabetes that involves the parameter α is unreliable. However, the parameter ω_0, the natural frequency of the system, was relatively insensitive to experimental error in measuring G. Thus we may regard a value of ω_0 as the basic descriptor of the response to a glucose tolerance test. For discussion purposes, using the corresponding natural period $T_0 = 2\pi/\omega_0$ is more convenient. The remarkable fact is that data from a variety of sources indicated that *a value of less than four hours for T_0* indicated normalcy, while appreciably more than four hours implied mild diabetes.

Remarks.

1. The usual period between meals in our culture is about four hours. This suggests the interesting possibility that sociological factors may also play a role in the blood glucose regulatory system.

2. We wish to emphasize that the model described above can only be used to diagnose mild diabetes or prediabetes, since we have assumed throughout that the deviation g of G from its optimal value G_0 is small. Very large deviations of G from G_0 usually indicate severe diabetes or diabetes insipidus, which is a disorder of the posterior lobe of the pituitary gland.

A serious shortcoming of this simplified model is that it sometimes yields a poor fit to the data in the time period three to five hours after ingestion of the glucose load. This indicates, of course, that variables such as epinephrine and glucagon play an important role in this time period. Thus these variables

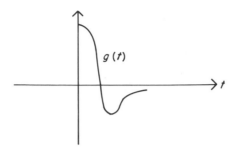

Figure 7.2. Graph of $g(t)$ if $\alpha^2 - \omega^2 > 0$

should be included as separate variables in our model, rather than being lumped together with insulin. In fact, evidence indicates that levels of epinephrine may rise dramatically during the recovery phase of the GTT response, when glucose levels have been lowered below fasting levels. This can also be seen directly from (9). If $\alpha^2 - \omega_0^2 > 0$, then $g(t)$ may have the form described in Figure 7.2. Note that $g(t)$ drops very rapidly from a fairly high value to a negative one. Quite conceivably, therefore, the body will interpret this as an extreme emergency and thereby secrete a large amount of epinephrine.

Medical researchers have long recognized the need of including epinephrine as a separate variable in any model of the blood glucose regulatory system. However, they were stymied by the fact that the concentration of epinephrine in the blood could not be reliably measured. Thus they had to assume, for all practical purposes, that the level of epinephrine remained constant during the course of a glucose tolerance test. This author has just been informed that researchers at Rhode Island Hospital have devised an accurate method of measuring the concentration of epinephrine in the blood. Thus we will be able to develop and test more accurate models of the blood glucose regulatory system. Hopefully, this will lead to more reliable criteria for the diagnosis of diabetes.

Exercises

1. Derive (10).

2. A patient arrives at the hospital after an overnight fast with a blood glucose concentration of 70 mg glucose/100 mL blood (milligrams per 100 milliliters). His blood glucose concentration 1 h, 2 h, and 3 h after fully absorbing a large amount of glucose is 95, 65, and 75 mg glucose/100 mL blood, respectively. Show that this patient is normal. *Hint:* Show that the time interval between two successive zeros of $G - G_0$ is one-half the natural period.

   According to a famous diabetologist, the blood glucose concentration of a non-diabetic who has just absorbed a large amount of glucose will be at or below the fasting level in 2 h, or less. Exercises 3 and 4 compare the diagnoses of this diabetologist with those of Ackerman *et al.*

3. The deviation $g(t)$ of a patient's blood glucose concentration from its optimal concentration satisfies the differential equation

$$\frac{d^2g}{dt^2} + 2\alpha\frac{dg}{dt} + \alpha^2g = 0$$

immediately after he fully absorbs a large amount of glucose. The time t is measured in minutes, so that the units of α are reciprocal minutes. Show that this patient is a diabetic if we use the criterion of the famous diabetologist. On the other hand, observe that this patient is normal according to Ackerman *et al.* if $\alpha > \pi/120(\text{min})$.

4. A patient's blood glucose concentration $G(t)$ satisfies the initial value problem

$$\frac{d^2G}{dt^2} + \frac{1}{20(\text{min})}\frac{dG}{dt} + \frac{1}{2500(\text{min})^2}G = \frac{1}{2500(\text{min})^2}75 \text{ mg glucose}/100 \text{ mL blood}$$

$$G(0) = 150 \text{ mg glucose}/100 \text{ mL blood,}$$

$$G'(0) = -\frac{e^{\sqrt{3}} + e^{-\sqrt{3}}}{e^{\sqrt{3}} - e^{-\sqrt{3}}}G(0)/(\text{min})$$

immediately after he fully absorbs a large amount of glucose. This patient's optimal blood glucose concentration is 75 mg glucose/100 mL blood. Show that this patient is a diabetic according to Ackerman *et al.* but is normal according to the famous diabetologist.

5. Consider the equation $ag'' + bg' + cg = 0$, with a, b, and c positive. Show that every solution of this equation approaches zero as t approaches infinity.

Reference

[1]   E. Ackerman, L. Gatewood, J. Rosevear, and G. Molnar, "Blood glucose regulation and diabetes," in *Concepts and Models of Biomathematics*, F. Heinmets, Ed. Marcel Dekker, 1969, ch. 4, pp. 131–156.

Notes for the Instructor

Objectives. A mathematical model is developed for the interaction of glucose and hormones, including insulin, in the blood. A simple criterion is developed for interpreting the results of a glucose tolerance test.

Prerequisites. Linear second-order homogeneous differential equations.

Time. The material can be covered in one or two lectures.

CHAPTER 8

Combat Models

Courtney S. Coleman*

Satietie of sleepe and love, satietie of ease,
Of musicke, can find place, yet harsh warre still must please
Past all these pleasures, even past these.[1]

1. Introduction

During the first World War, F. W. Lanchester outlined several tentative
mathematical models of the fledgling art of air warfare [9], [10]. These
models have since been extended to represent a variety of competitions
ranging from isolated battles to entire wars. We shall outline and "solve"
some simple models, commenting on a mixed conventional–guerrilla combat
such as Vietnam and studying in some depth the battle of Iwo Jima during
World War II.

2. Three Lanchester Combat Models

An x force and a y force are engaged in combat. Let $x(t)$ and $y(t)$ denote the
respective strengths of the forces at time t, where t is measured in days from
the start of the combat. It is not easy to quantify "strength," including as it

* Department of Mathematics, Harvey Mudd College, Claremont, CA 91711.

[1] From the translation of Homer's *Iliad* by George Chapman [6]. In contemporary English
rather than Chapman's poetic Elizabethan, the lines might be read as "*Men grow tired of sleep,
love, singing, and dancing sooner than of war.*" See [10].

does the numbers of combatants, their battle readiness, the nature and number of the weapons, the quality of the leadership, and a host of psychological and other intangible factors difficult even to describe, much less to turn into numbers. We shall take the easy way out and identify the strengths $x(t)$ and $y(t)$ with the numbers of combatants. (See Howes and Thrall [8] for another approach to the quantification of combat strength.)

We shall assume that $x(t)$ and $y(t)$ vary continuously and even differentiably as functions of time. This is, of course, an idealization of the true state of affairs, since the strengths must be integers and change only by integer amounts as time goes on. However, one might argue that when a strength is large an increase by one or two is infinitesimal compared to the total, and we might as well allow the strength to change by an arbitrarily small amount, not just integral, over small time spans. For example, if the American strength on Iwo Jima at 8:00 a.m. on the 18th day of the battle was 70,000 and at 9:00 a.m. it was 69,995, it seems reasonable to set the strength equal to 69,999.5833 $\cdots$ at 8:05 a.m. Once we have made the idealization to continuous functions, we might as well go all the way and "smooth" off any "corners" on the graphs of $x(t)$ and $y(t)$ versus t; thus we can reasonably take $x(t)$ and $y(t)$ to be continuous and differentiable.

Although we may not have a specific formula for $x(t)$, say, as a function of t, we may have a good deal of information about the *operational loss rate* (OLR) of the x force (i.e., the loss rate due to the inevitable diseases, desertions, and other noncombat mishaps), the *combat loss rate* (CLR) due to encounters with the y force, and the *reinforcement* (or *supply*) *rate* (RR). We shall assume that the net rate of change in $x(t)$ is given by

$$\frac{dx(t)}{dt} = -(\text{OLR} + \text{CLR}) + \text{RR}. \tag{1}$$

A similar equation applies to the y force. The problem is to find appropriate formulas for these rates for each force and then to analyze the solutions $x(t)$ and $y(t)$ of the respective differential equations to determine who "wins" the combat.

In the analysis which follows, we shall use a number of symbols which are listed here for convenience:

a, b, c, d, g, h	Nonnegative loss rate constants.
$P(t), Q(t)$	Reinforcement rates in numbers of combatants per day.
$x(t), y(t)$	Strengths of the opposing forces at time t.
x_0, y_0	Strengths at the start of combat.
t	Time in days from the start of combat.

We shall write out three Lanchestrian combat models using these symbols and then discuss the meaning of the models.

Conventional combat:

$$(\text{CONCOM})^2 \quad \begin{cases} \dfrac{dx(t)}{dt} = -ax(t) - by(t) + P(t) \\[2mm] \dfrac{dy(t)}{dt} = -cx(t) - dy(t) + Q(t). \end{cases}$$

Guerrilla combat:

$$(\text{GUERCOM}) \quad \begin{cases} \dfrac{dx(t)}{dt} = -ax(t) - gx(t)y(t) + P(t) \\[2mm] \dfrac{dy(t)}{dt} = -dy(t) - hx(t)y(t) + Q(t). \end{cases}$$

Mixed conventional–guerrilla combat:

$$(\text{VIETNAM}) \quad \begin{cases} \dfrac{dx(t)}{dt} = -ax(t) - gx(t)y(t) + P(t) \\[2mm] \dfrac{dy(t)}{dt} = -cx(t) - dy(t) + Q(t). \end{cases}$$

Each equation relates the time rate of change of the strength of a force to other rate terms and has the form of (1), as we shall now show. The reinforcement rates $P(t)$ and $Q(t)$ are almost self-evident. For an isolated force such a term would be zero. Ordinarily, one would expect a reinforcement rate to be nonnegative, although combatants could be withdrawn from battle, producing a negative reinforcement rate. The *operational loss rates* $-ax(t)$ and $-dy(t)$ yield *constant relative loss rates* (in the absence of combat and reinforcement):

$$\frac{dx/dt}{x} = -a \qquad \frac{dy/dt}{y} = -d. \tag{2}$$

Other loss rates are possible but these give a reasonably good match to most situations.

If the only terms appearing in the Lanchestrian models are reinforcement and operational loss rates, then no combat occurs, since neither side has any effect whatsoever on the other. It is the *interaction terms* $-by(t)$, $-cx(t)$, $-gx(t)y(t)$, and $-hx(t)y(t)$ which introduce actual combat into the models. Now, a conventional force operates in the open (comparatively speaking), and Lanchester assumes that every member of such a force is within kill range of the enemy. It is also assumed that, as soon as the conventional force

---

[2] The author dislikes acronyms such as CONCOM. Nevertheless, the military is the most prolific source of these nonwords, and it is fitting to use them in this article.

suffers a loss, fire is concentrated on the remaining combatants. Under these conditions, Lanchester proposes for a conventional x force a combat loss rate of the form $-by(t)$, where b is the *combat effectiveness coefficient* of the y force. Observe that the combat loss rate of the x force per individual combatant in the y force is given by

$$\frac{dx/dt}{y} = -b. \tag{3}$$

Thus b is a measure of the average effectiveness in combat of each member of the y-force. A similar explanation can be given for the term $-cx(t)$.

It is not a simple matter to calculate the combat effectiveness coefficients b and c. One way is to set

$$b = r_y p_y, \qquad c = r_x p_x, \tag{4}$$

where r_y and r_x are the respective *firing rates* (shots/combatant/day) of the y and the x force and p_y and p_x are the respective probabilities that a single shot kills an opponent. Observe that, according to (4), the y force determines b, the x force determines c. Sometimes an *ex post facto* battle analysis will reveal the values of b and c (see, e.g., the Iwo Jima model in Section 6).

These loss rates for conventional forces are linear; according to Lanchester, the combat loss rates of guerrilla forces are nonlinear. The argument is as follows. Let $x(t)$ denote the strength of a guerrilla force invisible to the opponent and occupying a fixed region R. The opponent fires into R but cannot know when a kill has been made. It seems plausible that the combat loss rate for the guerrilla force should then be proportional to its own numbers $x(t)$ in R; for the larger $x(t)$, the greater the probability that an opponents' shot will kill. The combat loss rate of the guerrilla x force is also proportional to the strength $y(t)$ of the opponents. Thus we have that the combat loss rate for the guerrillas is

$$-gx(t)y(t).$$

The *combat effectiveness coefficient* g for the opponent y of a guerrilla x force is somewhat more complicated to estimate than the coefficient b given by (4). We can still use the firing rate r_y, but p_y is no longer solely determined by the y force. Lanchester argues that, instead, the probability of a kill of a member of the guerrilla x force is directly proportional to the *area of effectiveness* A_{ry} of a single y shot and inversely proportional to the area A_x of the region R occupied by guerillas. The number A_{ry} is the area of the exposed part of the body of a single guerrilla combatant under cover. Thus plausible formulas for g and h are

$$g = r_y \frac{A_{ry}}{A_x} \quad \text{and} \quad h = r_x \frac{A_{rx}}{A_y}. \tag{5}$$

The crucial difference here is that the guerrilla force has some degree of control over its own combat loss constants, while this is not the case with a conventional force. Presumably, the guerrilla force will try to occupy as extensive a region as possible during combat, subject, of course, to the difficulties of having forces spread too thinly.

The three models, CONCOM, GUERCOM, and VIETNAM, are based on various combinations of these Lanchestrian supply and loss rates. *The reader should now look back at the models, see just how the various terms fit together, and justify the titles given* (Exercise 1). Note that the term "VIET-NAM" is not strictly accurate for a model of the quarter-century conflict in that country, since various forces used various modes of combat at various times. Nevertheless, the term and the model have some degree of significance, as we shall see below.

To test the validity of these models, data are needed from various battles. First, however, let us "solve" simplified (indeed, oversimplified) versions of CONCOM, GUERCOM, and VIETNAM.

3. Conventional Combat: The Square Law

Suppose a pair of isolated conventional forces are fighting in an idealized setting for which operational losses are nil. With no reinforcements and no operational losses on either side, CONCOM reduces to the simple linear system

$$\frac{dx}{dt} = -by \qquad (6a)$$

$$\frac{dy}{dt} = -cx. \qquad (6b)$$

Dividing (6a) by (6b) we have that

$$\frac{dy}{dx} = \frac{cx}{by}. \qquad (7)$$

Separating the variables in (7) and integrating, we obtain

$$b \int_{y_0}^{y(t)} y \, dy = c \int_{x_0}^{x(t)} x \, dx$$

$$b(y^2(t) - y_0^2) = c(x^2(t) - x_0^2). \qquad (8)$$

This quadratic relation between the opposing forces explains why the *linear* system (6) has acquired the anomalous name of the *square law model*.

Let K be the constant $by_0^2 - cx_0^2$. The graph of the equation

$$by^2 - cx^2 = K \qquad (9)$$

obtained from (8) is a hyperbola (a pair of straight lines if $K = 0$), and we shall call (9) the *hyperbolic law*. These hyperbolas are plotted in Figure 8.1

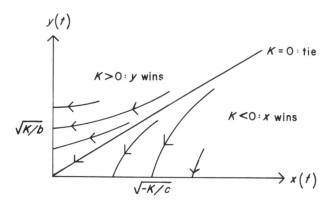

Figure 8.1. Hyperbolas of Square Law

for various values of K. We only look at the curves in the *strength quadrant*, $x \geqq 0$, $y \geqq 0$, for obvious reasons. The arrowheads on the curves indicate the direction of changing strengths as time passes. Since $dx/dt < 0$ and $dy/dt < 0$ whenever $x(t) > 0$ and $y(t) > 0$, the directions of the arrowheads are as indicated.

Who "wins" in such a combat? Let us say that one force wins if the other force vanishes first. For example, y wins if $K > 0$ since, according to (9), y never vanishes in this case, while the x force has been annihilated by the time $y(t)$ has decreased to $\sqrt{K/b}$. Thus the y force seeks to establish a combat setting in which $K > 0$, i.e., the y force wants the following inequality to hold,

$$by_0^2 > cx_0^2. \tag{10}$$

From (4) we see that (10) can be written as

$$\left(\frac{y_0}{x_0}\right)^2 > \frac{r_x}{r_y} \cdot \frac{p_x}{p_y}, \tag{11}$$

the condition for the predominance of the y force. Assuming that both forces are well-trained and in good combat condition, it is hard to see how either force can do much about the right side of (11). The square on the left side of (11) shows that changes in the *initial force ratio* y_0/x_0 are magnified quadratically. Clearly, the aim of the y force will be to increase the force ratio, while the opponents try to decrease the ratio. This is obvious, but what may not be obvious is the squaring effect indicated in (11).

This effect is just what makes achieving a favorable local force ratio so important. For example, a change from $y_0/x_0 = 1$ to $y_0/x_0 = 2$ results in a fourfold advantage for the y force. Of course, (11) is the inequality condition favoring the y force. The x force will try to reverse the inequality by increasing x_0.

Equation (9) only relates the strengths of the two forces to each other, not to the passage of time. Formulas for the variation of the strengths of the forces in time can be obtained from (6) by the following procedure. Differentiate (6a) and use (6b) to obtain

$$\frac{d^2 x}{dt^2} = -b\frac{dy}{dt} = bcx$$

or

$$\frac{d^2 x}{dt^2} - bcx = 0. \tag{12}$$

Using as initial conditions

$$x_0 = x(0), \qquad -by_0 = \frac{dx}{dt}\bigg|_{t=0},$$

the solution of the second-order, constant coefficient, linear, homogeneous ordinary differential equation (12) is given by

$$x(t) = x_0 \cosh \beta t - \gamma y_0 \sinh \beta t, \tag{13}$$

where $\beta = \sqrt{bc}$ and $\gamma = \sqrt{b/c}$. Similarly,

$$y(t) = y_0 \cosh \beta t - \frac{x_0}{\gamma} \sinh \beta t. \tag{14}$$

The graphs of (13) and (14) are sketched in Figure 8.2 in the particular case that $K > 0$ (i.e., $by_0^2 > cx_0^2$ or, equivalently, $\gamma y_0 > x_0$). Observe that for the victory of the y force it is not necessary that y_0 exceed x_0, but that $\gamma y_0 > x_0$.

4. Guerrilla Combat: The Linear Law

The dynamical equations modeling a pair of guerrilla forces in combat can be easily solved if, as above, there are no operational losses and no reinforcements on either side. Under these drastic conditions, GUERCOM reduces to the nonlinear quadratic system

$$\frac{dx}{dt} = -gxy \tag{15a}$$

$$\frac{dy}{dt} = -hxy. \tag{15b}$$

Dividing (15b) by (15a), we obtain the simple equation,

$$\frac{dy}{dx} = \frac{h}{g},$$

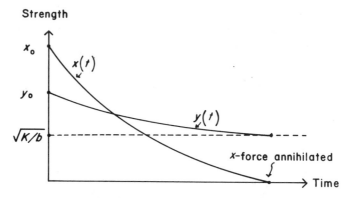

Figure 8.2. Strengths versus Time, $K > 0$. Curves are Sketched Only over the Time Span for Which Both Forces Still Exist

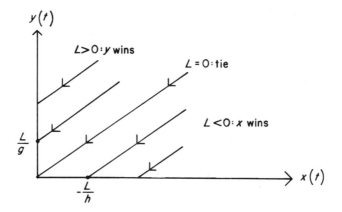

Figure 8.3. Linear Law

which can be integrated to give the relation

$$(\textit{linear law}) \qquad g(y(t) - y_0) = h(x(t) - x_0). \qquad (16)$$

The linear form of (16) explains why the nonlinear system (15) is often called the *linear combat law*. Rewriting (16), we have

$$gy - hx = L, \qquad (17)$$

where $L = gy_0 - hx_0$. The y force wins the combat if $L > 0$ and loses if $L < 0$. As before, the y force wins if the x force is annihilated while some y troops still remain active combatants. See Figure 8.3 for the linear graphs of (17).

Finding formulas relating the strengths to the time is not difficult, but we leave that to the reader (see Exercise 2). Now y wins if $L > 0$, i.e., if

$$\frac{y_0}{x_0} > \frac{h}{g}. \qquad (18)$$

Recalling from (5) one method for calculating g and h in this case of guerrilla combat, the win condition for y is that

$$\frac{y_0}{x_0} > \frac{r_x}{r_y} \cdot \frac{A_{rx}}{A_{ry}} \cdot \frac{A_x}{A_y}. \tag{19}$$

The precombat strategy for the y force involves attempting to maximize the force ratio y_0/x_0 and to minimize the reciprocal area ratio A_x/A_y. Probably neither side can do much to affect the ratios r_x/r_y and A_{rx}/A_{ry}. Observe that the left side of (19) is linear, whereas in CONCOM the corresponding term is squared [see (11)]. Thus changes in the ratio y_0/x_0 are *not* magnified by squaring. On the other hand, another term A_x/A_y now exists which can be altered; no such ratio exists in the CONCOM model. A better way of writing (19) might be as

$$\frac{A_y y_0}{A_x x_0} > \frac{r_x}{r_y} \cdot \frac{A_{rx}}{A_{ry}}. \tag{20}$$

Thus we see that the products $A_y y_0$ and $A_x x_0$ are the critical terms. We shall say no more about this model, since no significant recent wars have been exclusively of this type.

5. Vietnam: The Parabolic Law

In VIETNAM a guerrilla force opposes a conventional force. Once again, let us simplify matters and assume that no reinforcements arrive and no operational losses occur. In this case VIETNAM reduces to

$$\frac{dx}{dt} = -gxy \tag{21a}$$

$$\frac{dy}{dt} = -cx, \tag{21b}$$

where x denotes the guerrilla and y the conventional force. Dividing (21b) by (21a), we have that

$$\frac{dy}{dx} = \frac{c}{gy}; \tag{22}$$

integrating we have that

$$(parabolic\ law)\ gy^2(t) = 2cx(t) + M, \tag{23}$$

where $M = gy_0^2 - 2cx_0$. The guerrilla force wins if $M < 0$ and the conventional y force wins if $M > 0$. See Figure 8.4 for sketches of the parabolas defined by (23).

Experience seems to indicate that a conventional y force can defeat a

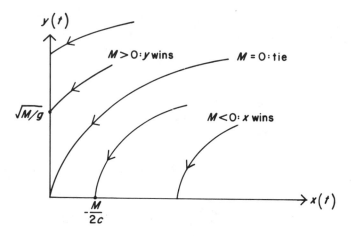

Figure 8.4. Parabolic Law

guerrilla x force only if the force ratio y_0/x_0 is significantly greater than one. Let us see what the parabolic law of (23) has to say about the force ratio y_0/x_0 needed to ensure a win by the conventional y force. Recall that such a win occurs if $M > 0$; i.e., if we have that

$$\left(\frac{y_0}{x_0}\right)^2 > \frac{2c}{g} \cdot \frac{1}{x_0} = 2 \cdot \frac{r_x}{r_y} \cdot \frac{A_x p_x}{A_{ry}} \cdot \frac{1}{x_0}, \tag{24}$$

where we have used (4) and (5) to evaluate c and g in terms of firing rates, kill regions, and the area of the region occupied by the guerrilla force. Let us assume that the firing rates r_x and r_y are approximately equal, so that $r_x/r_y \sim 1$. In addition, suppose that the probability that a shot by a guerrilla kills an opponent is $p_x = 0.1$ and that the vulnerable part of the body of a single guerrilla combatant under cover is 2 ft^2 ($A_{ry} = 2.0$). Thus we have that (24) reduces to

$$\left(\frac{y_0}{x_0}\right)^2 > \frac{0.1 A_x}{x_0}. \tag{25}$$

Now, guerrilla forces usually operate in relatively small units, so let us set $x_0 = 100$ and let us assign 1000 ft^2 to each guerrilla combatant, thereby setting $A_x = 100 \times 1000 = 100{,}000$. Under all these hypotheses, (25) reduces to

$$\left(\frac{y_0}{x_0}\right)^2 > \frac{0.1 \times 100{,}000}{100} = 100, \text{ or } \frac{y_0}{x_0} > 10. \tag{26}$$

Thus the simplified VIETNAM model indicates that the ratio of the conventional force to the guerrilla force must indeed be large if the guerrillas operate in small units from relatively large regions in which they can remain

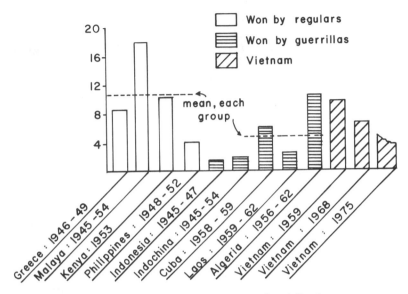

Figure 8.5. Mixed Guerrilla–Conventional Combats

hidden. Of course, what is important is the *local* ratio, although it might be assumed that this reflects the global ratio.

Now the war in South Viet Nam was not actually fought according to such simplified models. Sometimes the Viet Cong and the North Vietnamese fought in a conventional way (e.g., during the Tet offensive of February 1968), while the American and South Vietnamese forces did occasionally use guerrilla units. Nevertheless, both sides perceived the overall struggle in terms of a mixed conventional–guerrilla model, at least in a qualitative way.

According to (26), the conventional South Vietnamese and American Forces would have needed a tenfold force ratio over the guerrilla Viet Cong and North Vietnamese in order to win. What was the actual ratio? Deitchman, writing in 1962 [2], listed the average force ratios for 10 mixed conventional–guerrilla wars following World War II and included figures for the South Vietnam conflict up to that point (see Figure 8.5).[3]

From Figure 8.5 it would appear that the overall force ratio in such a conflict should favor the conventional force only if it is at least eight to one:

$$\frac{\text{conventional forces}}{\text{guerrilla forces}} > 8. \qquad (27)$$

Thus the data strengthen our assertion that (26) is approximately correct. Of course, Figure 8.5 is a bit misleading since no account is taken of rein-

---

[3] The last two columns have been added; the 1968 data are taken from *Encyclopaedia Britannica*, 15th ed., vol. 19, 1974, p. 130.

forcement rates or operational loss rates over a period of time; rather, only the average actual strengths over the period of the conflict are used.

As can be seen from Figure 8.5, by 1968 the ratio of conventional forces to guerrilla forces had declined to the point (6:1) where victory for the guerrilla forces was likely, assuming the validity of the parabolic law for the Vietnam combat and the numbers used above. This decline occurred in spite of massive American reinforcements. In fact, shortly after the Tet offensive, in which the Viet Cong and North Vietnamese won some local victories (but fighting mostly in a *conventional* manner), General Westmoreland requested an additional 206,000 troops from President Johnson. Could this number have been any help? The answer is "not much," as the following table of troop strengths indicates (again we assume a continuation of the mixed conventional–guerrilla type of conflict).

Forces in South Viet Nam, Spring 1968

Conventional Forces		*Guerrilla Forces*	
American	510,000	North Vietnamese[†]	50,000
South Vietnamese:		Viet Cong[†]	230,000
regulars	600,000		
South Vietnamese:			
local defense[†]	500,000		
Other allies	70,000		
Total	1,680,000	Total	280,000

$$\text{Force ratio}: \frac{1,680,000}{280,000} = 6$$

[†] Approximate

If President Johnson had sent the 206,000 troops as requested, the force ratio would have increased to

$$\frac{1,886,000}{280,000} \sim 6.7,$$

still not enough to have improved the situation much for the conventional forces. Moreover, the guerrilla forces would only have had to increase to 314,000 in order to maintain a ratio of 6:1. It was analyses such as this, coupled with the disquiet of the American people about the whole affair, that led President Johnson to seek a political solution to the Vietnamese conflict. He rejected Westmoreland's request and initiated the Paris peace talks, which eventually led to the American disengagement of 1973. The final victory of the Viet Cong and North Vietnamese came in April 1975. (Let us again emphasize that we have ignored many significant factors such as the bombing of North Vietnam and the effect of the terrorist campaigns.)

6. The Battle of Iwo Jima

My body shall not decay in the field
Unless we are avenged;
I will be born seven more times again
To take up arms against the foe.[4]

One of the fiercest battles of the Second World War was fought between Japanese and American troops on the 8 mi^2 volcanic island of Iwo Jima, 660 mi south of Tokyo. The island is actually in the Tokyo Prefecture and has always been considered a part of Japan itself. Partly for this reason, but more because of its value as a base for fighters attacking American aircraft on their way to and from bombing missions over Tokyo and other Japanese cities, the Japanese were determined to hold onto the island. Conversely, the Americans needed the island as a bomber base close to the Japanese islands. The American invasion began on February 19, 1945 after an extensive preinvasion bombardment. As was discovered later, this early bombardment had little effect on the Japanese strength since their forces were protected by numerous natural and man-made caves. The fighting on February 19—and indeed throughout the month-long combat—was intense, with high casualties on both sides. The Japanese forces had been ordered to hold the island at all costs, fighting to the last man if they had to, and that is just about what they did. The island was declared "secure" by the American forces on March 16, active fighting ceased on March 26, and the final mopping up of Japanese combatants hiding out in the caves was essentially complete by mid-May. (The last two Japanese holdouts surrendered in 1951!)

The bloody battle is mostly forgotten now. Every now and then, Kuribayashi's lines are reprinted, or the photograph of four American marines raising the U.S. flag on Iwo Jima's Mt. Suribachi reappears in a news magazine or collection of war photos. A transitory resurgence of interest in Iwo Jima will follow, but by and large the battle survives only in the memories of the families of the combatants and in the pages of military histories.

J. H. Engel [3] has used actual combat records to construct a Lanchestrian model of the battle, and we shall follow his analysis and use his data. During the conflict the American forces were reinforced, but the Japanese were not. Operational losses, at least on the American side, were negligible and can be ignored. One might argue that a VIETNAM model would be appropriate for the battle of Iwo Jima, with the Japanese taking cover in the caves as the guerrilla force. However, Engel shows that a CONCOM model can also be used:

---

[4] These lines by General Kuribayashi, the commander of the Japanese forces on Iwo Jima, were uttered shortly before he and the remnants of his army were killed in the bloody battle. See [4].

$$\text{(a)} \quad \frac{dx}{dt} = -by + P(t)$$

$$\text{(b)} \quad \frac{dy}{dt} = -cx,$$

where x denotes the strength of the American forces and y that of the Japanese. In order to keep the variables straight we shall use A and J, respectively, instead of x and y:

$$\frac{dA}{dt} = -bJ + P(t) \tag{28a}$$

$$\frac{dJ}{dt} = -cA. \tag{28b}$$

Let us solve (28) by quadratures. This can be done by *variation of parameters for systems* (see Exercise 7) or as follows. Differentiating (28b) and then using (28a), we have that

$$\frac{d^2 J}{dt^2} = -c\frac{dA}{dt} = bcJ - cP(t)$$

or

$$\frac{d^2 J}{dt^2} - bcJ = -cP(t). \tag{29}$$

Equation (29) is a second-order linear constant-coefficient ordinary differential equation with "driving force" $-cP(t)$. The technique of variation of parameters for a single second-order equation enables us to solve (29), obtaining (see (13))

$$J(t) = J_0 \cosh \beta t - \frac{A_0}{\gamma} \sinh \beta t - \frac{1}{\gamma} \int_0^t \sinh \beta(t - s) P(s)\, ds, \tag{30}$$

where J_0 and A_0 are the respective strengths of the Japanese and the American forces *on the island* just before the invasion (hence $A_0 = 0$), $\beta = \sqrt{bc}$, $\gamma = \sqrt{b/c}$. The term A_0 enters (30) because $dJ/dt_{t=0} = -cA_0$ by (28b).

Now we can determine $A(t)$ by calculating dJ/dt from (30) and then using (28b). From (30) we have that

$$\frac{dJ}{dt} = \beta J_0 \sinh \beta t - \frac{\beta A_0}{\gamma} \cosh \beta t$$

$$- \frac{1}{\gamma} \sinh \beta(t - t) P(t) - \frac{\beta}{\gamma} \int_0^t \cosh \beta(t - s) P(s)\, ds, \tag{31}$$

where we have used the Leibniz rule for differentiating an integral:

$$\frac{d}{dt} \int_a^t h(t, s)\, ds = h(t, t) + \int_a^t \frac{\partial}{\partial t} h(t, s)\, ds.$$

Since $\sinh 0 = 0$ and $\beta/\gamma = c$, (31) becomes

$$\frac{dJ}{dt} = \beta J_0 \sinh \beta t - cA_0 \cosh \beta t - c \int_0^t \cosh \beta(t - s)P(s)\,ds. \qquad (32)$$

Recalling that $A_0 = 0$ and using $A(t) = -(1/c)(dJ/dt)$ from (28b), we have that

$$A(t) = -\gamma J_0 \sinh \beta t + \int_0^t \cosh \beta(t - s)P(s)\,ds. \qquad (33)$$

Thus the Lanchestrian models for the variation in time of the strengths of the American and the Japanese forces lead to

$$A(t) = -\gamma J_0 \sinh \beta t + \int_0^t \cosh \beta(t - s)P(s)\,ds \qquad (34a)$$

$$J(t) = J_0 \cosh \beta t - \frac{1}{\gamma} \int_0^t \sinh \beta(t - s)P(s)\,ds. \qquad (34b)$$

These equations are, of course, simply (30) and (33) with $A_0 = 0$.

Let us compare these theoretical models with the actual data. We immediately run into a problem: no Japanese records exist! Apparently, the casualty lists kept by General Kuribayashi were destroyed in the battle itself, while whatever records were maintained in Tokyo were consumed in the fire bombings of the remaining five months of the war. Nevertheless a grisly circumstance does allow an initial count of the Japanese forces. No Japanese escaped from the island during the combat, no Japanese ship or aircraft landed on, or departed from, the island; every Japanese combatant was either killed or captured. Thus a body count and a prisoner count provide the data on the Japanese forces. The American forces kept daily records of their own casualties [11]; only the overall totals appear in the following table.

Casualties on Iwo Jima

	Total United States Casualties at Iwo Jima			
	Killed, Missing or Died of Wounds	Wounded	Combat Fatigue	Total
Marines	5,931	17,272	2,648	25,851
Navy units:				
Ships and air units	633	1,158		1,791
Medical corpsmen	195	529		724
Seabees	51	218		269
Doctors and dentists	2	12		14
Army units in battle	9	28		37
Grand Totals	6,821	19,217	2,648	28,686

Casualties on Iwo Jima (*contd.*)

Defense Forces	Japanese Casualties at Iwo Jima Prisoners		Killed
(Estimated) 21,000	Marine	216	20,000
	Army	867	
	Total	1,083	

From [12, p. 296]

Newcomb's figure of 21,000 for the Japanese forces is a little low, since he apparently did not count some of the living and the dead found in the caves in the final days. We shall use Engel's figure of 21,500. Thus we set $J_0 = 21,500$ in (34).

The American reinforcement rate $P(t)$ is known:

$$P(t) = \begin{cases} 54,000, & 0 \leq t < 1 \\ 0, & 1 \leq t < 2 \\ 6,000, & 2 \leq t < 3 \\ 0, & 3 \leq t < 5 \\ 13,000, & 5 \leq t < 6 \\ 0, & 6 \leq t < 36, \end{cases} \tag{35}$$

where we assume that the rate is constant each day and is measured in number of troops put ashore per day.

This leaves the computation of β and γ (that is, of b and c) yet to be carried out. Here we have to fudge a little. First of all, b and c cannot be the effectiveness coefficients of only the active combatants, but must be the average over all the active forces, both combat and support, since we have lumped all such forces together. Second, there is no way, *a priori*, to calculate even average values of such coefficients for a battle lasting so many days. We shall instead follow Engel [3] and ask whether or not we can use the daily American casualty lists [11], the reinforcement data of (35), and $J_0 = 21,500$ to determine possible values for b and c. From the casualty lists given by Morehouse and from (35), we can calculate $A_{act}(t)$, the actual number of American forces, for $t = 1, 2, \cdots, 36$. Then from (28b), we have by integration that

$$J(36) - J_0 = -c \int_0^{36} A_{act}(t)\, dt$$

$$= -c \sum_{t=1}^{36} A_{act}(t),$$

since we assume that $A_{act}(t)$ remains constant each day. From this, we have that

$$c = \frac{J_0 - J(36)}{\sum\limits_{t=1}^{36} A_{\text{act}}(t)} = \frac{21,500 - 0}{2,037,000} \sim 0.0106 \tag{36}$$

where Engel has calculated the denominator from the data given by Morehouse.

Now we need to estimate b. Using the value of c given in (36), we can find an approximation $J_{\text{app}}(t)$ to the actual value $J_{\text{act}}(t)$, which, of course, is only known for $t = 0$ and $t = 36$:

$$J_{\text{app}}(t) = 21,500 - 0.0106 \sum\limits_{k=1}^{t} A_{\text{act}}(k), \qquad t = 0, \cdots, 36, \tag{37}$$

where we have integrated (28b) from 0 to t and then replaced $\int_0^t A_{\text{act}}(s)\,ds$ by the summation in (37). As stated earlier, active combat came to an end on the 36th day; however, the fighting from the 28th to the 36th day was so sporadic that it seems dubious that our continuous model should be used. To find b, therefore, we shall use $t = 28$. (We had no choice but to use $t = 36$ to calculate c since only when all combat had ceased could there be any accurate count of the Japanese losses.) From (28a), we have by integration that

$$\begin{aligned} A(t) &= A_0 - b \int_0^t J_{\text{app}}(s)\,ds + \int_0^t P(s)\,ds \\ &= -b \sum\limits_{k=1}^{t} J_{\text{app}}(k) + \sum\limits_{k=1}^{t} P(k). \end{aligned} \tag{38}$$

Thus we can use (38) with $t = 28$ to find b:

$$b = \frac{\sum\limits_{k=1}^{28} P(k) - A(28)}{\sum\limits_{k=1}^{28} J_{\text{app}}(k)}. \tag{39}$$

For the quantities on the right-hand side of (39) we shall use the following numbers:

$\sum\limits_{k=1}^{28} P(k) = 73,000,$ from (35)

$A(28) = 52,735,$ taken by Engel from the data of Morehouse

$\sum\limits_{k=1}^{28} J_{\text{app}}(k) = 372,500$ Calculated by Engel using (37) and the data of Morehouse on $A_{\text{act}}(k)$.

We have that

$$b \sim 0.0544. \tag{40}$$

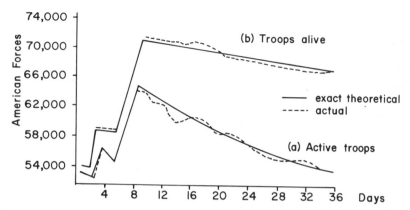

Figure 8.6. Comparison of Actual Troop Strength with Theoretical for the Battle of Iwo Jima (Adapted from Engel [3])

It should be pointed out once again that b has been calculated using (38) with $t = 28$ and using the actual day-to-day American strength (as used by Engel to calculate $\sum_{k=1}^{28} J_{app}(k)$). (Alternatively, $b(k)$ could have been calculated from (38) for $k = 1, \cdots, 28$, and then an average taken, but we have not done this.) We now ask just how $A_{act}(k)$ and $A(k)$ from (38) compare for $k = 0, 1, 2, \cdots$. Engel has plotted the two on a common graph, and we include an adaptation of the graph in Figure 8.6. Observe the upward jumps due to the reinforcements in the opening phase of the battle. The fit is remarkably good, although this is not too surprising given the way b and c were calculated. Thus it appears that a Lanchestrian model does indeed fit the battle of Iwo Jima. Of course, other models may do as well or better.

7. Mathematical Models of Combat: The Dark Side of Modeling

Models like these raise ethical questions about the uses made of mathematics. G. H. Hardy, the best known English "pure" mathematician of the first half of the 20th century, has spoken on this matter [5]:

So a real mathematician has his conscience clear; there is nothing to be set against any value his work may have; mathematics is ... a "harmless and innocent" occupation. The trivial mathematics on the other hand has many applications in war.

To Hardy, pure mathematics was deep, "real," "harmless and innocent," while it was "trivial" mathematics which could be applied. One wonders if it is really all that simple.

Mathematics is not a world entirely apart from the cultures and societies of the mathematicians who create or discover it. Since war and the preparation for war continue to be a principal preoccupation of humankind, it is not surprising that mathematics is applied to its study and analysis. Mathematics has mostly been used in the design of weaponry—viz., Archimedes and his catapults or the contemporary calculations of the paths of missiles. However, mathematical analysis can also be used in the planning of optimal strategy, supply, and tactics. Lanchester's curious models fit into the latter category of military uses of mathematics. See [13] for an intriguing mathematical model of the absurdities of an arms race.

The author has his own views on the ethics of applying mathematics to the study of war, and the reader undoubtedly does as well. Whatever these views may be, everyone must know that mathematics will continue to be applied to this grim subject.

Exercises

1. Explain each of the terms in CONCOM, GUERCOM, and VIETNAM. In VIETNAM, which is the guerrilla force, x or y?

2. Find all solutions $x(t)$ and $y(t)$ of (15) inside the strength quadrant, $x > 0$, $y > 0$. *Hint*: Solve (17) for y in terms of x and insert that expression for y into (15a). Then separate variables in (15a) and use partial fractions (or a table of integrals) to solve. Finally, express x as a function of t. Proceed similarly with (15b).

3. Apply variation of parameters to (29) and deduce (30).

4. Derive (33) from (32).

5. Suppose the simplified conventional combat law (6) applies with $x_0 = 15,000$, $y_0 = 20,000$, $b = 0.04$, $c = 0.5$. Suppose the y force has another 5000 troops which can be used as reinforcements but must be inserted into the battle all at once if at all. Give a complete analysis of the optimal point for bringing in the reinforcements. (There need not be a single answer here; you must define what you mean by "optimal." Use (9) with K determined by the initial 20,000, and then use (9) again with a new K determined by the value of $y(t)$ just after the reinforcements are brought in.)

6. Repeat 5, but with the linear law. Use (15) and (17) with $x_0 = 15,000$, $y_0 = 20,000$, $g = 0.04$, $h = 0.05$, y reinforcements of 5000.

7. Use variations of parameters for systems directly on (28) to obtain (34).

Project

The Battle of the Ardennes (or Battle of the Bulge) was the last major offensive of the German Wehrmacht during World War II. The battle began on December 16, 1944,

128

with a surprise German attack, which in the first six days created a bulge in the Allied lines 70 miles wide and 50 miles deep. By January 28, 1945 the lines had returned to their original configuration, and the battle was over. The table below summarizes the battle statistics, using mostly guesswork for the daily nonbattle casualty figures. There are no daily figures for the German battle casualties, just as there are none for the Japanese at Iwo Jima. Construct and solve a model for this combat. (A student team at Harvey Mudd College assembled the data in the table and constructed and solved a model of their own devising [7]. Their equations are

$$\frac{dA(t)}{dt} = -C_A(t) - bG(t) + R_A(t)$$

$$\frac{dG(t)}{dt} = -C_G(t) - cA(t) + R_G(t)$$

where they assume that the noncombat casualty rates $[-C_A(t)$ and $-C_G(t)]$ were independent of the size of the forces. You may prefer to construct another model. A suggestion: think of the first ten days of struggle as one battle, the remaining thirty-four days as another.)

References for the Project [*in addition to* [7]]:

H. M. Cole, *The Ardennes: Battle of the Bulge—The European Theater of Operations*, Office of the Chief of Military History, Department of the Army, Washington, DC, 1965.
J. S. D. Eisenhower, *The Bitter Woods*. New York: Putnam, 1969.
Encyclopedia Americana, vol. 29. New York: Americana, 1971, pp. 410–413.
Report of Operations (*Final After Action Report*), 12th Army Group, vol. I–XII.

Daily Statistics for American and German Forces

Day Number	Date	Rein- Forcements	American Forces Battle Casualties	Nonbattle Casualties	Force Strength	German Forces Rein- Forcements	Nonbattle Casualties
0		0	0	0	92,200	0	0
1	December 16	0	2500	1000	88,700	0	1000
2	17	0	6000	1000	81,700	24,400	1000
3	18	33,100	3750	1000	110,050	0	1000
4	19	13,200	3800	1000	118,450	14,400	1000
5	20	48,800	2750	1000	163,500	0	1000
6	21	26,100	1625	1000	186,975	0	1000
7	22	26,700	3500	1000	209,175	14,400	1000
8	23	24,400	1875	1000	230,700	0	1000
9	24	12,900	2575	1000	240,025	0	1000
10	25	13,000	2000	1000	250,025	34,000	1000
11	26	0	2000	1000	247,025	0	1000
12	27	12,200	1875	1000	256,350	0	1000
13	28	0	1250	1000	254,100	0	1000
14	29	0	1250	1000	251,850	0	1000
15	30	23,300	1000	1000	273,150	20,000	1000

Daily Statistics for American and German Forces (*contd.*)

Day Number	Date	American Forces Rein-Forcements	American Forces Battle Casualties	American Forces Nonbattle Casualties	Force Strength	German Forces Rein-Forcements	German Forces Nonbattle Casualties
16	31	10,300	1500	1000	280,950	0	1000
17	January 1	0	1125	1000	278,825	48,000	1000
18	2	0	875	1000	276,950	10,000	1000
19	3	8,500	1500	1000	282,950	0	1000
20	4	0	1500	1000	280,045	0	1000
21	5	0	1250	1000	278,200	0	1000
22	6	0	875	1000	276,325	0	1000
23	7	0	1375	1000	273,950	0	1000
24	8	0	1000	1000	271,950	0	1000
25	9	12,900	1250	1000	282,600	0	1000
26	January 10	0	1375	1000	280,225	0	1000
27	11	0	1000	1000	278,225	0	1000
28	12	0	1000	1000	276,225	0	1000
29	13	0	1500	1000	273,735	0	1000
30	14	0	1500	1000	271,225	0	1000
31	15	0	2000	1000	268,225	0	1000
32	16	0	1625	1000	265,600	0	1000
33	17	0	1000	1000	263,600	0	1000
34	18	0	1250	1000	261,350	0	1000
35	19	0	1125	1000	259,225	0	1000
36	20	0	875	1000	257,350	0	1000
37	21	0	1125	1000	255,225	0	1000
38	22	0	875	1000	253,350	0	1000
39	23	0	875	1000	251,475	0	1000
40	24	0	1000	1000	249,475	0	1000
41	25	0	875	1000	247,600	0	1000
42	26	0	750	1000	245,850	0	1000
43	27	0	875	1000	243,975	0	1000
44	28	0	500	1000	242,475	0	1000
		265,400				164,000	

Also for Germans:
  Initial attack force $t = 0$ $G(0) = 200,000$
  Battle Casualties   100,000
  Final Force Strength   220,000

References

[1] M. Braun, *Differential Equations and Their Applications*, 2nd ed.   New York: Springer-Verlag, 1978. The best book around on the elementary level. Braun introduces Richardson's theory [13] of the arms race on pp. 513–525.

[2] S. J. Deitchman, "A Lanchester model of guerrilla warfare," *Operations Res.*, vol. 10, pp. 818–827, 1962. A readable account of the topic with a brief introduction to Lanchester's laws.

[3] J. H. Engel, "A verification of Lanchester's law," *Operations Res.* (*J. Operations Res. Soc. Amer.*), vol. 2, pp. 163–171, 1954. The source of our analysis of the battle of Iwo Jima. Should be accessible to the undergraduate.

[4]   G. W. Garand and T. R. Strobridge, *History of United States Marine Corps Operations in World War II*, vol. 4 [*Western Pacific Operations*], Histor. Div. Hdqr. USMC, 1971. An extensive treatment of the planning and execution of the Iwo Jima operation. No mathematics at all.

[5]   G. H. Hardy, *A Mathematician's Apology*, 2nd ed.   New York: Cambridge Univ. Press, 1967, pp. 140–141. From the book jacket: " . . . a personal account by a distinguished mathematician of what mathematics meant to him as a man. Hardy discusses and illustrates the attractive force of mathematics. He dismisses its utility but describes its depth and beauty as a creative art."

[6]   Homer, *The Iliad*, G. Chapman, tr.   New York: Pantheon Books, 1956, V.1. The lines quoted are from Book XIII, lines 571–575.

[7]   G. J. Hueter, M. A. McClelland, L. A. Resner and M. G. Zevallos, "An application of the Lanchester model to the battle of the Ardennes," *Interface* vol. 5, no. 1, pp. 15–26, 1978. Available from Department of Mathematics, Harvey Mudd College, Claremont, CA 91711.

[8]   D. R. Howes and R. M. Thrall, "A theory of ideal linear weights for heterogeneous combat forces," *Naval Res. Logistics Quart.*, vol. 20, pp. 645–659, 1973. We have considered only homogeneous forces; Howes and Thrall take up the more realistic heterogeneous case using the techniques of linear algebra. Accessible to students at the end of a linear algebra course. Good bibliography.

[9]   F. W. Lanchester, *Aircraft in Warfare; The Dawn of the Fourth Arm*. Constable and Co., Ltd., 1916. This book is hard to find now but is the basic source.

[10] F. W. Lanchester, "Mathematics in warfare," in *The World of Mathematics*, vol. 4, J. R. Newman, Ed. New York: Simon and Schuster, 1956, pp. 2138–2157. This article is taken from Lanchester's book and is a readable introduction to his simplest models. Lanchester relates heterogeneous forces to the Pythagorean theorem!

[11] C. P. Morehouse, *The Iwo Jima Operation*, Histor. Div. Hdqr. USMC, 1946. Contains the daily American casualty lists for the battle; book is hard to find.

[12] R. F. Newcomb, *Iwo Jima*.   New York: Holt, Rinehart, and Winston, 1965. A fairly complete account of the battle; shorter than Garand and Strobridge (and not as incisive), not mathematical.

[13] L. R. Richardson, "Mathematics of war and foreign politics," in *The World of Mathematics*, vol. 4, J. R. Newman, Ed.   New York: Simon and Schuster, 1956, pp. 1240–1253. Richardson was a physicist with a Quaker background whose aversion to war led him to make a serious study of its causes. In this very readable essay he uses differential equations to model the arms race. (See also Braun [1].)

[14] H. K. Weiss, "Lanchester-type models of warfare," *Proc. 1st Internat. Congr. on Operations Research*, Baltimore, MD 1957, pp. 82–99. A good survey of the basic models and some of their extensions.

Notes for the Instructor

Objectives. The goal of this module is to show how mathematical models can be used to portray a battle. The ethics, not to mention the utility, of such an application are a matter of concern and might well be discussed in a class using this module.

Prerequisites. Separable differential equations, variation of parameters for a scalar second order equation, hyperbolic functions (or the equivalent exponentials), the Leibniz rule for differentiating an integral.

Time. The material can be covered in one or two lectures.

CHAPTER 9

Modeling Linear Systems by Frequency Response Methods

William F. Powers*

1. Introduction

In his classic book *Cybernetics* [1], Norbert Wiener viewed the human being from an information processing and control system point of view. Such a viewpoint has led to the development of communications and control system oriented mathematical models for human reactions and for physiological components of the human body. Much of this work has involved frequency response techniques, and this module will be concerned with the use of such techniques in mathematical modeling.

Briefly, frequency response techniques are based on the Fourier representations of functions. These representations describe a function in terms of the "amounts" of components with various frequencies that combine to produce the function. For example, if $f(t)$ has a Fourier series expansion

$$f(t) = \sum a_n \sin n t,$$

the different sine functions represent different frequencies, and the coefficients are the "weights" with which they are combined. (Nonperiodic functions f can be represented by Fourier integrals like

$$f(t) = \int a(x) \sin xt \, dx$$

employing a continuum of frequencies, weighted by $a(x)$.) Thus the dynamics of certain systems can be recovered from knowledge of their responses to sinusoidal inputs.

* Control Systems Department, Ford Motor Company, Dearborn, MI 48121.

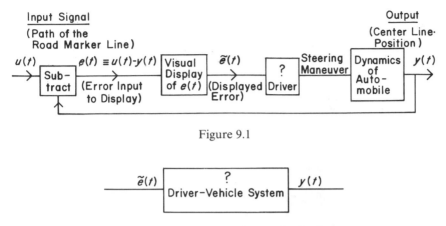

Figure 9.1

Figure 9.2. Input–Output System for Analysis

The major emphasis of the module is on modeling the human as an operator in a "compensatory task." More specifically, a mathematical model for the human operator is developed from data obtained from experiments in which the subject is supposed to null (i.e., eliminate) a visual error signal by manipulating a control system with known control dynamics. The visual error signal is random or "random appearing." An example of such a situation is the simulation of automobile steering. That is, suppose the difference between the center line of the car and a random (or random appearing) road marker line (the input) is displayed to the driver. This will cause the driver to turn the steering wheel, which in turn affects the path of the automobile and in turn the error signal. The dynamic characterists of the system that the subject controls are assumed known; the automobile's steering system in our example. A "feedback" diagram of this situation is shown in Figure 9.1. Approximate mathematical models can then be developed for the driver by frequency response analysis of error signal (programmed path for the road marker line minus the actual path) and the output (the resultant center line position of the automobile). That is, given measurements of the error viewed by the driver $\tilde{e}(t)$ and the output $y(t)$, one then applies frequency response methods to determine a model of the combined driver–vehicle system (see Figure 9.2). Since the mathematical model of the vehicle is known, the mathematical model of the driver may then be extracted from the combined model.

The type of modeling discussed above is an active research area, and a number of models developed by this approach have been used in applications. For example, such analyses have been used to show that the Saturn–Apollo space booster can be flown manually by the astronauts in a contingency situation [2], that trucks and buses larger than are currently permitted on highways may be uncontrollable in certain situations because of human operator limitations [3], and that inclusion of pilot models in aircraft design

Figure 9.3

improves the resultant aircraft performance [4]. In addition, this modeling technique has been employed in the generation of numerous physiological models, some of which are surveyed in [5]–[7].

The remainder of the module is as follows. In Section 2 the experimental aspects of the human operator modeling problem are discussed, along with the motivation for the utilization of frequency response techniques. In Section 3 a tutorial modeling problem which demonstrates the basic mathematics is presented. (If desired, the student may perform his own experiment and modeling problem.) In Section 4 data from actual human operator experiments are presented for use in modeling and interpretation by the student. Finally, in Section 5 extensions which involve more sophisticated mathematics are discussed briefly, along with appropriate references to the literature.

2. Modeling the Human Operator

Because of the large degree of variability among humans, one would not, at first glance, expect to be able to form a mathematical model for human behavior. However, for certain well-defined tasks which involve a human operator, good approximate mathematical models for the nominal (normal) operator of the task may be developed. Such models are then useful in the design of the equipment involved in the particular task. In this section we shall discuss some of the basic considerations involved in the development of human operator models for compensatory tasks (i.e., error nullification).

Assuming a well-defined task for the operator, one must devise an experiment to generate data useful for the development of the model. Figure 9.3 shows a typical laboratory setup for an experiment involving the modeling of a compensatory task. The human operator will move the stick in an attempt to null out the time-varying error signal displayed on the screen. A computer generates the error signal and also contains the dynamics of the physical system to be controlled (e.g., automobile, aircraft).

Since such an experiment involves human subjects, a number of considerations must be taken into account. Some of the major ones are described as follows.

(1) *Motivation:* The human subjects involved in the experiment should be motivated to approximately the same degree to perform to the best of their ability during the test.

(2) *Learning:* Typically, the longer one does the experiment, the better one's performance—up to a certain point. Such a period is a learning period which should not be part of the data track. Thus the subject usually does a number of trial experiments before data are collected.

(3) *Training:* Some subjects may be better than others at performing a compensatory task; e.g., a pilot is usually more skilled than a subject who is not a pilot. Such information should be noted since it will affect the parameters of the model.

(4) *Attention Span:* The experiment should be long enough to get a good data track but not so long that the subjects lose interest or full attention. Typical experiments (such as Figure 9.3) are on the order of 2–5 min, of which the initial and terminal portions are usually deleted because of learning and decreasing attention, respectively.

(5) *Physical and Psychological Conditions:* The general physical and psychological condition of the subject should be noted. (In this respect, such experiments can be used to determine the effects of, for example, drugs on the ability to do a compensatory task.)

(6) *Input Signals:* The displayed error signal should not be so simple that the subject can predict it from instant-to-instant and yet not so complicated that mathematical analysis is impossible.

The last consideration noted above, i.e., the mathematical character of the input signals, is critical since it will strongly influence the type of mathematics to be employed in the data analysis and model building. Let us consider some typical inputs and hypothesize the ability of the subject to follow them. If a constant error signal is displayed, the normal subject should have little trouble following it. Note that a constant signal is a sinusoidal signal with frequency equal to zero (i.e., $c = c \cos \omega t$ with $\omega = 0$). Next, suppose a sinusoidal signal of very low frequency is displayed. Again, the subject should be able to track the error signal. However, as the frequency of the error signal increases, the tracking error should increase, and finally a point will come when the subject can no longer track the signal even knowing what the signal is (i.e., a high-frequency sine wave). This frequency is a rough indicator of the "bandwidth" capability of the human operator (typically 5–10 rad/s). Since the human cannot physically track above that bandwidth, input signals containing frequencies higher than the bandwidth are of no use in the development of a mathematical model for the human's behavioral characteristics in tracking an error signal. This argument suggests that one possible classification of input signals is according to frequency. If this classification is adopted, then the admissible frequency range is $[O, BW]$, where $BW \equiv$ bandwidth.

Since frequency is a very important variable in many electrical and mechanical systems, numerous mathematical techniques and theories exist associated with the frequency characteristics of systems. Such approaches are usually referred to as "frequency domain" approaches.

If the system to be modeled is approximately linear, then there exists a number of frequency domain results which can be applied directly in its identification and modeling. However, one would expect that a human operator does not behave in a linear manner. In fact, one of the pioneers of human operator modeling, A. Tustin, used a simple experiment to show that the human operator is nonlinear. As will be shown in the next section, if a system is described by a linear differential equation with a sinusoidal forcing function of frequency ω, then certain interesting solutions (steady-state solutions) can contain sinusoidal components only of frequency ω. Tustin [8] employed an input of three sine waves of different frequencies in a human operator experiment and obtained an output containing frequencies other than the three input frequencies. This result implies that the human operator is nonlinear. However, Tustin also argued that for many tasks the human operator is near-linear and that an approximate linear model exists which accounts for the dominant features of the human operator's output. Because of this, frequency domain techniques of linear analysis have often been employed in the modeling of human operators.

In the next section basic *deterministic* techniques associated with frequency domain modeling will be discussed. For a complete discussion of techniques associated with human operator modeling, the theory of stochastic processes is necessary. Indeed, the class of input functions is frequently random, usually a sum of sine functions with random amplitudes and phases; typically, 10 or more frequencies are employed on the interval $[O, BW]$ so that the frequency appears random. However, even in such cases the final model is often a deterministic one, and knowledge of the mathematics in the next section allows a good understanding of the interplay between mathematics and the modeling. Stochastic process techniques are mainly used in the generation of data plots from the experiment and in the interpretation of the nearness-to-linearity of the model.

3. Properties and Techniques of Frequency Response Modeling

In this section a simple experiment will be described and used to illustrate mathematical models of the frequency-response type. For students with some background in engineering or physics, the governing equations for the

Figure 9.4. Experimental Setup

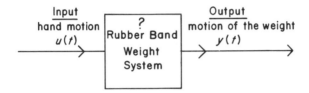

Figure 9.5. Input–Output Characterization of Modeling Problem

motion should be obvious by Newton's laws. However, even for students without such background, the experiment demonstrates a way of "discovering" the laws (or model) and the mathematical ideas behind frequency domain techniques.

3.1 Description of the Experiment

The following simple experiment will introduce the ideas and mathematical techniques associated with the application of frequency response plots to modeling linear or near-linear dynamic systems.

Experiment. String rubber bands together to a length of approximately 2 ft and tie a weight, e.g., a *heavy* bolt, to the end of the string of rubber bands. Our purpose is to model this system so that we will be able to predict the motion of the weight, i.e., the output, due to a time-varying hand motion at point a, i.e., the input (see Figures 9.4 and 9.5).

3.2 Models Based on Linear Differential Equations

A mathematical model which provides a useful approximate description of the system discussed is one involving a linear differential equation with constant coefficients. This choice of a model will be supported in what follows.

The differential equation to be used has the form

$$\frac{d^n y}{dt^n} + a_{n-1}\frac{d^{n-1}y}{dt^{n-1}} + \cdots + a_1\frac{dy}{dt} + a_0 y = u(t). \tag{1}$$

Such an equation is said to be *time invariant* since the coefficients on the left are independent of t. The function u is called the *forcing function* or the *input function*. A standard problem is to investigate the behavior of the system for various input functions. We shall be concerned in particular with input functions of the form $u(t) = \sin \omega t$ (or with linear combinations of such functions). The number ω is called the *frequency* of the input. If ω is small, the input is said to be a low-frequency input (e.g., 1 cycle/second = 2π rad/s = 1 Hz); if ω is large, the input is said to be a high frequency input (e.g., 10 c/s).

It will be assumed that the reader has a basic working knowledge of the theory of linear differential equations with constant coefficients and with Laplace transform methods for solving such equations. This material is contained in most textbooks on elementary differential equations, e.g., [20]–[23].

Let $y(t)$ be the solution of (1) with $y(0) = y'(0) = \cdots = y^{(n-1)}(0) = 0$, and take the Laplace transform of (1) to obtain the transformation equation

$$(s^n + a_{n-1}s^{n-1} + \cdots + a_1 s + a_0)V(s) = W(s)$$

where V and W are the Laplace transforms of y and u, respectively. If s is not a root of $s^n + a_{n-1}s^{n-1} + \cdots + a_1 s + a_0 = 0$, then

$$V(s) = \frac{W(s)}{s^n + \cdots + a_1 s + a_0} \tag{2}$$

or

$$V(s) = W(s) Y(s), \tag{2a}$$

with

$$Y(s) = (s^n + \cdots + a_1 s + a_0)^{-1}.$$

The function Y is called the *transfer function* for the equation (or the physical system modeled by the equation). Equation (2a) says that the Laplace transform of y may be found by multiplying the Laplace transform of u by the transfer function. The reader should note that the transfer function may be determined from (1) by inspection and that, conversely, the coefficients in (1) can be determined from the transfer function. We now consider some very important properties of linear time-invariant differential equations.

Definition. The differential equation

$$\frac{d^n y}{dt^n} + a_{n-1}\frac{d^{n-1}y}{dt^{n-1}} + \cdots + a_1\frac{dy}{dt} + a_0 y = 0$$

is said to be *stable* if all solutions y have the property that $\lim_{t\to\infty} y(t) = 0$.[1]

---

[1] Some authors say the equation is *asymptotically stable* if for every solution $\lim_{t\to\infty} y(t) = 0$.

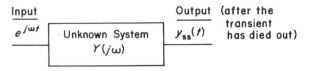

Figure 9.6. Frequency Response Identification of Unkown System Y

Definition. Equation (1) is said to be *stable* if the corresponding homogeneous equation is stable.

In what follows, j is used to denote $\sqrt{-1}$. This notation is common in engineering literature.

Theorem 1. *Consider an equation of the form* (1) *with real coefficients* $a_0, \cdots,$ a_{n-1} *and with* $u(t) = \exp(j\omega t)$.

a) *Equation* (1) *is stable if and only if all of the roots* $\lambda_1, \cdots, \lambda_n$ *of the characteristic equation* $s^n + a_{n-1}s^{n-1} + \cdots + a_1 s + a_0 = 0$ *have negative real parts.*

b) *If* Y *(the transfer function of* (1)*) is defined at* $j\omega$, *then the general solution of* (1) *can be written as* $y_g = y_T + y_{ss}$ *when* y_T *is the general solution of the homogeneous equation and* y_{ss} *is the steady-state solution of* (1) *defined by*

$$y_{ss}(t) = Y(j\omega)\exp(j\omega t). \tag{3}$$

If all the λ_k *are distinct, then* $y_T(t) = \sum_{k=1}^{n} c_k \exp(\lambda_k t)$.

c) *If* (1) *is stable, then* $\lim_{t\to\infty} y_T(t) = 0$ *and* y_g *is asymptotic to* y_{ss}. *The function* y_T *is called the* transient *part of* y_g.

The proof of this result is readily available in the textbook literature [21]. Alternatively, it (or parts of it) may be assigned as an exercise.

Part b) of this theorem illustrates an intimate relation between linear time-invariant dynamical systems and sinusoidal inputs: the steady-state solution of the equation modeling such a system is simply the product of the transfer function evaluated at $j\omega$ and the input. In most input–output analyses, the steady-state solution is of primary importance since the transient solutions die out when the system is stable.

Equation (3) is important in that it presents us with the key to the development of an experimental technique for the identification of the dynamics of a linear time-invariant system. Consider the input–output diagram in Figure 9.6. In this figure we see that, by proper choice of the class of input signals (always a critical decision in a modeling problem), we can construct the complex function $Y(j\omega)$ from (3). Then it is simply a matter of curve-fitting or parameter estimation to determine the transfer function $Y(s)$ and consequently the dynamics of the unknown system.

3.3 Frequency Response (Bode) Plots

Experiments involving frequency response methods for the modeling of an unknown system usually result in *frequency response* or *Bode plots*. These are plots of the magnitude and argument (phase) of the complex function $Y(j\omega)$ versus frequency (ω). More precisely, with $M(\omega) = |Y(j\omega)|$ and $N(\omega) = \arg(Y(j\omega))$ (so $Y(j\omega) = M(\omega)e^{jN(\omega)}$) the magnitude plot is the graph of $M(\omega)$ versus ω. It is common engineering practice to use a logarithmic scale $\log_{10} M(\omega)$ versus $\log_{10}\omega$ for the magnitude plot, and phase angle in degrees versus $\log_{10}\omega$ for the phase plot. $\log_{10}\omega$ is used in place of ω since this allows a much greater range of values to be included, and $\log_{10} M(\omega)$ is used since it makes graphical manipulations somewhat easier (the plot for a product of functions becomes the sum of the plots). Amplitude and phase plots for a typical second-order transfer function

$$Y_2(s) = \frac{1}{1 + 2as + s^2}, \qquad 0 < a < 1 \tag{4}$$

are shown in Figures 9.7 and 9.8. (See [9, ch. 15] for further examples.) After stating the following corollary to Theorem 1 (b), we will be ready to attack the rubber band–weight modeling problem.

Corollary. *Consider an equation of the form* (1) *with real coefficients and with* $u(t) = \sin\omega t$. *Then* $y_{ss}(t) = M(\omega)\sin(\omega t + N(\omega))$, *where* $Y(j\omega) = M(\omega)e^{jN(\omega)}$.

PROOF. The steady-state solution $y_{ss}(t)$ of $y^{(n)} + a_{n-1}y^{n-1} + \cdots + a_0 y = \sin\omega t$ is the imaginary part of the steady-state solution $\tilde{y}_{ss}(t)$ of

$$y^{(n)} + a_{n-1}y^{(n-1)} + \cdots + a_0 y = e^{j\omega t}.$$

Thus

$$\tilde{y}_{ss}(t) = Y(j\omega)e^{j\omega t} = M(\omega)e^{jN(\omega)}e^{j\omega t} = M(\omega)e^{j(\omega t + N(\omega))}$$

$$= M(\omega)[\cos(\omega t + N(\omega)) + j\sin(\omega t + N(\omega))]$$

and

$$y_{ss}(t) = M(\omega)\sin(\omega t + N(\omega)). \qquad \square$$

The meaning of this corollary is that if the input to the system is $\sin\omega t$, then the output has frequency ω, amplitude $M(\omega)$, and phase shift $N(\omega)$.

3.4 Linearity and Time Invariance

The mathematical results noted in Sections 3.2 and 3.3 are valid only for linear time-invariant differential equations. Since these results lead directly to a procedure for determining a model for the dynamics of such an unknown

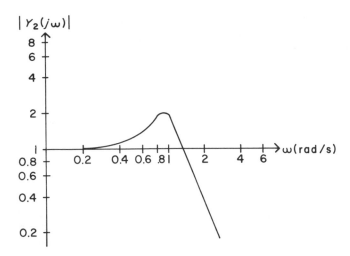

Figure 9.7. Magnitude Plot of Second-Order System (Equation (4) with $a = 0.25$). Note log coordinates.)

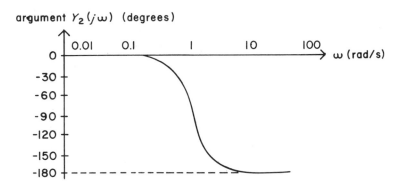

Figure 9.8. Phase Plot of Second-Order System (Equation (4) with $a = 0.25$)

system from input and output data, it would be convenient to have tests to check to see if these conditions (linearity and time invariance) are satisfied or approximately satisfied for certain regions of operation.

In many cases the question of time-invariance can be settled *a priori* by considering the system and the environment in which it operates. For example, in the rubber band–weight experiment none of the parameters of the system (e.g., weight, elasticity of the rubber band) vary with time. However, environmental parameters (e.g., wind) could vary with time. Naturally, one should attempt to perform this experiment in a uniform environment to eliminate the possibility of time-variable dynamics.

The question of linearity is more difficult. It is very rare for a physical system to be absolutely linear over a wide range of operating conditions. However, good results can be obtained with linear models if the physical system is near-linear over the range of operating conditions of interest (i.e., the system need not be near-linear for *all* parameter values). Systems describable by (1) have the property that, for each input–output pair (u_1, y_1) and (u_2, y_2),

(i) $u_1 + u_2$ is a legitimate input and the output corresponding to this input is $y_1 + y_2$;

(ii) for any constant α the input αu_1 is legitimate and the corresponding output is αy_1.

To test whether a system may be described by (1), one can apply various inputs and sums of inputs and observe the outputs. If i) or ii) are true over a certain frequency range of inputs, one may adopt as a working hypothesis that the system may be adequately so described. The corollary to Theorem 1 provides another test. The output to a sinusoidal input of frequency ω must also have frequency ω, although generally a phase shift occurs.

3.5 The Experimental Procedure

The corollary of Section 3.3 was formulated so that the mathematical results derived there could be applied directly to an especially convenient experimental input, $\sin \omega t$. To perform the experiment, attach one end of the rubber band string to your finger and have an aide to record observations.

1. Move your hand up and down very slowly, imagining that your hand is tracing a sine function of *constant, very low* frequency, e.g., one cycle in four or five seconds. The amplitude of the oscillation should be roughly 3–9 in and as constant as you can make it. The result should be that the weight moves just as your hand does; i.e., the amplitude is the same, and the phase change is nearly imperceptible. Plot these values on the frequency response plots, i.e., amplitude ratio $\equiv M(\omega) \approx 1$, phase $\equiv N(\omega) \approx 0°$ (where ω very small). The plotted points should correspond roughly to points ① on Figures 9.9 and 9.10.

2. Gradually increase the frequency of your hand motion, but keep the amplitude roughly constant. You will begin to detect a change in the relative motion of the weight with respect to the hand. Estimate roughly this frequency and the resultant amplitude ratio and phase lag. (The amplitude of the weight should be larger than the amplitude of the hand, and the weight should be lagging behind the hand motion.) Plot these points on the frequency response plots; they should correspond roughly to points ② on Figure 9.9 and 9.10.

3. Again increase the frequency (with constant amplitude of the hand) in discrete steps, and roughly estimate the frequency, amplitude ratio, and phase lag at the point where the weight has its largest amplitude; the motion of the weight at this point may be somewhat wild. (This is indicating the *region of resonance* of the system, i.e., the input frequency is close to the

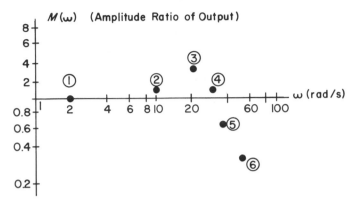

Figure 9.9. Amplitude Ratio of Position of Weight for $\sin \omega t$. Input by Hand

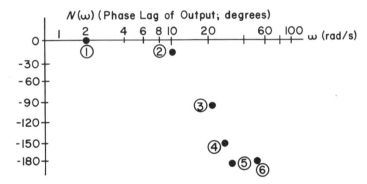

Figure 9.10. Phase Lag of Position of Weight with Respect to Position of Hand

natural frequency of the system.) Plot these points on the frequency response plots; they should correspond roughly to points ③ on Figures 9.9 and 9.10. If the experiment is done carefully, the phase shift should be almost exactly $-90°$.

4. Continue as above and plot points for higher and higher frequencies. One can clearly see that the amplitude of the oscillations decreases. However, it is difficult to see (with the eye) that the phase lag is increasing (because of the low-amplitude oscillation of the weight). The phase shift should approach $-180°$ as ω increases (points ④, ⑤, ⑥, on Figures 9.9 and 9.10).

Now that the frequency response plots are completed, the mathematical model of the system can be constructed. The following theorem is especially helpful in this connection.

Theorem 2. *Suppose* (1) *is stable and that Y is the transfer function. Then*
a) $\lim\limits_{\omega \to \infty} M(\omega) = \lim\limits_{\omega \to \infty} |Y(j\omega)| = 0.$

b) $\lim\limits_{\omega \to \infty} N(\omega) = \lim\limits_{\omega \to \infty} (\text{argument } Y(j\omega)) = -n\pi/2 \text{ (rad)}$

where n is the degree of (1).

PROOF. Since

$$\lim_{\omega \to \infty} \omega^n Y(j\omega) = j^{-n},$$

it follows that

$$|Y(j\omega)| \text{ is asymptotic to } |\omega^{-n}|,$$

from which we conclude that a) holds and also that

$$\arg(Y(j\omega)) \arg(j^{-n}) = n(-\pi/2)$$

from which we conclude b).   □

   This theorem gives us a mathematical basis for estimating the order of the differential equation which models our dynamical system *provided the high-frequency data are fairly accurate* (recall the difficulty we had in measuring the high-frequency phase angle).

   In our example, we estimate the order by determining the asymptotic value of the phase angle in Figure 9.10. This value is $-\pi = 2(-\pi/2)$ which implies that the dynamical system can be approximately modeled by a second-order differential equation.

   Note: It may be shown (as an exercise) that the slope of the magnitude plot as $\omega \to \infty$ also gives the order of the differential equation:

$$\lim_{\omega \to \infty} \frac{d\log_{10}|Y(j\omega)|}{d\log_{10}\omega} = -n.$$

   Now that we know a second-order differential equation provides an approximate model for the system, we can complete the model by using a least-squares curve fit of the magnitude plot and a parameter optimization routine. That is, we have deduced that the equation

$$\frac{d^2y}{dt^2} + a_1\frac{dy}{dt} + a_0y = u$$

is a good candidate for a mathematical model of the system and we need only determine the parameter values a_0, a_1 which give a good curve fit of the data in Figure 9.9.

   Note: Bode [10], [12] proved that, for stable differential equations of the form (1), the phase plot is uniquely defined by the magnitude plot. Thus the least-squares curve fit need only be done on the magnitude plot.

3.6 Determination of a_0, a_1 (Least Squares and Function Minimization)

At this point we know that the transfer function has the form

$$Y(s) = \frac{1}{s^2 + a_1s + a_0}$$

for some constants a_0 and a_1, and we have a set of m experimental measurements of the form (ω_i, M_i) where M_i is the observed amplitude for input ω_i. The problem now is to determine the constants a_0 and a_1 in such a way that $M(\omega) = |Y(j\omega)|$ fits the observed data in the best possible way, where "best" remains to be defined.

For each i we have an observed value M_i and a theoretical value $|Y(j\omega_i)|$. The difference $|Y(j\omega_i)| - M_i$ is the "error" in the ith measurement. We want to choose a_0 and a_1 in such a way that the sum of the squares of the errors is minimized. That is, we seek to minimize the function

$$f(a_0, a_1) = \sum_{i=1}^{m} (|Y(j\omega_i; a_0, a_1)| - M_i)^2$$

as a function of a_0 and a_1. The choice of least-square error as a measure of fit is a common one which can be supported on several grounds. There is also an element of arbitrary choice in selecting this over alternative measures.

If one suspects that the high-frequency data are not as reliable as the rest, one might wish to de-emphasize that data by minimizing instead the function

$$f(a_0, a_1) = \sum_{i=1}^{m} W_i(|Y(j\omega_i; a_0, a_1)| - M_i)^2 \tag{5}$$

where $W_1, W_2, \cdots, W_n$ are constant weighting parameters. The choice of weights will depend upon the particular data and experimental procedure.

Many canned computer subroutines are available for the determination of a_0 and a_1, given (5). Examples are FMCG, FMFP, and NEWT in the IBM Scientific Subroutine manual [13]. Alternatively, it is not difficult to implement algorithms based on either the gradient method or Newton's method.

The gradient method begins with a guess

$$\begin{bmatrix} a_0^{(0)} \\ a_1^{(0)} \end{bmatrix}$$

and constructs a sequence

$$\begin{bmatrix} a_0^{(n)} \\ a_1^{(n)} \end{bmatrix}$$

based on the formula

$$\begin{bmatrix} a_0^{(J+1)} \\ a_1^{(J+1)} \end{bmatrix} = \begin{bmatrix} a_0^{(J)} \\ a_1^{(J)} \end{bmatrix} - \alpha_J \begin{bmatrix} \dfrac{\partial f}{\partial a_0}(a_0^{(J)}, a_1^{(J)}) \\ \dfrac{\partial f}{\partial a_1}(a_0^{(J)}, a_1^{(J)}) \end{bmatrix}.$$

The positive scalar α_J is chosen at each step so as to lead to the greatest decrease in f possible at the Jth step. (See [14], [15] for details of both the theory behind the algorithms and methods for determining α_J.)

Newton's method is based on the iteration

$$
\begin{bmatrix} a_0^{(J+1)} \\ a_1^{(J+1)} \end{bmatrix} = \begin{bmatrix} a_0^{(J)} \\ a_1^{(J)} \end{bmatrix} - D_J^{-1} \begin{bmatrix} \dfrac{\partial f}{\partial a_0}(a_0^{(J)}, a_1^{(J)}) \\[2ex] \dfrac{\partial f}{\partial a_1}(a_0^{(J)}, a_1^{(J)}) \end{bmatrix}
$$

where

$$
D_J = \begin{bmatrix} \dfrac{\partial^2 f}{\partial a_0^2}(a_0^{(J)}, a_1^{(J)}) & \dfrac{\partial^2 f}{\partial a_0 \partial a_1}(a_0^{(J)}, a_1^{(J)}) \\[3ex] \dfrac{\partial^2 f}{\partial a_1 \partial a_0}(a_0^{(J)}, a_1^{(J)}) & \dfrac{\partial^2 f}{\partial a_1^2}(a_0^{(J)}, a_1^{(J)}) \end{bmatrix}.
$$

In each case, if all goes well, the sequence

$$
\begin{bmatrix} a_0^{(J)} \\ a_1^{(J)} \end{bmatrix}
$$

will converge to the values

$$
\begin{bmatrix} a_0^* \\ a_1^* \end{bmatrix}
$$

that minimize $f(a_0, a_1)$. There are advantages and disadvantages to each method.

1) If a poor initial estimate of the parameters is used in a Newton iteration, the sequence may not converge. If the estimate is good, convergence is usually very rapid.

2) The gradient method will always decrease the function $f(a_0, a_1)$ on successive iterates, but the rate of convergence may be intolerably slow.

The canned algorithms FMCG and FMFP use the conjugate gradient method and the Davidon–Fletcher–Powell formulas, respectively, which were designed to overcome the undesirable properties of the classical gradient and Newton methods. See [14] for the theory behind the algorithms and [15] for some representative simulations.

3.7 Model Verification

Let us now assume that optimal values $\tilde{a}_0$, $\tilde{a}_1$ of the parameters a_0 and a_1 have been determined. The following questions should be asked.

(1) Is the original assumption that the system is nearly linear a justifiable one? How can it be checked?
(2) What is the physical significance of $\tilde{a}_0$? How can the conjectured significance be checked?
(3) What is the physical significance of $\tilde{a}_1$? How can the conjectured significance be checked?

In model verification, one can only hope to develop *necessary conditions*, and in this sense, every model is tentative. For example, let us recall how one might answer the first question above. If the system is indeed linear, then the output for the sum of *any* two inputs $u_1 + u_2$ must be the same as the sum of the respective outputs for the separate inputs u_1 and u_2; and the output for the input αu ($\alpha = $ constant) must be equal to α multiplied by the output for the input u. Since only a finite number of inputs can ever be tested, we only accumulate necessary conditions for the validity of the model with such tests. In practice, one uses such tests to delineate regions of linear operation (e.g., our system is not truly linear, but for a certain frequency range it is very close to linear).

Finally, let us address the second and third questions raised at the beginning of this section, i.e., the physical characterization of a_0, a_1. The system consists of only two different elements: the rubber band string and the weight. Thus these must be the main contributors to the physical characterization of a_0, a_1. Also, the environment of the experiment must be considered, since it will affect to some degree the values of a_0, a_1. For example, suppose the same experiment was conducted under water; then the environment of the experiment might not allow us to determine an accurate model. This point is of major importance in the modeling of physiological systems in that the organ to be modeled cannot be isolated in a reasonable experimental environment.

The atmosphere will affect the values of a_0, a_1, but if the weight is heavy enough, this effect can be minimized. Assuming that a_0, a_1 are mainly determined by the rubber bands and weight, let us now determine how they are related. Our first thought might be that each parameter is influenced by only one of the components, e.g., a_0 is influenced only by the weight and a_1 is influenced only by the rubber band's elasticity. To check these hypotheses, the student can vary separately the weight and elasticity of the rubber bands and repeat the estimation of the parameters a_0, a_1. Actually, the system is essentially a spring-mass-damper system which is shown (idealized) in Figure 9.11. The usual model for such a system (based on Newton's laws, see [21, p. 2]) leads to the equation

$$m\ddot{y} + c\dot{y} + ky = 0 \qquad (6)$$

or

$$\ddot{y} + \frac{c}{m}\dot{y} + \frac{k}{m}y = 0, \qquad (7)$$

where m is the mass of the weight, c is the linear damping parameter of the rubber band, and k is the linear spring constant of the rubber band. Thus m is mainly associated with the weight, and c, k are associated with the rubber bands. The connection between our parameters a_0 and a_1 and the parameters of this system is

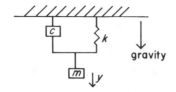

Figure 9.11. Spring-Mass-Damper Representation of System

$$a_0 = \frac{k}{m} \qquad a_1 = \frac{c}{m}. \tag{8}$$

Thus if enough experiments are conducted with various weights and rubber bands of various thicknesses, the data should indicate (roughly) these relationships.

4. Project: Human Operator Experimental Data

In this section data from an actual experiment will be presented for analysis and interpretation. Before presenting the data, one further mathematical property involving a nonlinearity present in all human operators is needed. This nonlinearity is called the *inherent time delay* or transport lag, and it is due to the fact that the human operator cannot react instantaneously to a stimulus. (This delay time is roughly 0.2–0.4 sec. for straightforward compensatory tasks.) Even though this is a nonlinearity, it may be incorported very easily into linear frequency domain analyses because of the following fact.

Theorem 3. *Suppose that the Laplace transform* W *of* u *exists for* $s \geq s_0$, *and define* y *by*

$$y(t) = \begin{cases} 0, & t \leq \tau \\ u(t - \tau), & t > \tau. \end{cases}$$

Then the Laplace transform $L[y]$ *is related to* W *by*

$$L[y](s) = e^{-\tau s} W(s), \qquad \text{for } s > s_0.$$

This is a standard fact concerning Laplace transforms. It means that if y is viewed as the output and u as the input, then y follows u exactly but is delayed by τ units of time.

If a block with a nonlinearity of this type is inserted into a feedback control loop (such as Figure 9.1), it presents little difficulty if frequency-domain techniques are employed. Indeed, if the transfer function Y is of the form $Y(s) = e^{-\tau s}$, then

$$|Y(j\omega)| = 1, \qquad \text{and} \qquad \arg Y(j\omega) = -\tau\omega.$$

That is, the magnitude of the frequency response is unaffected by $Y(s)$ whereas the phase is decreased by $\tau\omega$.

Experimental data. The data for this experiment were developed by Elkind [16] in one of the earliest comprehensive human operator modeling projects. The basic setup of the experiment is shown in Figure 9.12. This setup differs from Figure 9.1 in that no vehicle dynamics (or plant dynamics) are in the loop. Experiments with vehicle dynamics are presented in [2], [3], [8], [17], and [18]. This part of the experiment was mainly concerned with determining transfer function models of the human operator for inputs of various bandwidths, i.e., ranges of frequencies of the components. The magnitude and phase frequency response plots resulting from the experiment are shown in Figure 9.13.

The four curves on each plot correspond to four inputs with different bandwidths, with R.16 and R.64 representing the smallest and largest bandwidths, respectively. The input labeled R.64 has higher frequency components than the other inputs. (The designations, R.16, etc., refer to the fact that the input functions have rectangular power spectra with the specified cutoff frequency, 16 c/s in the case of R.16. See [17a] for the definitions and details.) The dynamics of the display are neglected, so the plots represent the transfer function of the human operator.

Points for Discussion and Analysis.

1. Discuss the relation between the human operator's response and the bandwidth of the input. (Rough answer: As the bandwidth increases, the magnitude of the response decreases and the lag in the phase decreases. Thus the operator makes smaller motions if the input has higher frequency components, and this probably aids in phase synchronization.)

2. What would be a reasonable transfer function for the data of Figure 9.13? Elkind proposed

$$Y_{P_1} = \frac{K e^{-\tau s}}{(Ts + 1)}. \tag{9}$$

Curve fitting *both* the magnitude and phase curves results in the following values [17b].

Input	$\tau(s)$	$T(s)$	$K(dB)$
R.16	0.64	4.55	34.5
R.24	0.264	3.18	31.5
R.40	0.214	1.27	22.5
R.64	0.183	0.58	15.0

(If computer routines are available, a parameter optimization scheme may be employed to solve for the parameters; otherwise, rough estimates can be made of the trends of τ, T, and K as functions of the input bandwidth.)

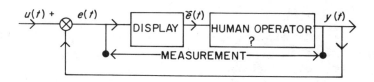

Figure 9.12. Experiment Setup

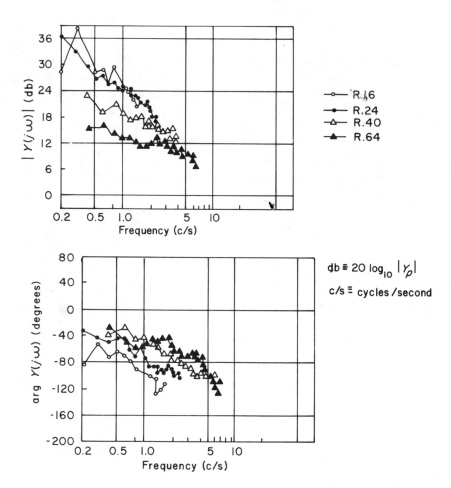

Figure 9.13. Experimentally Derived Frequency Response Plots [17b, fig. 29]

These parameter values imply that the "gain" (K) of the human operator decreases with increasing bandwidth, which agrees with intuition. The term $(Ts + 1)^{-1}$ is called a "first-order lag" since it causes an increase in the phase lag, and it has been attributed to a neuromuscular lag, as opposed to the sensory transport lag τ. Note that T approached 0 as the input bandwidth increased, which implies the human operator tends to behave more as a pure gain and pure transport time-delay with increasing bandwidth.

In [17] it is noted that if the model of (16) is inserted into a feedback control loop, the resultant feedback system is unstable at high frequencies. Thus the following model is proposed to represent a stable feedback system:

$$Y_{P_1} = \frac{K e^{-\tau_1 s}}{(T_0 s + 1)(T_1 s + 1)}. \tag{10}$$

With this model, the following values [17a], [17b] are obtained.

Input	τ_1(s)	T_0(s)	T_1(s)	K(dB)
R.16	0.110	4.55	0.531	34.5
R.24	0.104	3.18	0.161	31.5
R.40	0.133	1.27	0.081	22.5
R.64	0.150	0.58	0.033	15.0

These values are considered reasonable. Typically, $0.1 \leq \tau_1 \leq 0.2$ and is attributed mainly to the transport lag; T_1 is attributed to neuromuscular lag; and T_0 is a lag which is dependent upon what is being controlled and the bandwidth of the input. Further discussions are presented in [17] and [18].

5. Further Topics

The generation of the data presented in Section 4 involve random inputs and correlation methods. The resultant human operator models are more accurately labeled "random input describing functions," the theory of which is presented in [19]. However, they are treated like transfer functions in the analysis of control systems and in the interpretation of models. A student with knowledge of stochastic processes should be able to follow the presentation in [19] if more insight into the data generation process is desired.

Current research in this area includes, among other things, the development of models for more than a single task and the application of optimal control and state estimation theory to the development of models (which assumes the human operator is an optimal information processor and controller subject to inherent constraints). Current results are usually presented at the Annual Conference on Manual Control, which publishes a *Proceedings* released either by NASA, the United States Air Force, or a sponsoring university (e.g., see NASA SP-144, NASA SP-214).

References

[1]   †N. Wiener, *Cybernetics.*   Cambridge, MA: MIT Press, 1948, 1961 (paperback).
[2]   H. R. Jex *et al.*, "A study of fully-manual and augmented-manual control systems for the Saturn V booster using analytical pilot models," National Aeronautics and Space Administration, Rep. NASA CR-1079, 1968.
[3]   D. H. Weir and D. T. McRuer, "A theory for driver steering control of motor vehicles," *Highway Res. Rec.*, vol. 247, pp. 7–28, 1968.
[4]   †D. T. McRuer, "Development of pilot-in-the-loop analysis," *AIAA J. Aircraft*, vol. 10, no. 9, pp. 515–522, Sept. 1973.
[5]   †J. H. Milsum, *Biological Control Systems Analysis.*   New York: McGraw-Hill, 1966, pp. 242–252, 418–425.
[6]   G. A. Bekey, "Parameter estimation in biological systems: a survey," in *Identification and System Parameter Estimation*, P. Eykhoff, Ed.   New York: American Elsevier, 1973.
[7]   P. Eykhoff, *System Identification.*   New York: Wiley, 1974, ch. 14.
[8]   A. Tustin, "The nature of the operator's response in manual control and its importance for controller design," *J. Inst. Elec. Engineers* (England), vol. 94, Part IIA, no. 2, pp. 190–202, 1947.
[9]   ††J. J. DiStefano, A. R. Stubberud and I. J. Williams, *Schaum's Outline on Feedback and Control Systems.*   New York: McGraw-Hill, 1967.
[10] J. L. Melsa and D. G. Schultz, *Linear Control Systems.*   New York: McGraw-Hill, 1969.
[11] R. A. Struble, *Nonlinear Differential Equations.*   New York: McGraw-Hill, 1962.
[12] H. W. Bode, *Network Analysis and Feedback Amplifier Design.*   Princeton, NJ: Van Nostrand Reinhold, 1945.
[13] IBM System/360 Scientific Subroutine Package (360-CM-03X), Version III, IBM Technical Publication Dept., 1968.
[14] J. M. Ortega and W. C. Rheinboldt, *Iterative Solution of Nonlinear Equations in Several Variables.*   New York: Academic, 1970.
[15] †D. M. Himmelbrau, *Applied Nonlinear Programming.*   New York: McGraw-Hill 1972.
[16] J. I. Elkind, "Characteristics of simple manual control systems," MIT Lincoln Lab., Tech. Rep. 111, Apr. 6, 1956.
[17a] †D. T. McRuer and E. S. Krendel, "The human operator as a servo system element," *J. Franklin Inst.*, vol. 267, nos. 5 and 6, pp. 381–403 and 511–536, May and June, 1959.
[17b] USAF Wright Air Development Center Rep. 56–524, Oct. 1957. (The details for [17a] are contained in this report.)
[18a] †D. T. McRuer, D. Graham and E. S. Krendel, "Manual control of single-loop systems," *J. Franklin Inst.*, vol. 283, nos. 1 and 2, pp. 1–27 and 145–170, Jan. and Feb. 1967.
[18b] USAF Flight Dynamics Lab. Rep. AFFDL-TR-65-15, July 1965. (The details for [18a] are contained in this report.)
[19] †D. Graham and D. T. McRuer, *Analysis of Nonlinear Control Systems.*   New York: Dover, 1961, 1971 (paperback), ch. 6.
[20] A. Rabenstein, *Introduction to Ordinary Differential Equations.*   New York: Academic, 1972.

† indicates readily available background reference.

†† indicates essential reference.

[21] F. Brauer and J. A. Nohel, *Ordinary Differential Equations*. New York: Benjamin, 1967.
[22] W. E. Boyce and R. C. DiPrima, *Elementary Differential Equations and Boundary Value Problems*, 3rd ed. New York: Wiley, 1977.
[23] M. Braun, *Differential Equations and Their Applications*, 2nd ed. New York: Springer-Verlag, 1978.

Notes for the Instructor

Objectives: From a scientific point of view, this module is concerned with illustrating some of the problems that arise in the development of models of the performance of human subjects on certain tasks, such as steering. In particular, it demonstrates the use of frequency response techniques in mathematical modeling. Although the module is intended as a case study in modeling, Section 3 is basically self-contained and may be appropriate for use in a course on ordinary differential equations. A brief outline of the various sections is as follows.

Section 1: The basic type of human behavior to be modeled is described, and previous applications of these models are noted. This section is mainly descriptive.

Section 2: Some of the problems in obtaining data from experiments involving human subjects are discussed, as well as the basic experimental setup. The motivation for employing frequency response mathematical methods is also presented. Again, this section is mainly descriptive.

Section 3: A specific mathematical model is formulated and analyzed. Important results are contained in Theorem 1 (especially part c), Theorem 2, and the Corollary. Both the mathematical and the intuitive aspects of these results are helpful in understanding the model. It may be worthwhile for the instructor to familiarize the students with the basic aspects of frequency response plots beforehand. The manner in which Section 3.6 is handled will depend upon the time available; the goal is to demonstrate the roles of least squares and function minimization techniques in parameter estimation. In Section 3.7 the very difficult problem of model verification is considered. In addition to the specific technical questions which arise in verifying the model discussed in this module, one can raise general philosophical questions about the contribution of mathematical modeling at this point.

Section 4: This section contains data from actual human operator experiments. The instructor may use the data in a variety of ways; e.g., 1) give it to students with only an explanation of the experimental procedure and have them hypothesize the model, estimate the parameters of the model, and present arguments for and against the model with respect to intuition or 2) give the students the two models (9) and (10) along with the data and have them discuss the trends of the parameters with increasing bandwidth and determine the physical basis for each component in the models (this approach would not require any numerical procedures).

Section 5: Current research areas are briefly discussed and examples of references of recent developments are provided. (Such information would be of primary interest to students who wish to pursue the subject further.)

Prerequisites. Calculus, differential equations, Laplace transforms.

Time. Four or five lectures are needed to cover the material.

Number of Students on Project. Although the project could be done alone, it is recommended that either two or three students work together.

Subroutines (Optional). IBM Scientific Subroutines for Function Minimization (e.g., FMCG, FMFP, NEWT).

Potential Local Experts. Faculty in engineering.

PART IV

TRAFFIC MODELS

CHAPTER 10

How Long Should a Traffic Light Remain Amber?

Donald A. Drew*

1. The Problem and the Model

Let us consider the problem of calculating how long a traffic light should remain amber before turning red. Essentially, the amber cycle exists to allow vehicles in the intersection, or those too close to stop, to clear the intersection. Thus the light should remain amber long enough so that all drivers who cannot stop have a chance to pass through the intersection on the amber. A driver approaching an intersection should never be in the dilemma of being too close to stop safely and yet too far away to pass through the intersection before the red phase starts.

A "rule of thumb" exists for this calculation. Allow 1 s of amber for each 10 mi/h legal approach speed. Let us see if a theoretical calculation confirms this rule of thumb.

A driver approaching an intersection who gets an amber signal has a decision to make: whether to stop or to pass through the intersection. If he is traveling at the legal speed (or below it), he must have an adequate distance in which to stop, if he should decide to stop. If he should decide to pass through the intersection, he must have adequate time to pass completely through the intersection, including a time interval in which to decide to stop (reaction time) and the time it would take to drive the minimum distance needed to stop. Hopefully, drivers who see the amber soon enough will use the braking distance to stop their vehicles.

Thus the amber phase should be long enough to include a driver's reaction time, plus the time it would take him to drive through the intersection, plus

* Department of Mathematical Sciences, Rensselaer Polytechnic Institute, Troy, NY 12181.

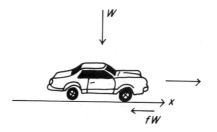

Figure 10.1. Friction Force

the time it would take him to drive the distance which would be required to brake his vehicle to a halt (the braking distance). Drivers having this much time would be able to stop safely within the braking distance.

If the legal speed is v_0, the width of the intersection is I, and the typical vehicle length is L, then the time to *clear* the intersection is $(I + L)/v_0$. (Note that the rear end of the vehicle must clear the intersection, thus the effective length of the intersection is $I + L$.)

Now, let us compute the braking distance. We note that the actual braking and stopping process is quite complicated, with drivers decelerating by first releasing the pressure on the accelerator, then stepping on the brake pedal with varying degrees of hardness (perhaps pumping it) until the vehicle stops. We shall bypass most of these processes; they are indeed difficult to model exactly.

We shall assume that the effect of braking the vehicle can be modeled by the introduction of a resisting frictional force upon the application of the brakes. Suppose W is the weight of the vehicle and f is the coefficient of friction. Then by *definition*, the braking force on the vehicle is fW, opposite to the direction of motion (see Figure 10.1). The distance traveled in stopping can be found by solving the differential equation of motion in a straight line subject to a constant force $-fW$. Thus

$$\frac{W}{g} \cdot \frac{d^2x}{dt^2} = -fW, \tag{1}$$

where g is the gravitational acceleration.

The correct conditions to impose on x are that at $t = 0$, $x = 0$, and $dx/dt = v_0$. The braking distance, then, is the distance traveled when $dx/dt = 0$.

2. The Solution

Integrating (1) subject to $dx/dt\,(0) = v_0$ gives

$$\frac{dx}{dt} = -fgt + v_0. \tag{2}$$

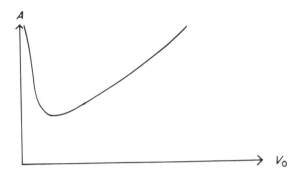

Figure 10.2. Amber Cycle versus Legal Speed

Hence the speed is zero when $t = t_b = v_0/fg$. Integrating (2) subject to $x(0) = 0$ gives

$$x = -fgt^2/2 + v_0 t. \tag{3}$$

The value of x when $t = t_b$ is

$$x(t_b) = D_b = \frac{v_0^2}{2fg}. \tag{4}$$

Caution with units must be used at this point: we are dealing with a distance D_b which is measured in feet by traffic engineers, and a speed v_0 which is usually measured in miles per hour. The reader should convert miles to feet before making any calculations.

Let us plunge ahead to computing the amber phase:

$$A = \frac{D_b + I + L}{v_0} + T,$$

where T is the driver reaction time. Thus

$$A = \frac{v_0}{2fg} + \frac{I + L}{v_0} + T.$$

If we sketch the graph of A versus v_0, it looks like Figure 10.2.

We shall assume that $T = 1$ s, $L = 15$ ft, and $I = 30$ ft. Moreover, we shall accept the word of a highway engineer that $f = 0.2$ is representative. (See Exercise 1.) The amber cycles for $v_0 = 30$, 40, and 50 mi/h are shown in Table 1, along with the rule of thumb values.

Table 1

v_0(mi/h)	A	Rule of Thumb
30	5.46 s	3 s
40	6.35 s	4 s
50	7.34 s	5 s

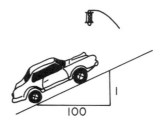

Figure 10.3

Note that the rule of thumb is consistently shorter than our predicted amber phases. This suggests that many intersections are designed so that it is quite possible that vehicles will be in the intersection when the light changes to red.

Even given adequate stopping time, many motorists will attempt to accelerate and try to beat the light. Most do not know (and some do not even care) when the light will turn red. A partial solution may lie in a "countdown" type of traffic light, where during the last few seconds of amber a countdown sequence of digits superimposed on the amber light warns the drivers exactly when the light will change. A system like this one was tried several years ago in Houston with some success in lowering accident rates. See Exercise 6.

Exercises

1. A vehicle traveling on dry, level pavement at 40 mi/h has the brakes applied. The vehicle travels 270 ft before stopping. Compute the coefficient of friction.

2. Suppose the intersection is such that all vehicles entering from the east are traveling up a grade of 1 ft/100 ft (Figure 10.3). Compute the braking distance for this case. *Hint:* The normal force is reduced, and an additional stopping force due to gravity is present. What is the distance for vehicles approaching from the west?

3. With better tires and brakes, the realistic value of f can become as great as 0.4. Make a table of amber cycles for $v_0 = 30, 40$, and 50 mi/h, for this value of f. These values are less conservative than the values in Table 1. Comment on the desirability of designing stop lights using more conservative versus less conservative models.

4. In actuality, drivers do not all approach the intersection at the same speed v_0. Assume that the drivers approach at speeds between $v_0 - (\Delta v/2)$ and $v_0 + (\Delta v/2)$ with equal probability. Calculate the average amber cycle based on this information. *Hint:* $\bar{A} = \dfrac{1}{\Delta v} \displaystyle\int_{v_0 - \Delta v/2}^{v_0 + \Delta v/2} A(v_0)\, dv_0$. Why?

5. What is the value of the friction f such that the rule of thumb is exactly correct at $v_0 = 40$ mi/h? Make a table of amber cycles similar to Table I for this value of f.

6. Can you think of other ways to reduce accidents at intersections?

Reference

[1]   D. R. Drew, *Traffic Flow—Theory and Control.*   New York: McGraw-Hill, 1968.

Notes for the Instructor

Objectives. Elementary integration is used to compute the stopping distance for a vehicle. The relation of the stopping distance to the amber cycle is discussed.

Prerequisites. A knowledge of elementary mechanics, friction, and calculus.

Time. The module can be covered in one lecture.

CHAPTER 11
Queue Length at a Traffic Light via Flow Theory

Donald A. Drew*

1. The Problem: A Simple Model

Let us consider the following design problem. Suppose we have a road which, under normal operation, has a traffic flow q of 1000 veh/h at a flow concentration k of 20 veh/mi. There are two intersections on this road, 0.2 mi miles apart. We wish to install a stoplight at one of these intersections. Our primary interest is whether the queue which forms at the downstream light backs up to the upstream intersection (see Figure 11.1). Clearly, if the downstream queue does interfere with the operation of the upstream intersection, then other considerations must be made, possibly installing a synchronized light at the upstream intersection.

The modeler's first task is to understand the problem and to be reasonably sure that enough information is given to get a solution. If the problem statement is not clear, or needed information is obviously missing, the modeler is obligated to confront the proposer (a highway engineer, perhaps) for additional information or clarification. When we do this with our traffic problem, we get the important information that the light will be green for 2 min out of a 3 min cycle. (We have already *assumed* that what happens at the upstream intersection is unimportant.)

Let us try to answer the question using a simplified model. (A simplified model is one in which we neglect many things at once in the hope of obtaining a quantitative estimate that will be useful to the proposer and will also help us in checking the estimates obtained from better models.) Let us assume

* Department of Mathematical Sciences, Rensselaer Polytechnic Institute, Troy, NY 12181.

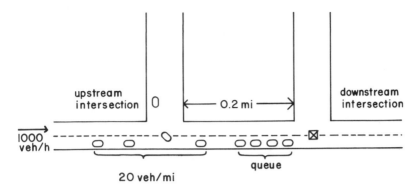

Figure 11.1. Diagram of Traffic Intersections

that we add vehicles to the queue at the rate at which they arrive at the downstream intersection. Since that flow rate is 1000 veh/h, 16 2/3 vehicles will arrive at the intersection in 1 min.[1] If the light is red, then these vehicles must have queued for that 1-min period. The resulting queue has 16 2/3 vehicles. The length can be calculated from knowing how much distance is occupied by each vehicle in the queueing mode. Our proposer gives us a slightly different number, the jam concentration $k_j = 257$ veh/mi. The above situation is not unusual; often the exact information we want must be extracted from other data. Often the data are not in quite as easy a form to reckon with as this model. For example, in a fluid mechanical problem, we may need to know a value of the pressure. Our proposer may suggest that the data are contained in a book of stream tables.

In our problem, 257 veh/mi translates into $1/257 \cong 0.00389$ mi/veh, so that 16 2/3 vehicles occupies 0.065 mi, well short of the 0.2 mi separating the intersections. Based on this information, we then suggest to the proposer that the lights at one intersection will probably not affect the other intersection. Notice, also, that we have not used one piece of information, the concentration of the flowing stream.

Our proposer returns at a later date and informs us that he has installed the light and the queue is much longer than 16 2/3 vehicles at most times when the flow rate is 1000 veh/h. In fact, he has observed that the length of the queue depends on the concentration. When the concentration reaches about 175 veh/mi, the queue is very long and does interfere with the upstream intersection.

---

[1] Since it seems that a vehicle is either in the queue or not, we should now interpret what it means to have 16 2/3 vehicles in the queue. If we go to a stoplight and take data on the problem, we may find that the recorded queue lengths are 16, 17, 17, 16, 18, and 16 vehicles during six randomly selected cycles of the light. The average of these observations is 16 2/3. Thus our predicted queue length of 16 2/3 is an average queue length.

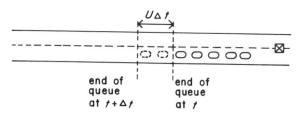

Figure 11.2. Rate at Which Vehicles are Added to Queue

2. A Better Model

One fact which we ignored in our first analysis is that the rear of the queue moves backward as vehicles are added. Essentially, we assumed that the queued vehicles occupy no distance on the highway. In fact, as vehicles are added to the queue, the rear end of the queue moves upstream and therefore intersects traffic at a much faster rate than traffic flows past an arbitrary point on the highway.

From a modeling point of view, we have made a simplification which turns out to be fatal (that is, it gave an answer which was so incorrect that our proposer is probably unhappy). Let us now consider a more detailed analysis which will account for the upstream movement of the rear of the queue.

If U is the relative speed at which the rear of the queue moves backward, then the rate at which vehicles are added to the queue can be expressed two ways. First, the rate is Uk_j. To see this, look at the number of vehicles added to the queue between the times t and $t + \Delta t$ (Figure 11.2). The length of queue added is $U\Delta t$. The number of vehicles added to the queue, then, is $k_j U\Delta t$. Thus the rate at which vehicles are added is Uk_j.

The rate at which vehicles are added to the queue can also be expressed by

$$q + Uk. \tag{1}$$

To see this, consider the *flowing* vehicles at time t and time $t + \Delta t$.

In Figure 11.3b, those vehicles which have flowed through the location of the end of the queue at the instant $t + \Delta t$ during the time interval t to $t + \Delta t$ are added to the queue: this contribution is $q\Delta t$. In addition, those vehicles which lie between the end of the queue at time t and the end of the queue at time $t + \Delta t$ are also added to the queue. This contribution is $kU\Delta t$. Thus the total rate is $q + kU$.

Equating the two rates of addition gives

$$U = \frac{q}{k_j - k}. \tag{2}$$

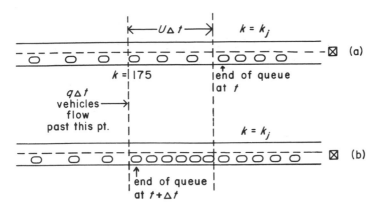

Figure 11.3. Queues at time t and time $t + \Delta t$.

(We note that the calculation which we did before essentially used this result with $k = 0$.)

If we assume that the queue grows for one minute (the length of the red cycle), then its length will be

$$U1 = \frac{q}{k_j - k} 1 \tag{3}$$

which for the heavier operating conditions with $k = 175$ veh/mi is about 0.203 mi. Thus at this heavier concentration we expect the intersections to interact. We might suggest that a light be installed at the upstream intersection and that the light cycles be synchronized so that they complement each other, with the upstream light releasing vehicles so that they arrive at the downstream light as it enters its green cycle.

We have probably satisfied the proposer with our last answer that the intersections will interact. The analysis is still incomplete, however. With a queue of vehicles 0.2 mi long, it is a poor assumption that the queue grows for only the interval of the red cycle. At the end of the red cycle, the lead car begins to move across the intersection, followed a short time later by the second car, and so on. Our proposer tells us that the flow rate q_s for this starting process has been observed to be 1500 veh/h, at a concentration k_s of 50 veh/mi. If the speed of this starting wave is U_s, then an analysis like the previous one gives

$$U_s = \frac{q_s}{k_j - k_s}, \tag{4}$$

which has a value of about 0.12 mi/min, as compared with the speed of the rear of the queue of about 0.20 mi/min. Thus we note that *the queue never really dissipates*; instead, the wave of vehicles starting up travels too slowly to catch up to the rear of the queue (see Fig. 11.4).

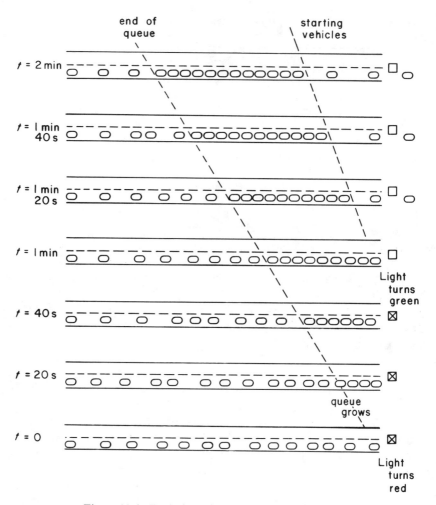

Figure 11.4. Evolution of Queue, with Starting Wave

Exercises

1. Compute the length of the queue for the same light at operating conditions given by
   a) $q = 500$ veh/h, $k = 20$ veh/mi;
   b) $q = 500$ veh/h, $k = 200$ veh/mi;
   c) $q = 1500$ veh/h, $k = 50$ veh/mi.

2. In the above analysis, we have ignored the recycling of the light. Describe the growth of the queues obtained when the light is red for one minute, green for two minutes, red for one minute, and green for two minutes. Use the values of $q = 1000$ veh/h, $k = 175$ veh/mi, $k_j = 257$ veh/mi, $q_s = 1500$ veh/h, and $k_s = 150$ veh/mi.

3. Suppose that we can regulate the concentration and flow upstream, subject to the *Greenshields* relation $q = u_f k(1 - k/k_j)$, where u_f is the free stream flow speed, which we take to be 50 mi/h. Find the value of k at which $U = U_s$, so that the starting wave moves at the same speed as the queue grows. Use $q_s = 1500$, $k_s = 150$.

References

The interested reader can find many approaches to queueing in the literature. The references below are listed in order of increasing difficulty.

[1]   F. A. Haight, *Mathematical Theories of Traffic Flow*.   New York: Academic, 1963.
[2]   D. R. Drew, *Traffic Flow—Theory and Control*.   New York: McGraw-Hill, 1968.
[3]   D. C. Gazis, *Traffic Science*.   New York: Wiley–Interscience, 1974, ch. 2.
[4]   D. L. Gerlough and M. J. Huber, "Traffic flow theory—A monograph," Special Rep. 165, Traffic Res. Board, *National Research Council*, Washington, D C 1975, ch. 8.

Notes for the Instructor

Objectives. Basic conservation and flow ideas are illustrated by considering the rate at which vehicles are added to a queue.

Prerequisites. No calculus is assumed, but the student should be able to handle concepts of speed and number density.

Time. The material can be covered in one lecture.

CHAPTER 12
Car-Following Models

Robert L. Baker, Jr.*

1. Introduction

The automobile is a pervasive feature of modern technological societies despite its accompanying problems of pollution, accidents, and congestion. In the last 25 years a vast amount of literature has been published which we might classify as "traffic science." This science has attempted to understand through modeling and data gathering the traffic processes and how to modify, optimize, and control them.

In this module, the reader is introduced to one subfield of traffic science known as car-following theory (CFT). CFT models the case of single-lane (no passing) relatively dense traffic on long straight highways. While this situation might seem unrealistic to city drivers, it is a situation which frequently does occur for long stretches of the interstate highway system and for shorter stretches of tunnels. CFT has gone through three overlapping stages: initial development (1953–1961, approximately), refinement and testing against experiment (1961–1968), and extension to automatically controlled vehicles (1968–present). CFT is relevant to modern technology since it is employed in virtually all studies of application of automatic control systems to high-speed ground transportation systems.

The minimum mathematical background required of the reader of this module is calculus, differential equations, and the operational calculus of the Laplace transform. To do further reading in some of the references, it would be necessary to have some knowledge of complex variable theory, particularly how it relates to the inversion integral and stability of linear

* IIT Research Institute, 132 Holiday Court, Annapolis, MD 21401.

systems. The author often teaches a junior level, two-semester sequence called engineering mathematics which includes Laplace transforms (and complex variables); students in this sequence would be adequately prepared to read this module. In selecting what aspects of CFT to present, the author's main criterion was that the mathematics be accessible to junior level technically oriented students, not necessarily math majors—a restriction that requires that only the beginnings of the subject be presented here. In addition to the appeal resulting from CFT's agreement with empirical evidence, this is a pleasing application of Laplace transform theory to a problem other than the electrical network and mass–spring systems which occur so frequently in textbooks.

Many of the original papers in CFT appear in the journal *Operations Research* which is probably available in many undergraduate libraries. A reasonably selfcontained chapter on CFT can be found in Drew [4, pp. 330–354]. An exhaustive 140-article bibliography appears in Wilhelm and Schmidt [16], but this publication may be difficult to locate at a school without a civil engineering major.

The author has included some problems for further work by the student to allow some "hands on" experience. Most of the problems can be done by mimicking and extending the results presented in the module. Numerical solutions of the differential equations for specific initial conditions would be very valuable but would require further background of the student than is expected to just read the module.

2. Complexity of Actual Car-Following; Simplicity of Car-Following Models

The situation we wish to model is n vehicles traveling in a straight line with no passing allowed. Any driver knows that the real situation is incredibly complex, with many mechanical and psychological factors entering into the behavior of each vehicle. For example, the speed of a car on a highway is often limited by signs, by the known presence of law enforcement officials, or by the driver's desire to "make time" or perhaps to have a leisurely, scenic drive. All cars are not alike mechanically, some having better acceleration at high speeds than others. Then, too, drivers have differing degrees of awareness of what is happening around them: some drivers are only aware of what their own cars are doing, some have a good idea of the situation several cars ahead and behind, and some are more concerned with the other occupants of the car than with driving.

It is obvious that we cannot hope to incorporate all of the factors mentioned above into a model. What we shall have to do to get a model which is mathematically tractable is to assume homogeneity, i.e., that all drivers in the line drive the same cars which are in the same mechanical condition,

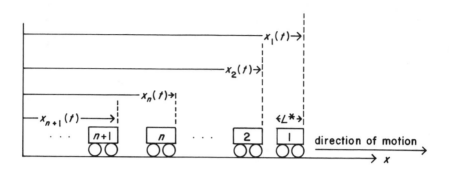

Figure 12.1. Geometry of Car-Following Models

that all drivers react the same way to a given situation, etc. Furthermore, our models will be described in terms of speed, acceleration, and separation distances, quantities which are very common in physical problems and relatively easy to measure.

3. Pipes' Model

Pipes [14] introduced a model for a line of traffic in which all drivers obey the California Vehicle Code statement: "A good rule for following another vehicle at a safe distance is to allow yourself the length of a car (about fifteen feet) for every ten miles per hour (14.67 ft/sec) you are traveling." This rule establishes a useful time constant $T^* \equiv 15$ ft/14.67 ft/s $= 1.02$ s, which is so close to one second that we set $T^* = 1$ s in the following calculations. There are several reasons for introducing Pipes' model: 1) historically, it was the first car-following model; 2) it makes substantial use of Laplace transforms; 3) it is an example of a model which required substantial modification to agree with reality—due to the safe distance assumptions of the model, no two vehicles ever collide.

Suppose that a line of traffic looks like Figure 12.1. If we let $L^* = 15$ ft be the length of the vehicles and b^* be a prescribed legal distance at rest, then $x_n(t)$ and $x_{n+1}(t)$ must satisfy

$$x_n(t) = x_{n+1}(t) + b^* + (L^*/14.67)v_{n+1}(t) + L^* \tag{1}$$

where $v_{n+1}(t)$ is the speed of the $(n + 1)$st vehicle. (Strictly speaking, the vehicle code would imply that at rest when $v_{n+1}(t) = 0$, $n = 1, 2, 3, \cdots$, each vehicle touches the bumper of the one ahead, so that $x_n - x_{n+1} = L^*$.

Pipes modified the code slightly to allow for some extra separation $b*$ at rest, so that the separation of one car length for each 10 mi/h would be measured from a point $b*$ ft behind the rear bumper of the car ahead. The introduction of $b*$, which is a constant, makes no difference in the following discussion where (1) is differentiated.) Using the definition of $T*$ and differentiating (1) with respect to time gives

$$v_n(t) = v_{n+1}(t) + T* \frac{dv_{n+1}(t)}{dt}, \tag{2}$$

which, after setting $T* = 1$ s, can be arranged into

$$a_{n+1}(t) \equiv \frac{dv_{n+1}(t)}{dt} = (v_n(t) - v_{n+1}(t)), \qquad n = 1, 2, 3, \cdots. \tag{3}$$

This is the first car-following equation we have met; it says that at any instant of time t the driver of a following car (subscript $n + 1$) adjusts his acceleration to be exactly equal in magnitude to the relative velocity between his vehicle and the vehicle directly ahead (subscript n). If the lead car is going faster, then the trailing car will accelerate, and vice versa. (In the language of the next section, (3) may be viewed as response = sensitivity × stimulus, where the stimulus is the relative velocity $v_n(t) - v_{n+1}(t)$, the sensitivity is $1/T* = 1\,\mathrm{s}^{-1}$, and the undelayed response is the acceleration $a_{n+1}(t)$.)

We now solve the set of equations (3) for $v_{n+1}(t)$ using a slight modification of the usual Laplace transform

$$\mathscr{L}\{v(t)\} \equiv \int_0^\infty e^{-st} v(t)\, dt. \tag{4}$$

The modification is to define s times the Laplace transform as the new quantity

$$V(s) \equiv L\{v(t)\} \equiv s\mathscr{L}\{v(t)\}, \tag{5}$$

which might be called the *s-multiplied Laplace transform.*[1] Applying (5) to (3) and using

$$L\left\{\frac{dv}{dt}\right\} = sV(s) - sv(0) \tag{6}$$

(verify using $\mathscr{L}\{df/dt\} = s\mathscr{L}\{f\} - f(0)$), we have

$$(s + 1)V_{n+1} = V_n + sv_{n+1}(0), \qquad n = 1, 2, 3, \cdots. \tag{7}$$

We shall now simulate the situation where the line of vehicles is at rest waiting for a green light at $t = 0$, so that

$$v_{n+1}(0) = 0, \qquad n = 1, 2, 3, \cdots, \tag{8}$$

---

[1] See Section 10.2 for a short table of *s*-multiplied Laplace transforms.

and then when the light changes for $t > 0$ the motion of the lead car is known, say $v_1(t) = g(t)$ for $t > 0$, so that $V_1(s) = L\{v_1(t)\} = L\{g(t)\}$ is known. In this case

$$V_{n+1}(s) = V_n(s)/(s + 1), \qquad n = 1, 2, 3, \cdots \tag{9}$$

and

$$V_{n+1}(s) = V_1(s)/(s + 1)^n, \qquad n = 1, 2, 3, \cdots \tag{10}$$

by successive substitution. This last set of equations relates the transform of the velocity of each vehicle to the known transform of the first vehicle. A formal solution for the velocity of each car in terms of the motion of the first car is then given by

$$v_{n+1}(t) = L^{-1}\{V_{n+1}(s)\} = L^{-1}\{V_1(s)/(s + 1)^n\}, \qquad n = 1, 2, 3, \cdots. \tag{11}$$

We shall use the convolution theorem, which states that if $\phi_1(s) = L\{f_1(t)\}$ and $\phi_2(s) = L\{f_2(t)\}$, then

$$L^{-1}\{\phi_1(s)\phi_2(s)/s\} = \int_0^t f_1(u)f_2(t - u)\, du, \tag{12}$$

to compute the inverse transform in (11). If we let

$$\phi_1(s) \equiv s/(s + 1)^n \tag{13}$$

and

$$\phi_2(s) \equiv V_1(s) = L\{v_1(t)\}, \tag{14}$$

and use the fact that

$$L^{-1}\{\phi_1(s)\} = \frac{t^{n-1}}{(n - 1)!}e^{-t} \tag{15}$$

from Laplace transform tables, then

$$v_{n+1}(t) = \frac{1}{(n - 1)!} \int_0^t u^{n-1} \exp(-u)v_1(t - u)\, du. \tag{16}$$

Given a specific function $v_1(t) = g(t)$, the integral in (16) may or may not be easy to carry out. In certain cases it may be easier to go back to (13) and (14) and redefine ϕ_1 and ϕ_2.

3.1 Instantaneous Acceleration of Leading Vehicle

Consider the physically unrealizable case that the lead vehicle is standing still at $t = 0$ and acquires a constant cruising speed v_c for $t > 0$, as illustrated in Figure 12.2. The mathematics gained from doing this example will allow us to discuss a more realistic situation later. In this case, then, $v_1(t - u) = v_c$ for $0 < u \le t, t > 0$, and by (16)

Figure 12.2. Instantaneous Acceleration of Leading Vehicle

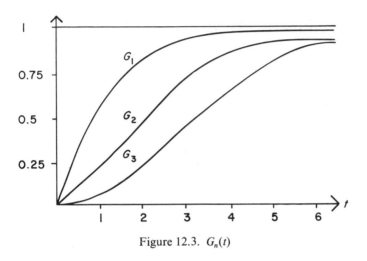

Figure 12.3. $G_n(t)$

$$v_{n+1}(t) = \frac{v_c}{(n-1)!} \int_0^t u^{n-1} \exp(-u)\, du. \tag{17}$$

If we define

$$G_n(t) \equiv \frac{1}{(n-1)!} \int_0^t u^{n-1} \exp(-u)\, du \tag{18}$$

the integral in (18) may be evaluated by successive integrations by parts to give

$$G_n(t) = 1 - e^{-t}\left(1 + \frac{t}{1!} + \frac{t^2}{2!} + \cdots + \frac{t^{n-1}}{(n-1)!}\right). \tag{19}$$

The function $G_n(t)$ has the obvious properties that $G_n(0) = 0$, $G_n(\infty) = 1$ for n any finite, positive integer, and it also may be verified that

$$G_n(t) - G_{n+1}(t) = e^{-t}\frac{t^n}{n!} = \frac{d}{dt}G_{n+1}(t). \tag{20}$$

The functions $G_n(t)$ can be easily tabulated by computer, and it may be established that

$$G_1(t) > G_2(t) > G_3(t) > \cdots > G_n(t) > \cdots$$

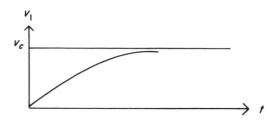

Figure 12.4. Gradual Acceleration to Constant Velocity

for $0 < t < \infty$. The general shape of the first few G_n are illustrated in Figure 12.3 and the velocities

$$v_{n+1}(t) = v_c G_n(t), \qquad n = 1, 2, 3, \cdots \tag{21}$$

are just proportional to these curves.

The accelerations of the vehicles are, with the help of (20),

$$a_{n+1}(t) = \frac{d}{dt} v_{n+1}(t) = v_c \frac{d}{dt} G_n(t) = v_c \frac{e^{-t}t^{n-1}}{(n-1)!}. \tag{22}$$

The coordinates of the vehicles at time $t > 0$ are

$$x_{n+1}(t) = x_{n+1}(0) + \int_0^t v_{n+1}(u)\, du = x_{n+1}(0) + v_c \int_0^t G_n(u)\, du \tag{23}$$

where the integrals could be approximated numerically. It is obvious from Figure 12.3 that

$$\int_0^t G_1(u)\, du > \int_0^t G_2(u)\, du > \cdots > \int_0^t G_n(u)\, du > \cdots$$

for $t > 0$, so that $x_i(t) \neq x_j(t)$, $i \neq j$, when the vehicles are initially separated by the postulated distance $b^* + L^*$, and hence no possibility of collision exists.

3.2 Gradual Acceleration of Leading Vehicle

It is more realistic to assume a gradual acceleration of the leading car from $v_1(0) = 0$ to v_c such as

$$v_1(t) = v_c(1 - \exp(-t)) \tag{24}$$

which is illustrated in Figure 12.4. Substituting (24) into (16) we get

$$v_{n+1}(t) = \frac{v_c}{(n-1)!} \int_0^t u^{n-1} \exp(-u)(1 - \exp(-(t-u)))\, du \tag{25}$$

which becomes

$$v_{n+1}(t) = v_c \left(G_n(t) - e^{-t} \frac{t^n}{n!} \right), \tag{26}$$

so that

$$v_{n+1}(t) = v_c G_{n+1}(t), \qquad n = 1, 2, 3, \cdots. \tag{27}$$

Once again, the velocities are proportional to the G_n curves.

4. Linear Car-Following with Delays [11, 4]

One of the objections to the simple model of the last section is that no accidents ever occur. If the lead car gradually accelerates, then all other vehicles gradually accelerate in turn and no vehicle ever "rear-ends" the vehicle ahead. Even under the extreme case described in Exercise 3 of instantaneous deceleration of the lead vehicle from $v_1 = v_c > 0$ to $v_1 = 0$, all vehicles behind the lead car are able to produce the required deceleration to obey the vehicle code and thus stop without collision. The fallacy in the model is that no driver can react instantaneously to the relative velocity difference he observes.

The basic assumption of car-following theory put in psychological terms is that a driver's response is directly proportional to a stimulus. The response, however, does not happen immediately at the time of stimulation, but only after a time delay T, i.e.,

$$\text{response}|_{t+T} = \text{sensitivity} \times \text{stimulus}|_t. \tag{28}$$

For example, we know that if a driver sees a situation which stimulates him to decelerate, a time delay occurs before he actually applies the brakes and an additional delay occurs before the action of the brake pedal is fully transmitted to the wheels. If a vehicle is traveling at 60 mi/h, then during a period of 0.5 s the vehicle travels about 44 ft, a not inconsequential distance in some accident situations.

Our basic mathematical model, which may be interpreted as (3) with T and λ added, is

$$a_{n+1}(t + T) \equiv \frac{dv_{n+1}(t + T)}{dt} = \lambda(v_n(t) - v_{n+1}(t)), \qquad n = 1, 2, 3, \cdots \tag{29}$$

where T is the driver response delay (assumed the same for each driver in the line) and λ is a sensitivity coefficient. In the psychological terms of (28) the stimulus to the $(n + 1)$st driver is the relative velocity $v_n - v_{n+1}$ he observes at time t and his response is his acceleration a_{n+1} at the later time $t + T$. If the sensitivity coefficient is assumed to be a simple constant, λ_0 say, (29) is known as the *linear* car-following model. In more complicated models λ is assumed to be a function of some feature of the problem, such as the spacing $x_n - x_{n+1}$ between successive vehicles.

We shall now solve the linear car-following equations, i.e., (29) with $\lambda = \lambda_0$, with the help of Laplace transforms.[2] We shall need the result

$$\mathcal{L}\left\{\frac{df}{dt}(t + T)\right\} = e^{sT}(s\mathcal{L}\{f\} - f(0))^3 \tag{30}$$

which may be verified from the standard table entries. Now if we define

$$V_n(s) \equiv \mathcal{L}\{v_n(t)\} \tag{31}$$

(note that this is slightly different from the last section where $V_n(s) \equiv L\{v_n(t)\}$) and then transform (29), we obtain

$$e^{sT}(sV_{n+1}(s) - v_{n+1}(0)) = \lambda_0(V_n(s) - V_{n+1}(s)). \tag{32}$$

Through algebraic manipulation, the last equation may be rearranged to yield

$$V_{n+1}(s) = \frac{\lambda_0 e^{-sT}}{s + \lambda_0 e^{-sT}} V_n(s) + \frac{v_{n+1}(0)}{s + \lambda_0 e^{-sT}}, \qquad n = 1, 2, 3, \cdots. \tag{33}$$

Equation (33) can be used to find the motion of any vehicle in the line of traffic if the motion of the lead car is given. As an illustration, we consider again the case where the line is at rest at $t = 0$ and the lead car instantaneously accelerates to a constant velocity $v_1(t) = v_c$ for $t > 0$. In this case

$$v_n(0) = 0, \qquad n = 1, 2, 3, \cdots \tag{34}$$

and

$$V_1(s) = v_c/s. \tag{35}$$

Substituting (34) and (35) into (33) and then using successive substitutions gives

$$V_{n+1}(s) = \left(\frac{\lambda_0 e^{-sT}}{s + \lambda_0 e^{-sT}}\right)^n \frac{v_c}{s}. \tag{36}$$

It is difficult and complicated to invert (36) for $v_{n+1} = \mathcal{L}^{-1}\{V_{n+1}(s)\}$ for arbitrary n, so we shall consider first the case $n = 1$, when

$$\frac{V_2(s)}{v_c} = \frac{\lambda_0 e^{-sT}}{s + \lambda_0 e^{-sT}}\left(\frac{1}{s}\right). \tag{37}$$

To invert (37) we shall employ the standard shifting theorem

$$\mathcal{L}\{f(t - a)h(t - a)\} = e^{-as}\mathcal{L}\{f(t)\}, \tag{38}$$

---

[2] Now we use only the traditional Laplace transform $\mathcal{L}$, not the modified form L of the previous section.

[3] This formula holds only if $df(t)/dt = 0$, $0 \le t \le T$, which is the case in the applications below.

where

$$h(t - a) = \begin{cases} 0, & t - a < 0 \\ 1, & t - a > 0 \end{cases}$$

is the Heaviside function, and the series expansion

$$1/(1 + x) = 1 - x + x^2 - x^3 + x^4 - + \cdots, \qquad |x| < 1. \qquad (39)$$

First we rewrite (37) in the form

$$\frac{V_2(s)}{v_c} = \frac{\lambda_0 e^{-sT}}{s^2} \left(\frac{1}{1 + \dfrac{\lambda_0 e^{-sT}}{s}} \right) \qquad (40)$$

and then, using (39) with the assumption that s is large enough so that $|\lambda_0 e^{-sT}/s| < 1$, we find that

$$\frac{V_2}{v_c} = \frac{\lambda_0 e^{-sT}}{s^2} \left(1 - \frac{\lambda_0 e^{-sT}}{s} + \frac{\lambda_0^2 e^{-2sT}}{s^2} - \frac{\lambda_0^3 e^{-3sT}}{s^3} + - \cdots \right). \qquad (41)$$

Now we invert each term separately using (38) and the fact that $\mathcal{L}^{-1}\{1/s^n\} = t^{n-1}/(n - 1)!$ to arrive at

$$\frac{v_2(t)}{v_c} = \lambda_0(t - T)h(t - T) - \frac{\lambda_0^2}{2!}(t - 2T)^2 h(t - 2T)$$

$$(42)$$

$$+ \frac{\lambda_0^3}{3!}(t - 3T)^3 h(t - 3T) - \frac{\lambda_0^4}{4!}(t - 4T)^4 h(t - 4T) + \cdots$$

or, in more compact notation,

$$\frac{v_2(t)}{v_c} = \sum_{k=1}^{\infty} (-1)^{k+1} \frac{\lambda_0^k}{k!}(t - kT)^k h(t - kT). \qquad (43)$$

It is also possible to express (42), (43) in the less formidable notation

$$\frac{v_2(t)}{v_c} = \begin{cases} 0, & 0 < t < T \\ \lambda_0(t - T), & T < t < 2T \\ \lambda_0(t - T) - \dfrac{\lambda_0^2}{2!}(t - 2T)^2, & 2T < t < 3T. \\ \quad \vdots \end{cases} \qquad (44)$$

The acceleration $a_2(t)$ is obtained by differentiation of (42)–(44) and the distance traversed is obtained by integration, i.e.,

$$x_2(t) = x_2(0) + \int_0^t v_2(u) \, du. \qquad (45)$$

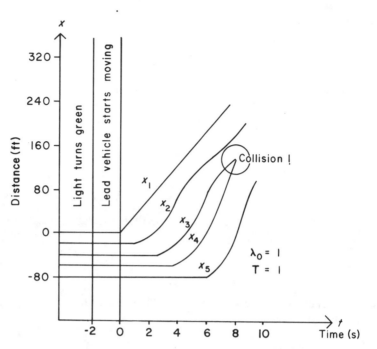

Figure 12.5. Motion of First Five Vehicles under Linear Car-Following when First Vehicle Executes Instantaneous Acceleration to 40 ft/s

In the Exercises suggestions are given to arrive at the solution for $v_3(t)$; each successive $v_{n+1}(t)$ can be calculated separately by the same basic techniques.

We end this section with an example from Drew [4] which illustrates the solutions just obtained. A line of cars is queued at a red light and 2 s after the light turns green (this time can be taken as $t = 0$), the lead vehicle instantaneously acquires a velocity of 40 ft/s. The reaction time $T = 1$ s and $\lambda_0 = 1/T = 1$ s^{-1}, so that $\lambda_0 T = 1$ (which lies in the asymptotically unstable zone discussed in Section 6). For the purposes of defining a collision, the car lengths can be neglected, and so a collision occurs if $x_n(t) = x_{n+1}(t)$ for some n. Figure 12.5 indicates a collision between cars 3 and 4 some 130 ft from the stopline.

5. Incorporation of Pipes' Model into Linear Car-Following

Chandler *et al.* [3] also studied the California Vehicle Code in the context of the linear car-following model. They began with (1) in the form

$$x_n(t) - x_{n+1}(t) - b^* - L^* - T^* v_{n+1}(t) = 0 \qquad (46)$$

and assumed that a fluctuation in the behavior of the lead car would, as a result of response lags, cause (46) to be violated. If we define

$$\delta_{n+1}(t) \equiv x_n(t) - x_{n+1}(t) - b^* - L^* - T^*v_{n+1}(t) \qquad (47)$$

and $\delta_{n+1}(t) > 0$, the $(n + 1)$st driver would accelerate in order to recover the equality in (46); if $\delta_{n+1}(t) < 0$, he would decelerate. So $\delta_{n+1}(t)$ can be viewed as the stimulus to the $(n + 1)$st driver who responds with an acceleration a_{n+1} at a time $t + T$ later with sensitivity λ_0. Thus the equation for the model is

$$a_{n+1}(t + T) = \lambda_0(x_n(t) - x_{n+1}(t) - b^* - L^* - T^*v_{n+1}(t)). \qquad (48)$$

6. Stability of Linear Car-Following Theory

Ultimately, the choice of one mathematical model over another depends on how well they describe actual phenomena. One extremely important occurrence in actual car-following situations is the collision of one vehicle with another.

A related mathematical topic is the stability of the linear car-following equations (29). The general question asked in stability studies is "What is the response of a system to a perturbation of the system?" If the response decays, the system is said to be stable; if the response is pure oscillatory or amplified the system is said to be unstable. Unfortunately, the mathematics of the stability analysis of linear car-following is tedious and hence has been placed in Appendix 3; we present here only a summary of the results.

"Local stability" in the car-following context asks, "What is the nature of the response of following cars to a maneuver of the lead car as t becomes large?" Specifically, suppose that before $t = 0$ all cars are traveling at the same velocity u; then the first driver accelerates and decelerates in some continuous fashion, for $0 < t < \bar{t}$, returning to his original velocity u for $t > \bar{t}$. If any of the following drivers must alternately accelerate and decelerate, with larger and larger magnitudes, their motion is erratic and certainly unstable. The stability results, whether expressed in terms of any of the three responses $a_{n+1}(t)$, $v_{n+1}(t)$, or $x_n(t) - x_{n+1}(t)$ where a_1, v_1, and x_1 refer to the lead car, are

(i) for $\lambda_0 T \leq 1/e$ (stable)—nonoscillatory, damped response for all followers ($n + 1 = 2, 3, 4, \cdots$);
(ii) for $1/e < \lambda_0 T < \pi/2$ (stable)—oscillatory, damped response for second car ($n + 1 = 2$); oscillatory response with amplitude which increases at first but is eventually damped for remaining cars ($n + 1 = 3, 4, \cdots$);
(iii) for $\lambda_0 T = \pi/2$ (unstable)—pure oscillatory response for second car ($n + 1 = 2$), oscillatory response with increasing amplitude for remaining cars ($n + 1 = 3, 4, \cdots$);

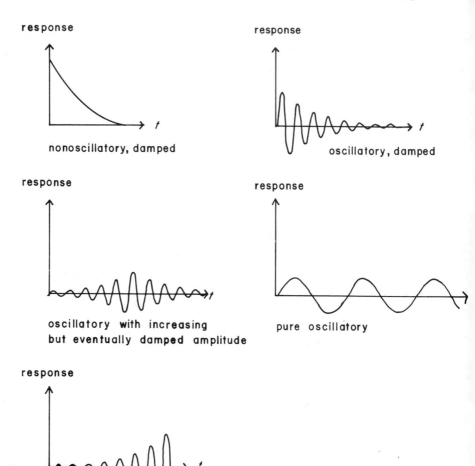

Figure 12.6. Types of Response in Local Stability Analysis

(iv) for $\lambda_0 T > \pi/2$ (unstable)—oscillatory response with increasing amplitude for all followers ($n + 1 = 2, 3, 4, \cdots$).

See Figure 12.6 for further explanation of the above terms.

The question of whether a collision occurs is not the same as the question of stability. It can be established that a line of traffic moving with velocity u under linear car-following control is safe from collisions due to the acceleration—deceleration of the lead car already mentioned if $\lambda_0 T \leq 1/e$ and the initial spacing is greater than u/λ_0. It is only the nonoscillatory case which ensures safety, however, because an oscillation, even though eventually damped, could cause a collision. More specific information about the initial spacing between the cars and the maneuver executed by the lead car is needed, in general, to discuss the possibility of a collision. However, one can generalize to say that $\lambda_0 T \le 1/e$ represents the safest response and $\lambda_0 T > \pi/2$ represents the most hazardous following condition for a line of cars for large times.

"Asymptotic stability" deals with a long line of cars and asks, "When does a fluctuation in the motion of the car in front get amplified as it propagates down the line?" Asymptotic stability is concerned with the behavior of the cars for large n, while local stability worries about responses for large t. It has been established that the stability dividing line of the fluctuation occurs at $\lambda_0 T = 1/2$, with attenuation if $\lambda_0 T < 1/2$ and amplification if $\lambda_0 T > 1/2$.

It thus appears that the crucial factor in both types of stability is the product of the response sensitivity with response time lag, $\lambda_0 T$. At first, thought, one might think that in order to produce a stable response, a large sensitivity λ_0 would be appropriate. For a system with a fixed response time lag T, this is not necessarily the case, because the response does not occur at the time t when the observation is made, but at the later time $t + T$ when the system has changed. The danger exists that a large response applied at the wrong time may overcompensate for small deviations from a desired goal and, instead of reducing the deviations, make them larger. It may be demonstrated (see comments at end of Section 10.3) that Pipes' model is both locally and asymptotically stable.

7. Nonlinear Car-Following Laws

Several modifications of the linear car-following model, known collectively as nonlinear car-following models, modify the assumption that the response sensitivity λ is a constant and the same for all drivers. A useful definition in this discussion is the car spacing $P_n(t)$ between two successive cars

$$P_n(t) \equiv x_n(t) - x_{n+1}(t). \tag{49}$$

Step function law. Assume the step function for λ:

$$\lambda = \lambda(P_n) = \begin{cases} \alpha, & \text{if } 0 < P_n \le P^* \\ \beta, & \text{if } P_n > P^*, \alpha \text{ and } \beta \text{ are constants.} \end{cases} \tag{50}$$

Under this law, each driver in the line will exhibit one of two possible sensitivities. Presumably, one would want to study only the case $\alpha > \beta$, since most drivers would react with a larger sensitivity if the spacing to the car ahead of them were small. In an extreme case where a driver's reaction is a panic one when close to the car in front but more subtle when further away, α might be chosen as 2β or 3β.

Reciprocal spacing law. Let

$$\lambda = \lambda(P_n) = \propto_0/P_n = \propto_0/(x_n(t) - x_{n+1}(t)), \tag{51}$$

where $\propto_0$ is a constant known as the characteristic speed. Each driver responds in inverse proportion to his spacing. This law is fundamentally different from any previous law because in a line of drivers it is now possible for each driver after the first to exhibit an individualized sensitivity different from that of any other driver. The reciprocal spacing law assumes added importance because it leads to agreement with both experimentally and theoretically determined relations between traffic flow and concentration (see the next section).

Edie's law. Edie [5] suggested that at lower densities the sensitivity would be proportional to the speed of the car at reaction time and inversely proportional to the square of the spacing, so that

$$\lambda = \lambda_0 \frac{dx_{n+1}(t + T)}{dt} \bigg/ (x_n(t) - x_{n+1}(t))^2. \tag{52}$$

More general versions of (52),

$$\lambda = \lambda_0 \left(\frac{dx_{n+1}(t + T)}{dt} \right)^m \bigg/ (x_n(t) - x_{n+1}(t))^l, \qquad m, l > 0, \tag{53}$$

were studied by Gazis *et al.* [8]. Their main conclusion was that comparisons with empirical data did not show a clear superiority of any one nonlinear model, although they did show the necessity of a nonlinear model rather than a linear one.

In general, researchers have had to turn to computer solutions of the nonlinear models. Certainly, Laplace transform techniques will not work; for example, Edie's model

$$\frac{d^2 x_{n+1}(t + T)}{dt^2} = \lambda_0 \frac{\left(\dfrac{dx_{n+1}(t + T)}{dt} \right)\left(\dfrac{dx_n(t)}{dt} - \dfrac{dx_{n+1}(t)}{dt} \right)}{(x_n(t) - x_{n+1}(t))^2} \tag{54}$$

is nonlinear, and transform techniques work only for linear systems.

8. Steady-State Flow

In the cases of constant sensitivity and reciprocal spacing sensitivity, it is easy to calculate the steady-state velocity u and traffic flow q (e.g., in cars/hour) as functions of the concentration k (e.g., in cars/mile). On the highway one measures the flow of traffic q by counting the number of cars which pass a point in a reasonably long time, and the concentration k by counting the number of vehicles on a reasonable length of road. To define a steady-

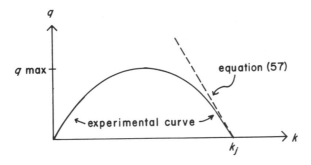

Figure 12.7. Schematic Diagram of Flow Versus Concentration

state we imagine that at some time $t = 0$ all cars are traveling at some common speed u_0 and that then the lead car accelerates or decelerates in some fashion to a new speed u. Because of the velocity control of the following cars, assume that a steady-state is eventually reached as t becomes large in which each car moves with speed u, provided, of course, that all stability conditions necessary to avoid collision are satisfied.

Consider first the constant sensitivity model $\lambda = \lambda_0$ in (29). We integrate this equation from $t = 0$ to some large time t which yields

$$u \equiv v_{n+1}(t + T) = \lambda_0(x_n(t) - x_{n+1}(t)) + c_1. \tag{55}$$

In the steady state we can neglect the time lag T, since all adjustments necessary have been made to produce constant spacing and constant speed; consequently, u may be identified with $v_{n+1}(t + T)$. Now the steady-state spacing $x_n - x_{n+1}$ is just the reciprocal of the concentration k, and the constant c_1 is determined by the condition that $u = 0$ when $x_n - x_{n+1} = 1/k_j$, where k_j is the jam concentration (see Appendix 1 for the definition). Therefore,

$$u = \lambda_0\left(\frac{1}{k} - \frac{1}{k_j}\right), \tag{56}$$

and, since $q = ku$,

$$q = \lambda_0\left(1 - \frac{k}{k_j}\right). \tag{57}$$

Experimentally, it has been established that the q versus k curve has the shape given in Figure 12.7. It seems intuitive that at low concentrations drivers would not attempt to "follow" the car ahead. It is not surprising, therefore, that (57) does not agree with the experimental result that the flow-concentration curve passes through the origin.

Next we treat the reciprocal spacing model

$$\frac{dv_{n+1}(t + T)}{dt} = \frac{\alpha_0}{(x_n(t) - x_{n+1}(t))}\left(\frac{dx_n(t)}{dt} - \frac{dx_{n+1}(t)}{dt}\right). \tag{58}$$

Integration of both sides of this equation as above yields

$$u \equiv v_{n+1}(t + T) = \alpha_0 \ln(x_n(t) - x_{n+1}(t)) + c_2, \qquad (59)$$

where c_2 is determined by the condition that the velocity of a car approaches zero as its spacing approaches the effective length $l \equiv 1/k_j$ of the vehicles (see Appendix 1 for further explanation of l). Thus $c_2 = -\alpha_0 \ln(l)$ and since the steady-state concentration k is just the reciprocal of the spacing $x_n - x_{n+1}$,

$$u = \alpha_0 \ln(k_j/k); \qquad (60)$$

and the steady-state flow is again given by ku, so

$$q = \alpha_0 k \ln(k_j/k). \qquad (61)$$

Equation (61) *does* have the general shape of the experimental flow versus concentration curve in Figure 12.7. Equations (56). (57), (60), and (61) could be classified as equations of state, since they are analogous to the state equations in thermodynamics which relate the macroscopic quantities pressure, temperature, density, etc.

Greenberg [9] derived the equations

$$u = c \ln(k_j/k), \qquad q = ck \ln(k_j/k), \qquad (62)$$

where c is the velocity for optimum flow q_{max}, based on a *completely different* fluid flow model of traffic. If one is willing to accept the somewhat astonishing identification of c, a characteristic of the total flow along a highway, with the parameter α_0, a measure of how a driver actually drives along that highway, two totally different approaches lead to the same equation of state. This circumstance has been partially responsible for the popularity of the reciprocal spacing model over some of the other models.

Values of α_0 and c have been estimated from experiments done in three New York tunnels; the methods and results are reported in Herman and Potts [11]. The results are tabulated as follows.

	α_0 (mi/h)	c (mi/h)
Lincoln Tunnel	20	17
Holland Tunnel	18	15
Queens–Midtown Tunnel	~22	25

These authors report on several experiments resulting in plots of λ versus $1/\bar{p}$ ($\bar{p}$ = average of $x_n - x_{n+1}$ over the line of cars). The data are adequately described by a least-squares fit of the straight line $\lambda = \alpha_0/\bar{p}$, and the values of α_0 are those occurring above. (If there are k experimentally determined values of λ, λ_1, λ_2, $\cdots$, λ_k, then a least-squares fit of $\alpha_0/\bar{p}$ minimizes the sum of the squares of the distances $\sum_{i=1}^{k}(\lambda_i - \alpha_0/\bar{p})^2$.)

9. Conclusion

Of necessity, only a brief introduction to CFT has been presented in this module. While more complex models may not present conceptual difficulties, the mathematics of understanding them often goes beyond the realm of the average junior level undergraduate. Wilhelm and Schmidt [16] mention that at least 30 vehicle-following models existed as of 1973. We mention several extensions of the ideas in this chapter just to stimulate the reader's appetite:

(1) Tuck [15] derives a set of two-dimensional CFT-type models to deal with convoys of ships and suggests possible extension to study bird flights, military marching formations, and aircraft flight;
(2) Newell [13] studied the question of different sensitivities for acceleration and deceleration (brake lights indicate deceleration, but no corresponding apparatus indicates acceleration);
(3) Lee [12] included a memory function to represent past behavior;
(4) Bexelius [1] studied control of n cars.

Many problems remain for future researchers. One of the potentially fertile areas is the application of CFT to environments other than the highway. More complex situations, such as dual lane highways with passing deserve study, and, as is the case with many models, a gap exists between the mathematical solution and data gathered on the highway. The hope is that through repeated feedback and the consequent updating of the model, driving safety could be improved.

10. Appendix

10.1 List of Symbols

T^*	$= 1$ s, California Vehicle Code constant.
L^*	$= 15$ ft, length of a vehicle in Pipes' model.
b^*	Prescribed legal separation distance at rest (in ft).
$x_n(t)$	Coordinate of the front of the nth vehicle in an arbitrary Cartesian reference frame (e.g., in ft).
$v_n(t)$	$= dx_n(t)/dt$, speed of the nth vehicle (e.g., in ft/s).
$a_n(t)$	$= dv_n(t)/dt = d^2 x_n(t)/dt^2$, acceleration of the nth vehicle (e.g., in ft/s^2).
$P_n(t)$	$= x_n(t) - x_{n+1}(t)$, car spacing, the distance from $(n+1)$st front bumper to nth front bumper (e.g., in ft).
$\mathscr{L}\{f(t)\}$	$= \int_0^\infty e^{-st} f(t)\, dt$, Laplace transform of $f(t)$.
$L\{f(t)\}$	$= s\mathscr{L}\{f(t)\}$, s-multiplied Laplace transform of $f(t)$.
$V_n(s)$	$= L\{v_n(t)\}$ in Section 3; $\mathscr{L}\{v_n(t)\}$ in Section 4.

λ	Sensitivity coefficient; λ_0 is used if λ is constant.
q	Flow, the number of cars passing a given point on the highway in unit time (e.g., in cars/h).
k	Concentration or density, the number of cars per unit length of highway at any instant (e.g., in cars/mi).
k_j	Jam concentration, the number of cars per unit length of highway when the traffic is jammed, i.e., all vehicles have been forced to stop (e.g., in cars/mi).
c	Optimum speed, the speed when the flow is a maximum (e.g., in mi/h).
u	Steady-state speed; common speed of a line of vehicles no longer adjusting to an initial fluctuation of the lead vehicle (e.g., in mi/h).
l	Effective length of vehicles in steady-state theory, usually defined as $1/k_j$, where k_j is determined by fitting a given equation of state $q = q(k)$ to experimental data. In the paper by Chandler *et al.* [3] $k_j = 228$ cars/mi which implies $l = 23.1$ ft.

10.2 Useful Laplace Transforms

Pipes' Model $f(t)$	*s-Multiplied Laplace Transform* $\phi(s) = L\{f(t)\}$ $= s \int_0^\infty e^{-st} f(t)\, dt$
$U(t) = 1$	1
t	$1/s$
$h(t - T)f(t - T)$	$e^{-sT}\phi(s)$
$G_n(t)$, see (19)	$1/(s + 1)^n$
$\Phi_n(t) \equiv \dfrac{dG_n(t)}{dt} = \dfrac{e^{-t}t^{n-1}}{(n-1)!}$, see (15), (20)	$s/(s + 1)^n$
$\displaystyle\int_0^t G_n(u)\, du$	$1/s(s + 1)^n$
$\begin{cases} t/t_0, & 0 \le t \le t_0 \\ 1, & t > t_0. \end{cases}$	$(1 - e^{-t_0 s})/t_0 s.$

Linear Car-Following $f(t)$	*Laplace Transform* $\psi(s) = \mathcal{L}\{f(t)\} = \displaystyle\int_0^\infty e^{-st} f(t)\, dt$
$\dfrac{df(t + T)}{dt}$	$e^{sT}(s\psi(s) - f(0))$, if $df(t)/dt = 0, 0 \le t \le T$
$f(t - T)h(t - T)$, see (38)	$e^{-sT}\psi(s)$
$\dfrac{t^{n-1}}{(n-1)!}.$	$1/s^n.$

10.3 Mathematics of Stability

(This discussion is a modification of the ideas of Donald Drew, Department of Mathematical Sciences, Rensselaer Polytechnic Institute, Troy, NY 12181).

In this Appendix we derive the stability results for the linear car-following model.

$$\frac{dv_{n+1}(t + T)}{dt} = \lambda_0(v_n(t) - v_{n+1}(t)), \qquad n = 1, 2, 3, \cdots \qquad (A1)$$

which are summarized in Section 6.

Let us consider asymptotic stability first. We want to study the amplitude of a disturbance as it propagates down the line of cars. Suppose that the velocity of the first car is proportional to $\cos(\omega t)$. We are going to set

$$v_1(t) = b_1 e^{i\omega t} = b_1(\cos \omega t + i \sin \omega t), \qquad (A2)$$

which is a complex valued quantity. Actually, we are interested only in the real part of (A2), but because the equations of (A1) are linear, we can recover the real quantities of interest at any point in the calculation by taking the real part of the equations. If we assume that the solutions to (A1) subject to the motion of the lead car given in (A2) have the form

$$v_n(t) = b_n e^{i\omega t}, \qquad (A3)$$

then

$$b_{n+1}(i\omega e^{i\omega T} + \lambda_0) = \lambda_0 b_n, \qquad n = 1, 2, 3, \cdots \qquad (A4)$$

and by successive substitutions

$$b_{n+1} = \frac{b_1}{\left(1 + \dfrac{i\omega e^{i\omega T}}{\lambda_0}\right)^n}. \qquad (A5)$$

The question of asymptotic stability or instability, then, becomes a question of whether $\left|1 + (i\omega e^{i\omega T}/\lambda_0)\right|$ is greater than one (stable) or less than one (unstable). Since

$$\left|1 + \frac{i\omega e^{i\omega T}}{\lambda_0}\right|^2 = \left|\left(1 - \frac{\omega}{\lambda_0}\sin \omega T\right) + i\frac{\omega}{\lambda_0}\cos \omega T\right|^2, \qquad (A6)$$

$$\left|1 + \frac{i\omega}{\lambda_0}e^{i\omega T}\right|^2 = 1 - \frac{2\omega}{\lambda_0}\sin \omega T + \frac{\omega^2}{\lambda_0^2}, \qquad (A7)$$

and so

$$\left|1 + \frac{i\omega}{\lambda_0}e^{i\omega T}\right| < 1 \quad \text{if} \quad \frac{\omega^2}{\lambda_0^2} < \frac{2\omega}{\lambda_0}\sin \omega T. \qquad (A8)$$

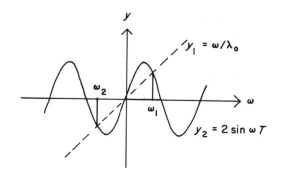

Figure 12.8. Unstable Case

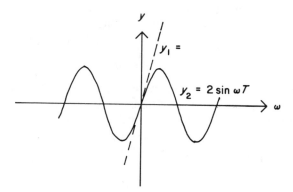

Figure 12.9. Stable Case

If $\omega/\lambda_0 > 0$, then (A8) becomes

$$\frac{\omega}{\lambda_0} < 2 \sin \omega T. \tag{A9}$$

and if $\omega/\lambda_0 < 0$, (A8) becomes

$$\frac{\omega}{\lambda_0} > 2 \sin \omega T. \tag{A10}$$

Plotting both $y_1 = \omega/\lambda_0$ and $y_2 = 2 \sin \omega T$ versus ω gives Figures 12.8 and 12.9. In Figure 12.8, for values of ω in $(0, \omega_1)$ or $(\omega_2, 0)$, disturbances will grow in n. In Figure 12.9, no values of ω lead to growing disturbances.

The case separating the unstable and stable cases is when the line $y_1 = \omega/\lambda_0$ is tangent to the curve $y_2 = 2 \sin \omega T$ at the origin. We compute the slopes at the origin $dy_1/d\omega|_{\omega=0} = 1/\lambda_0$ and $dy_2/d\omega|_{\omega=0} = 2T$. Thus the model will be asymptotically stable to disturbances of all frequencies if $1/\lambda_0 > 2T$ or $\lambda_0 T < 1/2$. For $\lambda_0 T > 1/2$, the model is asymptotically unstable.

Let us now consider the question of local stability for the second car only. The equation of motion for the second car is

$$\frac{dv_2(t + T)}{dt} + \lambda_0 v_2(t) = \lambda_0 v_1(t). \tag{A11}$$

The general solution of (A11) is the homogeneous solution plus a particular solution. We assume that $v_1(t)$ is given and that a stable particular solution $v_2^P(t)$ exists and has been found. Since $v_2(t) = v_2^P(t) + v_2^h(t)$, where $v_2^h(t)$ is the homogeneous solution, $v_2^h(t)$ can be looked at as a perturbation to the particular solution, and local stability theory asks whether this perturbation grows or decays with time. Obviously, $v_2^h(t)$ satisfies

$$\frac{dv_2^h(t + T)}{dt} + \lambda_0 v_2^h(t) = 0 \tag{A12}$$

and substitution of the guess

$$v_2^h(t) = e^{\mu t} \tag{A13}$$

yields the characteristic equation

$$\mu e^{\mu T} + \lambda_0 = 0. \tag{A14}$$

If we let $\mu T = \sigma$ and $\tau = \lambda_0 T$, we have the transcendental equation

$$\sigma e^\sigma + \tau = 0, \tag{A15}$$

which is reasonably difficult to analyze for $\sigma = \sigma(\tau)$.
    We first look for real σ. We write

$$\sigma = -\tau e^{-\sigma}. \tag{A16}$$

Sketching both $y_1 = \sigma$ and $y_2 = -\tau e^{-\sigma}$ in Figure 12.10 shows that

(i) no values of σ satisfy (A16) if $\tau > \tau_1$;
(ii) one value of σ satisfies (A16) if $\tau = \tau_1$ (this is the case pictured in Figure 12.10);
(iii) two values of σ satisfy (A16) if $0 < \tau < \tau_1$.

Let us calculate τ_1. The condition $y_1 = \sigma$ is tangent to $y_2 = -\tau_1 e^{-\sigma}$ is

$$\frac{dy_1}{d\sigma} = \frac{dy_2}{d\sigma} \quad \text{or} \quad 1 = \tau_1 e^{-\sigma} \tag{A17}$$

and

$$y_1 = y_2 \quad \text{or} \quad \sigma = -\tau_1 e^{-\sigma}. \tag{A18}$$

Thus $\sigma = -1$ and $\tau_1 = 1/e$. Thus for $0 < \tau = \lambda_0 T < 1/e$, two real negative values of σ exist which satisfy (A16); both solutions of (A13) have exponentially decaying character. For $\lambda_0 T = 1/e$, one exponentially decaying solution of (A13) exists.

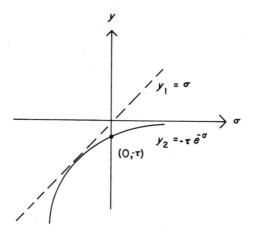

Figure 12.10. Solution of Equation (A16)

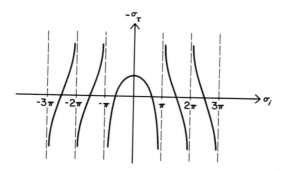

Figure 12.11. Equation (A23)

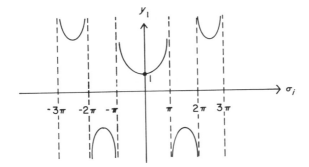

Figure 12.12. Equation (A24)

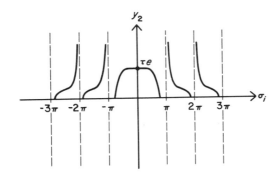

Figure 12.13. Equation (A25)

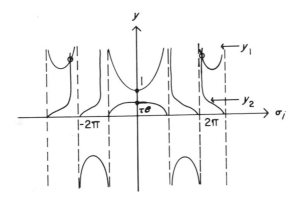

Figure 12.14. Equations (A24) and (A25) for $\tau e < 1$

Now let us consider σ complex. We write $\sigma = \sigma_r + i\sigma_i$, and so (A15) becomes

$$(\sigma_r + i\sigma_i)\, e^{\sigma_r}(\cos \sigma_i + i \sin \sigma_i) + \tau = 0. \tag{A19}$$

Equating real and imaginary parts to zero, we obtain

$$\sigma_r\, e^{\sigma_r} \cos \sigma_i - \sigma_i\, e^{\sigma_r} \sin \sigma_i + \tau = 0 \tag{A20}$$

$$\sigma_i\, e^{\sigma_r} \cos \sigma_i + \sigma_r\, e^{\sigma_r} \sin \sigma_i = 0. \tag{A21}$$

Substituting (A21) into (A20) gives

$$\frac{\sigma_r}{\cos \sigma_i} + \tau e^{-\sigma_r} = 0 \tag{A22}$$

and (A21) may be rearranged into

$$\frac{-\sigma_r}{\cos \sigma_i} = \frac{\sigma_i}{\sin \sigma_i} \tag{A23}$$

which is sketched in Figure 12.11. To analyze (A22) and (A23), we plot

$$y_1(\sigma_i) = \frac{-\sigma_r}{\cos \sigma_i} = \frac{\sigma_i}{\sin \sigma_i} \tag{A24}$$

in Figure 12.12 and

$$y_2(\sigma_i) = \tau e^{-\sigma_r} = \tau \exp\left(\frac{\sigma_i \cos \sigma_i}{\sin \sigma_i}\right) \tag{A25}$$

in Figure 12.13 and look for intersections of the two graphs.

Let us first consider $\tau e < 1$. Figure 12.14 does not explicitly show the two real roots we found before. It does show a sequence of roots (two are shown) near $\sigma_i = \pm 2\pi k$, $k = 1, 2, 3, \cdots$. For the two roots near $\pm 2\pi$, σ_r is negative and large in magnitude (see Figure 12.11). The solutions corresponding to these roots die out very rapidly, even though they do oscillate. Superposed on any of the exponentially decaying solutions, they are insignificant after a very short time.

Now let us consider $\tau e > 1$. No real roots exist to contribute exponentially decaying solutions by our previous analysis. Two complex roots exist between $-\pi < \sigma_i < \pi$. Because $\sigma_i \neq 0$, these roots correspond to oscillatory solutions. To see whether these solutions grow in time, we must look at σ_r. If τ is such that $|\sigma_i| < \pi/2$, $\sigma_r < 0$, and the solutions decay. If τ is such that $|\sigma_i| > \pi/2$, the two solutions grow. Let us attempt to find τ_2 such that $\sigma_i = \pi/2$ is a solution of (A20) and (A21). From (A21), we find $\sigma_r = 0$, and from (A20), $\tau_2 = \pi/2$. See Figure 12.15

Let us summarize our results. If

(i) $0 < \tau = \lambda_0 T \leq 1/e$, the solution of (A12) will decay, with only minor oscillations;

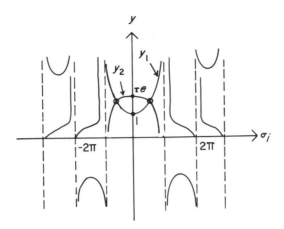

Figure 12.15. Equations (A24) and (A25) for $\tau e > 1$

(ii) $1/e < \tau = \lambda_0 T < \pi/2$, the solution of (A12) will oscillate with decaying amplitude;
(iii) $\tau = \lambda_0 T > \pi/2$, the solution of (A12) will oscillate with increasing amplitude.

Thus the motion of the second car is locally stable if $\tau = \lambda_0 T < \pi/2$, and the stability is strongest if $\lambda_0 T \leq 1/e$.

Incidentally, by setting $T = 0$ and $\lambda_0 = 1$ in (A7) and (A14), we see that Pipes' model, which is car-following without delay ($T = 0$) and sensitivity $\lambda_0 = 1/T^* = 1$, is both asymptotically and locally stable.

Exercises

Pipes' Model

1. Solve for the velocities $v_{n+1}(t)$, $n = 1, 2, 3, \cdots$ in the case that the lead vehicle moves with a constant acceleration from $t = 0$ to $t = t_0$ and then cruises at constant speed so that

$$v_1(t) = \begin{cases} v_c[t/t_0], & 0 \leq t \leq t_0 \\ v_c, & t > t_0. \end{cases}$$

Try to get some graphical feel for the solutions $v_{n+1}(t)$.
   Solution:

$$v_{n+1}(t) = \begin{cases} \dfrac{v_c}{t_0} \displaystyle\int_0^t G_n(u)\,du, & 0 < t < t_0 \\[4mm] \dfrac{v_c}{t_0}\left[\displaystyle\int_0^t G_n(u)\,du - \displaystyle\int_0^{t-t_0} G_n(u)\,du\right], & t > t_0. \end{cases}$$

2. Return to (7) and reformulate the equation for the case that at $t = 0$ all vehicles are moving at a constant velocity v_c and then for $t > 0$ the lead vehicle decelerates. Obtain an equation for $V_{n+1}(s)$ in terms of $V_1(s)$ analogous to (10).

   Solution:

$$V_{n+1} = \frac{V_1}{(s + 1)^n} + sv_c \left(\sum_{k=1}^{n} \frac{1}{(s + 1)^k} \right).$$

3. Show that the instantaneous deceleration of the lead vehicle represented by

$$v_1(t) = \begin{cases} v_c, & t = 0 \\ 0, & t > 0 \end{cases}$$

leads to

$$v_{n+1}(t) = v_c \sum_{k=1}^{n} \Phi_k(t)$$

   (see Section 10.2 in the Appendix) and then graph $v_1, v_2, v_3, v_4,$ and v_5.

Linear Car-Following Model

4. Differentiate both sides of

$$\frac{1}{1 + x} = 1 - x + x^2 - x^3 + x^4 - x^5 + - \cdots$$

   to obtain a series for $1/(1 + x)^2$ and use the result to show that

$$\frac{v_3(t)}{v_c} = \frac{\lambda_0^2}{2!}(t - 2T)^2 h(t - 2T) - \frac{2\lambda_0^3}{3!}(t - 3T)^3 h(t - 3T)$$

$$+ \frac{3\lambda_0^4}{4!}(t - 4T)^4 h(t - 4T) - + \cdots.$$

Steady-State

5. Assume the model

$$\frac{d}{dt} v_{n+1}(t + T) = \alpha_0 \frac{v_n(t) - v_{n+1}(t)}{[x_n(t) - x_{n+1}(t)]^m}, \qquad m > 1.$$

   (If $m = 1$, this is the reciprocal spacing model.) Show that the solution of this equation is

$$v_{n+1}(t + T) = (m - 1)^{-1}\alpha_0\{l^{-(m-1)} - [x_n(t) - x_{n+1}(t)]^{-(m-1)}\}$$

   and that the steady-state velocity is

$$u = (m - 1)^{-1}\alpha_0[k_j^{-(m-1)} - k^{-(m-1)}].$$

Model Forming

6. Write down the differential equations for a CFT model in which the stimulus to the $(n + 1)$st driver is his spacing from the nth driver and his response is his acceleration after a delay T. Assume constant sensitivity.

Solution:

$$\frac{d^2}{dt^2} x_{n+1}(t + T) = \lambda_0(x_n(t) - x_{n+1}(t)), \qquad n = 1, 2, 3, \cdots.$$

7. Write down the differential equations for linear car-following with delay if the driver's stimulus is a linear combination of the relative velocity and the departure of the spacing from a desired value D, where the sensitivities for each stimulus are constant but different.

Solution:

$$\frac{d^2 x_{n+1}(t + T)}{dt^2} = \lambda_1 \left[\frac{dx_n(t)}{dt} - \frac{dx_{n+1}(t)}{dt} \right] + \lambda_2 [x_n(t) - x_{n+1}(t) - D],$$

$$n = 1, 2, 3, \cdots.$$

Emergency Control Model

8. Suppose that when two cars come closer than some critical spacing X, the following car stops obeying linear CFT with delay T and slams on his brakes after delay T with the maximum deceleration mechanically possible, say $-\beta$ with $\beta > 0$. Use the Heaviside function

$$h(t) = \begin{cases} 1, & t > 0 \\ 0, & t < 0 \end{cases}$$

and the deviation of X from the car spacing at time t

$$Z_n \equiv X - (x_n(t) - x_{n+1}(t))$$

to write the differential equations for this model.

Solution:

$$\frac{dv_{n+1}(t + T)}{dt} = \lambda_0 [v_n(t) - v_{n+1}(t)][1 - h(Z_n)] - \beta h(Z_n), \qquad n = 1, 2, 3, \cdots.$$

Forward–Backward Control Model

9. Write down the differential equations for a linear CFT with delay if a driver is stimulated by the relative velocity 1) between his car and the one in front and 2) between his car and the one behind. State conditions on the sensitivities which would reflect a stronger reaction to the car in front.

Solution:

$$\frac{dv_{n+1}(t + T)}{dt} = \lambda_1 [v_n(t) - v_{n+1}(t)] + \lambda_2 [v_{n+2}(t) - v_{n+1}(t)], \qquad n = 1, 2, 3, \cdots.$$

$$\lambda_1 > \lambda_2 > 0.$$

Traffic Light Simulation

10. Extend the results of Section 4 and Exercise 4 to compute $v_4(t)/v_c$ and $v_5(t)/v_c$, and then try to reproduce the results of Figure 12.5. The cars are initially separated by 20 ft. It will probably be easier if integration of the velocities to obtain the coordinates $x_n(t)$ is carried out on the computer.

Stability
11. After reading Section 10.3 in the Appendix, what can you conclude about stability for the case $\lambda_0 T = \pi/2$?

References

[1]   S. Bexelius, "An extended model for car-following," *Transportation Res.*, vol. 2, pp. 13–21, 1968.
[2]   E. A. Bender and L. P. Neuwirth, "Traffic flow: Laplace transforms," *Amer. Math. Monthly*, vol. 80., pp. 417–423, 1973. Readable by students; develops some details of stability theory.
[3]   R. E. Chandler, R. Herman, and E. W. Montroll, "Traffic dynamics: Studies in car-following," *Operations Res.*, vol. 6, pp. 165–184, 1958. First paper on stability theory; develops asymptotic stability for a number of models; higher level than present module, but accessible to better students with some complex variable background.
[4]   D. R. Drew, *Traffic Flow Theory and Control.* New York: McGraw-Hill, 1968. A book on many aspects of traffic science besides car-following; lots of practical examples, many topics accessible to undergraduates.
[5]   L. C. Edie, "Car-following and steady-state theory for non-congested traffic," *Operations Res.*, vol. 9, no. 1, pp. 66–76, 1961.
[6]   Denos C. Gazis, "Traffic flow and control: Theory and applications," *Amer. Scientist*, vol. 60, no. 4, pp. 415–424, 1972. A survey article on traffic science, like a *Scientific American* article.
[7]   D. C. Gazis, R. Herman, and R. B. Potts, "Car-following theory of steady state traffic flow," *Operations Res.*, vol. 7, no. 4, pp. 499–505, 1959. Very readable; uses only algebra and simple integration.
[8]   D. C. Gazis, R. Herman and R. W. Rothery, "Nonlinear follow-the-leader models of traffic flow," *Operations Res.*, vol. 9, no. 4, pp. 545–567, 1961.
[9]   H. Greenberg, "An analysis of traffic flow," *Operations Res.*, vol. 7, no. 1, pp. 79–85, 1959.
[10] R. Herman, *et al.*, "Traffic dynamics: Analysis of stability in car-following," *Operations Res.*, vol. 7, pp. 86–106, 1959. The definitive article on stability theory: not accessible to average undergraduates.
[11] Robert Herman and R. B. Potts, Single-Lane Traffic Theory and Experiment, *Theory of Traffic Flow.* Amsterdam: Elsevier, 1961, pp. 120–146. Very readable summary of car-following without too many technical details; good source of experimental results.
[12] G. Lee, "A generalization of linear car-following," *Operations Res.*, vol. 14, no. 4, pp. 595–606, 1966.
[13] G. F. Newell, "Theories of instability in dense highway traffic," *J. Operations Res. of Japan*, vol. 5, pp. 9–54, 1962.
[14] L. A. Pipes, "An operational analysis of traffic dynamics," *J. Appl. Phys.* vol. 24, no. 3, pp. 274–281, 1953. The original paper on car-following; easily read by those with background to read this module.
[15] E. Tuck, "Stability of following in two dimensions," *Operations Res.*, vol. 9, no. 4, pp. 479–495, 1961.
[16] W. E. Wilhelm and J. W. Schmidt, "Review of car-following theory," *Transportation Engineering J. of ASCE*, vol. 99, no. TE4, pp. 923–931, 1973. A bibliographic survey; easily read by students.

Notes for the Instructor

Objectives. Laplace transform techniques are used to study the differential equations and differential-delay equations of car-following models. A heuristic discussion of stability is included, with mathematical details left to an appendix. Steady-state equations for flow versus concentration are derived, using elementary integration, and their significance in relation to other traffic theories and empirical results is discussed. Pitfalls: be careful to distinguish between L and $\mathscr{L}$; $\mathscr{L}$ is the usual Laplace transform, while $L \equiv s\mathscr{L}$. Do not confuse instability with collisions; they are separate but related issues.

Prerequisites. Calculus, differential equations using Laplace transforms, simple kinematics of straight line motion.

Time. Three lectures should suffice to cover the material, if the students also read the module.

CHAPTER 13

Equilibrium Speed Distributions

Donald A. Drew*

1. Speed Distributions

One of the fundamental processes in traffic flow, which leads to delays and frustration for individual drivers is that of overtaking (that is, approaching from behind) a slower vehicle. Trucks, busses, and sightseers ("Sunday drivers") can cause substantial delays to a driver trying to move quickly from one place to another on the standard two-lane, two-way highway encountered in so much of the United States.

We wish to consider the processes of overtaking and passing on a long two-lane, two-way road. In order to study conditions leading to a faster vehicle overtaking a slower vehicle, we must have a mechanism of handling the different speeds of different vehicles. The concept which we use is that of number density function in speed space. (Our speed space is sometimes called phase space.)

We shall approach this concept through some examples. Suppose we count the number of vehicles on the road with speeds between 0 and 10 mi/h, the number between 10 and 20 mi/h, and so on. If we plot these data as a bar graph, we will have Figure 13.1(a). This is the correct type of information; however, it is still too crude.

Let us also take the data on the number of vehicles with speeds between 0 and 5 mi/h, between 5 and 10 mi/h, and so on. These data are shown in Figure 13.1(b). We note that the number of vehicles between 20 and 30 mi/h in Figure 13.1(b) is the sum of the number with speed between 20 and 25 mi/h and the number with speed between 25 and 30 mi/h.

* Department of Mathematical Sciences, Rensselaer Polytechnic Institute, Troy, NY 12181.

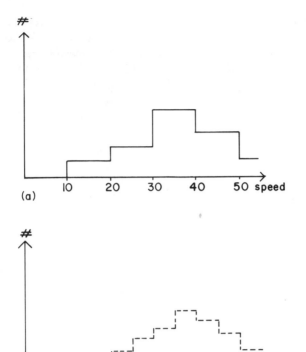

Figure 13.1

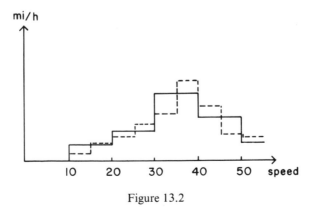

Figure 13.2

It is difficult to see that the data plotted in Figure 13.1 represent the same set of vehicles on the same road. The two sets of data look more similar if we plot the *densities*, that is, the number *per unit speed versus* the speed. To obtain the number density from Figure 13.1 we divide the number in each interval of speed by the length of that interval. The number densities corresponding to Figure 13.1(a) and (b) are plotted in Figure 13.2.

We are not restricted to 5-mi/h intervals in plotting our number density data; we could use 1-mi/h intervals, 0.5-mi/h intervals, 0.1-mi/h intervals, and so on. Using smaller intervals gives more detail about the speed space structure. Thus the smaller the interval, the better.

We note that to obtain the number of vehicles n_i in the ith speed interval from the density f_i we need only multiply by the length of that interval, Δu_i:

$$n_i = f_i \Delta u_i. \tag{1}$$

If we wish to compute the number of vehicles n with speeds between u_a and u_b, we must sum the number in each subinterval n_i

$$
\begin{aligned}
n &= \sum n_i \\
&= \sum f_i \Delta u_i,
\end{aligned}
\tag{2}
$$

where the summation is over subintervals lying between u_a and u_b.

We wish to work with a mathematical abstraction of the discrete number distributions described above. Suppose $f(u)$ is the *continuous* density distribution of vehicles at speed u, obtained by abstracting the number density process above by letting $\Delta u_i \to 0$. Then if u_a and u_b are any two speeds, the number of vehicles on the road with speeds between u_a and u_b is given by

$$n = \int_{u_a}^{u_b} f(u)\, du. \tag{3}$$

In general, the density distribution will change in time. Also, if we look at different subsections of the road, we expect to see different distributions, depending on local disturbances, like the presence of a truck, or a curve in the highway. These time and space dependences could be considered; however, for the sake of simplicity, we shall assume that the density f depends on speed alone and is independent of space x and time t.

2. Overtaking

Now let us turn our attention to the processes which shape $f(u)$, specifically, overtaking and passing. When we refer to overtaking, we mean catching up to a slower vehicle ahead. Passing refers to pulling out in the lane of on-

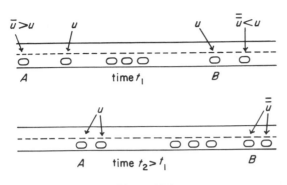

Figure 13.3

coming traffic, getting ahead of the slower vehicle, and then returning to a more desirable speed.

Let us consider the vehicles operating at speed u. The overtaking of vehicles changes the number of vehicles at speed u in two ways. First, a vehicle operating at a higher speed $\bar{u} > u$ may overtake a vehicle operating at speed u, causing an addition to the number of vehicles operating at speed u. These vehicles which are slowed down are added to *queues*, or lines of vehicles following other vehicles. (The vehicle may, of course, be the first vehicle in the queue.) See Figure 13.3. The second way in which overtaking can change the number of vehicles operating at speed u is that a vehicle operating at speed u may overtake a vehicle operating at a lower speed $\bar{u} < u$, thus causing a loss from the number of vehicles operating at speed u. See Figure 13.3.

Let us compute the rate of change θ of number of vehicles due to overtaking. Assume that there is a vehicle at time $t = 0$ at point x traveling at speed u. Let us compute the number of vehicles overtaking this given vehicle in time interval Δt. We note that all vehicles which have speed $\bar{u} > u$ and are in the interval $x - (\bar{u} - u)\Delta t$ to x will overtake the vehicle at x in the interval Δt. Assuming that the vehicles are distributed uniformly over a road of length L, the number of vehicles having speed $\bar{u} > u$ (per unit speed) in the interval $x - (\bar{u} - u)\Delta t$ to x is $[(\bar{u} - u)\Delta t/L]f(\bar{u})$, where L is the length of the road. The total number of such vehicles is

$$\int_u^\infty \frac{(\bar{u} - u)\Delta t}{L} f(\bar{u})\, d\bar{u}. \tag{4}$$

Thus the rate at which vehicles overtake a given vehicle at speed u, location x, is the number divided by the time interval Δt, so that the rate is

$$\int_u^\infty \frac{(\bar{u} - u)f(\bar{u})}{L}\, d\bar{u}. \tag{5}$$

To obtain the rate of gain G of vehicles at speed u due to overtaking, we must multiply this quantity by the probability of finding a vehicle at x at speed u. In fact, the probability of finding a vehicle at x at speed u is $f(u)/N \cdot l/L$, where N is the total number of vehicles on the road, l is the length of a vehicle, and l/L is the probability of a given vehicle being at x. Thus the rate of gain of vehicles at speed u (per unit speed) due to overtaking is

$$G = K \int_u^\infty (\bar{u} - u) f(u) f(\bar{u}) \, d\bar{u} \tag{6}$$

where $K = l/NL^2$.

We might comment on the fact that the integration extends to $\bar{u} = \infty$. We wish to include all vehicles with speed greater than u in the integration. In any reasonable physical situation, we expect $f(\bar{u}) = 0$ for $\bar{u} > u^*$, where u^* is the mechanical speed limit of a vehicle. By taking the upper limit of integration to be infinite, we bypass the uncertainty of the exact value of u^*.

Let us now compute the rate of loss $\mathscr{L}$ of vehicles from speed u due to their overtaking a slower vehicle. Let us again consider a vehicle at x traveling at speed u and consider the number of vehicles having speed $\bar{\bar{u}} < u$ (per unit speed) in the interval x to $x + (u - \bar{\bar{u}})\Delta t$. That number is

$$\frac{(u - \bar{\bar{u}})\Delta t f(\bar{\bar{u}})}{L}. \tag{7}$$

The total number of such vehicles is

$$\int_0^u \frac{(u - \bar{\bar{u}})\Delta t f(\bar{\bar{u}})}{L} \, d\bar{\bar{u}}, \tag{8}$$

and as before, the rate of loss due to overtaking for such vehicles is

$$\mathscr{L} = K \int_0^u (u - \bar{\bar{u}}) f(\bar{\bar{u}}) f(u) \, d\bar{\bar{u}}. \tag{9}$$

Combining the rates of gain and loss to get a net rate of change θ of $f(u)$ (per unit speed) due to overtaking gives

$$\theta = G - \mathscr{L} = K \int_0^\infty (\bar{u} - u) f(\bar{u}) f(u) \, d\bar{u}. \tag{10}$$

3. Passing

Now let us turn our attention to an analysis of the passing process. First we shall assume that a fraction p of those overtaking will pass instantaneously, essentially not interacting with the overtaken vehicle. This results in a loss of vehicles from speed u, and we have a negative rate due to instantaneous passing, $P_i = -p\theta$.

In addition, another complex passing process will take place depending

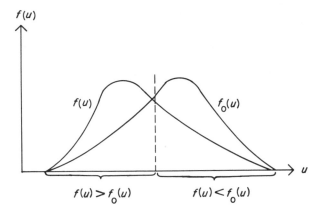

Figure 13.4

on the concentration of oncoming traffic, their position in the queue, and their general desire to move faster. This process seems impossible to model in a rational way. However, we shall bypass this difficulty by assuming a *phenomenological* model, essentially a model which does roughly what we expect. We expect that if the drivers were able to remain out of the queues (a result of overtaking), they would attain a distribution of desired speeds equal to $f_0(u)$. We assume that if more vehicles are actually traveling at a given speed u than the number of drivers who desire to do so, then drivers will be passing. Therefore, if $f(u) > f_0(u)$, a net loss of vehicles from speed u will occur due to passing. On the other hand, if fewer drivers are traveling at speed u than the number of drivers who desire to do so, then some of the drivers who do pass will accelerate to attain this speed. Thus if $f(u) < f_0(u)$, there will be a net gain of vehicles at speed u due to passing. In fact, we assume that if more drivers are not at their desired speed, then more passing will occur. See Figure 13.4.

Thus the rate of change in $f(u)$ (per unit speed) due to this noninstantaneous passing is

$$P_n = \lambda(f_0(u) - f(u)). \tag{11}$$

We note that λ has dimension (time)$^{-1}$. The quantity $1/\lambda$ is often referred to as a *relaxation* time, and this passing process is called *relaxation*.

4. Equilibrium Distributions

Since we are assuming equilibrium, the net rate of change of $f(u)$ vanishes, and hence the sum of the rates due to overtaking and to passing is zero:

$$\theta + P_i + P_n = 0 = (1 - p)K \int_0^\infty (\bar{u} - u)f(\bar{u})f(u)\,d\bar{u} + \lambda(f_0(u) - f(u)).$$

$$(12)$$

Equation (12) gives a balance between overtaking and passing. Solving for $f(u)$ gives

$$f(u) = \frac{f_0(u)}{1 - \dfrac{(1-P)K}{\lambda} \displaystyle\int_0^\infty (\bar{u} - u)f(\bar{u})\,d\bar{u}}.$$

$$(13)$$

Let us attempt some qualitative calculations using (13). We note that for

$$\frac{(1-p)K}{\lambda} \int_0^\infty (\bar{u} - u)f(\bar{u})\,d\bar{u} < 0 \tag{14}$$

we shall have

$$f(u) < f_0(u). \tag{15}$$

Since $p < 1$, $K > 0$, and $\lambda > 0$, (14) is equivalent to

$$u > \frac{\displaystyle\int_0^\infty \bar{u}f(\bar{u})\,d\bar{u}}{\displaystyle\int_0^\infty f(\bar{u})\,d\bar{u}} \equiv u_a \tag{16}$$

where u_a is defined to be the average speed of vehicles on the highway.

For $u < u_a$, we have $f(u) > f_0(u)$. Thus one property of our solution (13) is that more vehicles are traveling at speeds less than u_a than desire to be at that speed, while fewer vehicles are traveling at speeds greater than u_a than so desire. This seems to be an acceptable prediction.

A big problem in attempting to use (13) to make predictions is that to do so, we must determine a suitable form for $f_0(u)$, and suitable values for p and λ. Clearly, all of these quantities depend on the density of traffic, which is determined by N, the number of vehicles on the road.

On the other hand, by guessing at reasonable values for λ and p and taking a somewhat arbitrary form for $f_0(u)$, we can make some predictions for the flow-concentration relationship, which is a fundamental part of traffic flow theory. See Exercise 2 for a quite simple calculation of this sort.

The ideas involved in this module were first introduced in order to obtain an idea of how the velocity distribution evolved in time as a function of location along the road. The partial differential equation governing the distribution $f(t, x, u)$ is much like the Boltzmann equation studied in statistical mechanics. This analogy stimulates both the traffic theoretician and the statistical mechanician. Interesting predictions have been made about traffic flow from this Boltzmann-like approach; see the monograph by Prigogine and Herman. Some mathematical sophistication is required, however.

Exercises

1. Suppose $f(u) = Be^{-(u-u_1)^2/\sigma^2}$, where B and σ are positive numbers, and u_1 is a reference speed.
   a) Sketch this distribution.
   b) Calculate

$$u_a = \frac{\displaystyle\int_0^\infty uf(u)\,du}{\displaystyle\int_0^\infty f(u)\,du}.$$

   Hint: You will need an integral table or the fact that $\int_0^\infty e^{-x^2}\,dx = \sqrt{\pi}/2$.

2. Let

$$f_0(u) = \begin{cases} C(u - u_1)(u_2 - u), & \text{for } u_1 \le u \le u_2, \\ 0, & \text{for } u > u_2 \text{ or } u < u_1, \end{cases}$$

   where C is a constant.
   a) Sketch this desired speed distribution.
   b) Compute $N = \int_0^\infty f_0(u)\,du$.
   c) Assume that $p = 0.5$, $\lambda = 0.01/\text{second}$, $l = 15$ ft, $C = 500{,}000$ vehicles/$(\text{mi/h})^3$, $u_1 = 20$ mi/h, $u_2 = 40$ mi/h, and $L = 20$ mi.
      (i) Compute an approximation to u_a using a power series expansion of $f(u)$ in terms of $(1 - p)K/\lambda$.
      (ii) Using this approximation to u_a, find $f(u)$.
   d) Again, using $p = 0.05$, $\lambda = 0.01$, $l = 15$ ft, $u_1 = 20$ mi/h, $u_2 = 40$ mi/h, and $L = 20$ mi, but leaving C as arbitrary, find an approximation to $f(u)$ in terms of C. Using b, relate C to N, and sketch a graph of q versus N. This is essentially the flow-concentration diagram.

References

[1] I. Prigogine and R. Herman, *Kinetic Theory of Vehicular Traffic.* New York: Elsevier, 1971. This book is the basis of the analysis in this module. However, they treat nonsteady and nonhomogeneous distributions, and hence the prerequisite for reading the book is some knowledge of Boltzmann-like partial differential equations.

Summaries of the Prigogine–Herman approach can be found in several sources:

[2] D. C. Gazis, *Traffic Science.* New York: Wiley-Interscience, 1974.
[3] D. L. Gerlough and M. J. Huber, *Traffic Flow Theory—A Monograph*, special Report 165, Traffic Research Board, National Research Council, Washington, DC, 1975.

Notes for the Instructor

Objectives. This module discusses vehicle passing and overtaking on long roads. Balance concepts are emphasized in the velocity space, an example not usually met in ordinary applications problems.

Prerequisites. Calculus, probability density, and some sophistication with integrals.

Time. Two lectures should be enough to cover the material.

CHAPTER 14
Traffic Flow Theory

Donald A. Drew*

1. Basic Equations

Let us derive the basic conservation equation for traffic flow. We consider the flow of vehicles on a long road where the features of the flow we wish to calculate, such as bottlenecks, etc., are long compared with the average distances between vehicles. Let $n(x, x + \Delta x, t)$ denote the number of vehicles between point x and point $x + \Delta x$ on the road at time t (see Figure 14.1). We shall *assume* that $k(x, t)$ exists such that for any x, Δx, and t,

$$n(x, x + \Delta x, t) = \int_{x}^{x+\Delta x} k(\hat{x}, t)\, d\hat{x}. \tag{1}$$

We note that, by the fundamental theorem of calculus,

$$k(x, t) = \lim_{\Delta x \to 0} \frac{n(x, x + \Delta x, t)}{\Delta x}$$

if k is continuous. We shall assume that we can adequately model the situations of interest with the assumption that k is continuous.

In terms of infinitesimals, k is the number of vehicles per unit length in the infinitesimal length between x and $x + \Delta x$ at time t. Empirical values of k can be determined from aerial photographs of roads. We select some "small" (infinitesimal) length Δx, count the vehicles between x and $x + \Delta x$, and divide by Δx.

Now let us define the flow rate $q(x, t)$. The flow rate q is simply the rate at which vehicles pass point x at time t. The total number Q crossing point x between time t and time $t + \Delta t$ is then given by

* Department of Mathematical Sciences, Rensselaer Polytechnic Institute, Troy, NY 12181.

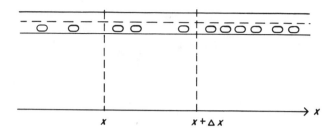

Figure 14.1. Traffic Situation at Some Time t

$$Q(x, t, t + \Delta t) = \int_t^{t+\Delta t} q(x, \hat{t}) \, d\hat{t}. \tag{2}$$

Again, by the fundamental theorem of calculus, we have

$$q(x, t) = \lim_{\Delta t \to 0} \frac{Q(x, t, t + \Delta t)}{\Delta t}. \tag{3}$$

Empirical values of q can be obtained by clocked counters which keep a time record of the vehicles crossing point x. We select some "small" Δt, count the vehicles crossing x between t and $t + \Delta t$, and divide by Δt.

Let us now consider the balance or conservation of vehicles in the road. Let us isolate a segment of the road lying between points x and $x + \Delta x$ and look at the rate of change of the number of vehicles in this segment.

The balance law which applies here is that no vehicles are created or destroyed (neglecting collisions!) in this segment. Thus conservation of vehicles requires that *the rate of increase of the number of vehicles between x and $x + \Delta x$ is equal to the rate at which vehicles flow in minus the rate at which they flow out.* Thus for any time instant t

$$\frac{d}{dt} \int_x^{x+\Delta x} k(\hat{x}, t) \, d\hat{x} = q(x, t) - q(x + \Delta x, t). \tag{4}$$

This is the fundamental conservation law (balance law) for the segment of road between x and $x + \Delta x$; it is a statement about the balance we would see in a snapshot of the road (see Figure 14.1). We note that

$$\frac{d}{dt} \int_x^{x+\Delta x} k(\hat{x}, t) \, d\hat{x} = \int_x^{x+\Delta x} \frac{\partial k}{\partial t}(\hat{x}, t) \, d\hat{x}. \tag{5}$$

Let us now divide by Δx and let $\Delta x \to 0$. We have

$$\lim_{\Delta x \to 0} \frac{1}{\Delta x} \int_x^{x+\Delta x} \frac{\partial k}{\partial t}(\hat{x}, t) \, d\hat{x} = \lim_{\Delta x \to 0} \frac{q(x, t) - q(x + \Delta x, t)}{\Delta x}. \tag{6}$$

By the fundamental theorem of calculus, the limit on the left is precisely $(\partial k/\partial t)(x, t)$, while by the definition of partial derivative, the limit of the quotient on the right is $-(\partial q/\partial x)(x, t)$. Thus we arrive at the fundamental

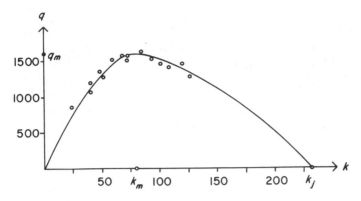

Figure 14.2. Flow Versus Concentration

balance law in differential form

$$\frac{\partial k}{\partial t} + \frac{\partial q}{\partial x} = 0. \tag{7}$$

This equation tells us how the concentration k changes in time at each x from the flow q. To predict how k changes, then, knowledge of another variable q is required.

To see that this equation is qualitatively correct, consider the following thought experiment. If $\partial q/\partial x < 0$, then q is decreasing in a neighborhood of the location x, and $q(x + \varepsilon, t)$ is less than $q(x - \varepsilon, t)$ for some small positive number ε. Thus the flow out of this section of road is less than the flow in, and hence k must increase in time near that location. Mathematically, this is expressed by (7).

We note that we have one equation (7) for two unknown functions k and q. Thus our system is *underdetermined*. A little more thought should convince us that this underdetermination is necessary at this stage. After all, we have in no essential way used the fact that we are modeling *vehicular traffic*. The concentration and flow could just as well be concentration and flow of a pollutant in a river, or of heat in a bar, or of electrons in a wire, or almost anything which flows in a one-dimensional situation.

We need more equations which reflect the peculiarities of vehicular traffic. These equations may be balance equations (perhaps an equation for $d/dt \int_x^{x+\Delta x} q(\hat{x}, t) \, d\hat{x}$) or maybe simply some abstraction of empirical data pertaining to the physical situation at hand. If we use empirical data, then those data contain (we hope!) the essential constitution of the physical situation. Such a relation is called a *constitutive equation*. For the traffic flow problem, we have much data of the form flow rate plotted against concentration (q versus k), as shown in Figure 14.2 (see Exercise 1).

Thus we assume that $q = q(k)$. It is noteworthy that the flow of vehicles increases with increasing k for k small, while it decreases to zero as k

---

[1] The number k_j is the *jam concentration*, the number of cars per unit length of highway when the traffic is jammed and nothing moves.

approaches the jam concentration k_j (here just slightly more than 225 veh/mi.[1] The maximum flow rate of 1500 veh/h occurs at about 75 veh/m.

Many different forms of $q(k)$ have been fitted to the data. They range from simple forms having the above general features to others which fit the data very accurately. One of the simpler models for $q(k)$ is *Greenshields'* model given by $q = u_f k(1 - k/k_j)$, where u_f is the (empirical) free speed of the road (the speed at which a vehicle would travel if it were alone on the highway), and k_j is the jam concentration (see Exercise 2).

2. Propagation of a Disturbance

Let us consider the evolution of the traffic concentration k on a long road. First let us assume that k is a function of x and t and that q is a function of k (Figure 14.2). Then (7) can be rewritten by applying the chain rule to calculate $\partial q(k)/\partial x$:

$$\frac{\partial k}{\partial t} + \frac{dq}{dk}\frac{\partial k}{\partial x} = 0. \tag{8}$$

Now, let us consider a curve $x = x(t)$ in the xt plane on which k is constant. Such a curve is called a *characteristic* of (8) and satisfies the implicit relation,

$$k(x(t), t) = \text{constant.} \tag{9}$$

The function $x(t)$ also satisfies the differential equation obtained by differentiating (9) with respect to t,

$$\frac{\partial k}{\partial x}\frac{dx}{dt} + \frac{\partial k}{\partial t} = 0, \tag{10}$$

where once again we have used the chain rule. Along a characteristic, $k(x, t)$ satisfies both (8) and (10). Hence we must have

$$dx/dt = dq/dk \tag{11}$$

along a characteristic. Since k is constant on $x = x(t)$, so is dq/dk, and we can immediately integrate (11) with respect to t, obtaining

$$x = (dq/dk)t + x_0 \tag{12}$$

as the equation of the family of characteristics. Since k is constant on each of these curves, each curve is a straight line. Note that we assume that $dq/dk > 0$ throughout.

If we know the value of k at x_0 at $t = 0$, the value of k at each point on the line $x = (dq/dk)t + x_0$ is the same as it is at x_0. However, in terms of x and t, $x_0 = x - (dq/dk)t$. Thus

$$k(x, t) = k(x_0, 0) = k\left(x - \frac{dq}{dk}t, 0\right), \tag{13}$$

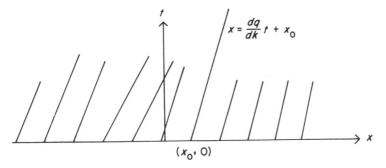

Figure 14.3. Curves of Constant k (Characteristics) (Slopes of lines are given by dq/dk for particular value of k associated with given line.)

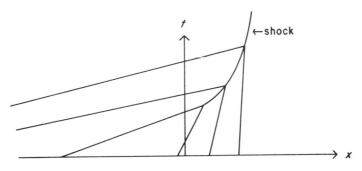

Figure 14.4. Propagation of Discontinuity

Thus if we supply the concentration of vehicles k at time $t = 0$, the initial time, the solution is determined by the relation

$$k(x, t) = k\left(x - \frac{dq}{dk}t, 0\right) \tag{14}$$

—almost. Consider the lines emanating from the neighborhood of $x = 0$ in Figure 14.3. If these lines are extended, they will cross. At a point where they cross, the equation predicts two *different* values of k. Physically, this cannot happen. Thus the partial differential equation cannot be valid everywhere along any two of the characteristic lines which cross at some point (x_1, t_1). A little thought suggests that somewhere on at least one of those characteristic lines, the solution must be discontinuous. We shall discuss the location of the discontinuity shortly. Before that, we note that the set of (x, t) where the discontinuity occurs must be such that each given line must be separated from all others which would intersect it by a discontinuity. Otherwise, two different value of k would be predicted for the point of intersection. This suggests that the discontinuity must be more than a mere point, that it must be a curve in $x - t$ space (see Figure 14.4). Such a curve of discontinuities is called a *shock*.

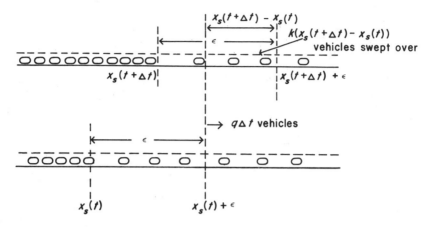

Figure 14.5. Outflow from Segment at $x_s(t) + \varepsilon$

3. Shocks

To obtain the condition valid at the shock, we must rederive the balance condition *without* the assumptions of differentiability which we made there. Consider a section of roadway which lies between $x_s - \varepsilon$ and $x_s + \varepsilon$, where x_s is the position of the shock and ε is some small positive distance. Since the shock moves in time, this section of roadway will change in time. We must account for this movement when we compute the inflow and outflow of vehicles from the segment.

First, we note that since the segment of road is small, essentially no vehicles will be found on it, and hence the rate of inflow to this segment must equal the rate of outflow from it.

Let us compute the rate of outflow from the segment of road through the end at $x_s(t) + \varepsilon$. Consider the road at two times t and $t + \Delta t$ where Δt is small (see Figure 14.5). The total number of vehicles which have flowed out of the segment between time t and time $t + \Delta t$ can be computed by considering the flow through the location $x_s(t) + \varepsilon$ during this time and subtracting those which did not make it out of the segment due to the movement of the end to position $x_s(t + \Delta t) + \varepsilon$.

Thus the number of vehicles flowing out of the segment is given by

$$q(x_s(t) + \varepsilon, t)\Delta t - k(x_s(t) + \varepsilon, t)(x_s(t + \Delta t) - x_s(t)). \qquad (15)$$

(Note that we use the flow rate and concentrations evaluated at time t. This approximation is not critical; we could have used other representative values of the time t.)

To compute the rate at which vehicles flow out, we must divide by Δt and let $\Delta t \to 0$. The rate of outflux of vehicles is then

$$q(x_s + \varepsilon, t) - k(x_s + \varepsilon, t)\frac{dx_s}{dt}(t). \tag{16}$$

A similar calculation of the inflow at $x_s - \varepsilon$ gives

$$q(x_s - \varepsilon, t) - k(x_s - \varepsilon, t)\frac{dx_s}{dt}(t). \tag{17}$$

Since no vehicles are created or destroyed in this segment, the difference between these two flow rates must equal the ráte of accumulation of vehicles in the segment from $x_s - \varepsilon$ to $x_s + \varepsilon$. If we let $\varepsilon \to 0$, we expect no vehicles to accumulate, and thus the flow rate in (16) must equal the flow rate out of (17). Thus we have

$$[\![q]\!] - [\![k]\!]\frac{dx_s}{dt} = 0, \tag{18}$$

where

$$[\![f]\!] = \lim_{\varepsilon \to 0}[f(x_s + \varepsilon, t) - f(x_s - \varepsilon, t)]$$

is the *jump* in a function f across the shock.

Thus the velocity of the shock dx_s/dt is given by $[q(k_2) - q(k_1)]/(k_2 - k_1)$, where k_2 and k_1 are the concentrations ahead of and behind the shockwave. We note that $[q(k_2) - q(k_1)]/(k_2 - k_1)$ is the slope of the chord connecting the points $(k_1, q(k_1))$ and $(k_2, q(k_2))$ in the flow-concentration diagram (Figure 14.2).

If we use Greenshields' relation for $q(k)$, we find that

$$\frac{q(k_2) - q(k_1)}{k_2 - k_1} = \frac{1}{2}\left[\frac{dq}{dk}(k_1) + \frac{dq}{dk}(k_2)\right],$$

so that dx_s/dt, the velocity of the shock, is the *average* of the slopes of the characteristics which meet at the shock. Using this rule and practicing a bit, it becomes possible to sketch the characteristics and the shocks for relatively complex traffic flows.

For example, let us consider the propagation of a traffic "hump." If the hump is as shown in Figure 14.6(b), with characteristics as sketched in Figure 14.6(a), we see that a shock must form somewhere around $x = 0$ and persist, intersecting pairs of characteristics at the average of their slopes.

The situation shown in Figure 14.6 corresponds to low concentration flows with $k < k_m$, where k_m is the concentration corresponding to maximum flow. If we consider concentrations greater than k_m, the shock will propagate backward. See Exercise 3.

We should also point out that dx_s/dt is the velocity of propagation of the shock and is not related to the velocities of individual vehicles. The average speed of the traffic defined by $u = q/k$ is always positive for $0 < k < k_j$. The shock speed, on the other hand, can be positive or negative, depending on the two concentrations on either side.

214

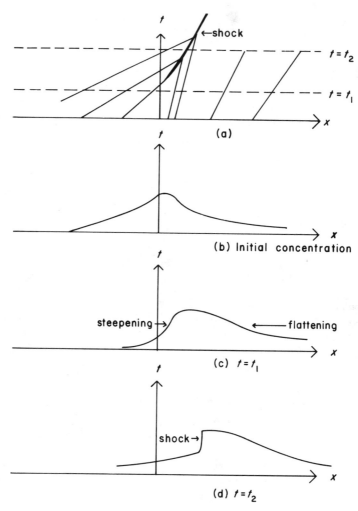

Figure 14.6. Development of Shock.

Exercises

1. Given the following data,

k (veh/mi)	$\frac{q}{k}$ (mi/h)	q (veh/h)
33	31	1023
43	26	1018
43	27	1061
48	23	1104
50	26	1300
92	12	1104
96	12	1112
98	11	1078
103	10	1030
106	10	1060
107	10	1070
110	8	880
110	9	990
114	9	1026
118	9	1062
119	9	1071
119	9	1071
121	9	1089
134	8	1072
135	8	1080
137	8	1096

use least squares to fit:
a) Greenshields' model

$$q = u_f k(1 - k/k_j);$$

b) Greenberg's model

$$q = u_m \ln(k_j/k).$$

That is, in a) choose u_f and k_j to minimize $\sum_{i=1}^{N} [q_i - u_f k_i (1 - k_i/k_j)]^2$, where (k_i, q_i) is an entry in the data table.

2. Compute the maximum flow rate for Greenshields' model. At what concentration does the maximum flow occur?

3. Consider the propagation of the traffic situation shown in Figure 14.7, where $k > k_m$. The characteristics for small t are given in Figure 14.7(b). Use your intuition about the formation of shocks to predict when a shock will form, and how it will propagate.

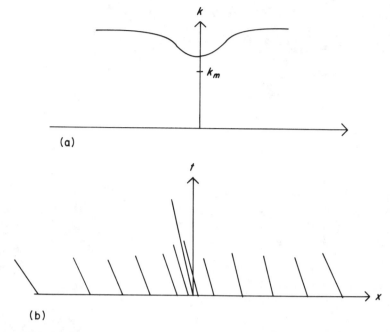

Figure 14.7. (a) Concentration at $t = 0$; (b) Characteristics Corresponding to Concentration in (a).

References

[1] F. A. Haight, *Mathematical Theories of Traffic Flow.* New York: Academic, 1963. This book is nicely mathematical, quite general, and not too hard to read. It is somewhat dated.

[2] D. R. Drew, *Traffic Flow Theory and Control.* New York: McGraw-Hill, 1968. No relation. A very general, readable book. It has many simple but important calculations pertaining to different aspects of highway design.

[3] D. C. Gazis, *Traffic Science.* New York: Wiley-Interscience, 1974. Chapter 1, written by L. C. Edie, deals with flow theories and is quite up to date.

[4] D. L. Gerlough and M. J. Huber, *Traffic Flow Theory, A Monograph*, Special Report 165, Traffic Research Board, National Research Council, Washington, D.C, 1975. Everything you always wanted to know about traffic flow theory—and more. This is an expensive paperbound monograph which synthesizes and reports, in a single document, the present state of knowledge in traffic flow theory. Not for children.

Notes for the Instructor

Objectives. This module introduces the fundamental balance idea necessary to derive the kinematic conservation equation. A discussion of constitutive equations for traffic flow is given. To illustrate the complexities of the model (and the physical situation), characteristics of first-order partial differential equations are derived and used from first principles. The modeling ideas are the main emphasis of this module.

Prerequisites. Multivariable calculus and differential equations and some perserverance should gain much understanding of modeling from this module.

Time. The module can be covered in three lectures.

PART V

INTERACTING SPECIES: STEADY STATES OF NONLINEAR SYSTEMS

CHAPTER 15
Why the Percentage of Sharks Caught in the Mediterranean Sea Rose Dramatically during World War I

Martin Braun*

In the mid-1920's the Italian biologist Umberto D'Ancona was studying variations in the population of various species of fish that interact with each other. In the course of his research, he came across some data on percentages-of-total-catch of several species of fish that were brought into different Mediterranean ports in the years that spanned World War I. In particular, the data gave the percentage-of-total-catch of selachians (sharks, skates, rays, etc.) which are not very desirable as food fish. The data for the port of Fiume, Italy, during the years 1914–1923 is as follows:

1914	1915	1916	1917	1918	1919	1920	1921	1922	1923
11.9%	21.4%	22.1%	21.2%	36.4%	27.3%	16.0%	15.9%	14.8%	10.7%.

D'Ancona was puzzled by the very large increase in the percentage of selachians during the period of the war. Obviously, he reasoned, the increase in the percentage of selachians was due to the greatly reduced level of fishing during this period, but how does the intensity of fishing affect the fish populations? The answer to this question was of great concern to D'Ancona in his research on the struggle for existence between competing species. It was also of concern to the fishing industry, since it would have obvious implications for the way fishing should be done.

    What distinguishes the selachians from the food fish is that the selachians are predators, while the food fish are their prey; the selachians depend on the food fish for their survival. At first, D'Ancona thought that this accounted for the large increase of selachians during the war. Since the level of fishing was greatly reduced during this period, more prey was available to the selachians, who therefore thrived and multiplied rapidly. However, this

* Department of Mathematics, Queens College, Flushing, NY 11367.

explanation does not hold any water since food fish were also more abundant during this period. D'Ancona's theory only shows that more selachians can be found when the level of fishing is reduced; it does not explain why a reduced level of fishing is more beneficial to the predators than to their prey.

After exhausting all possible biological explanations of this phenomenon, D'Ancona turned to his colleague, the famous Italian mathematician Vito Volterra in the hope that Volterra would formulate a mathematical model of the growth of the selachians and their prey, the food fish, and that this model would provide the answer to D'Ancona's question. Volterra began his analysis of this problem by separating all the fish into the prey population $x(t)$ and the predator population $y(t)$. Then he reasoned that the food fish do not compete very intensively among themselves for their food supply, since this is plentiful and the fish population is not very dense. Hence, in the absence of the selachians, the food fish would grow according to the Malthusian law of population growth $\dot{x} = ax$, for some positive constant a. Next, reasoned Volterra, the number of contacts per unit time between predators and prey is bxy, for some positive constant b. Hence $\dot{x} = ax - bxy$. Similarly, Volterra concluded that the predators have a natural rate of decrease cy proportional to their present number, and that they also increase at a rate dxy proportional to their present number y and their food supply x. Thus

$$\frac{dx}{dt} = ax - bxy \qquad \frac{dy}{dt} = -cy + dxy. \qquad (1)$$

The system of equations (1) governs the interaction of the selachians and food fish in the absence of fishing. We will carefully analyze this system and derive several interesting properties of its solutions. Then we will include the effect of fishing in our model, and show why a reduced level of fishing is more beneficial to the selachians than to the food fish. In fact, we will derive the surprising result that a reduced level of fishing is actually harmful to the food fish.

Observe first that (1) has two equilibrium solutions $x(t) = 0, y(t) = 0$ and $x(t) = c/d, y(t) = a/b$. The first equilibrium solution, of course, is of no interest to us. This system also has the family of solutions $x(t) = x_0 e^{at}$, $y(t) = 0$ and $x(t) = 0, y(t) = y_0 e^{-ct}$. Thus both the x and y axes are orbits of (1). This implies that every solution $x(t), y(t)$ of (1) which starts in the first quadrant $x > 0, y > 0$ at time $t = t_0$ will remain there for all future time $t \geq t_0$.

The orbits of (1) for $x, y \neq 0$ are the solution curves of the first-order equation

$$\frac{dy}{dx} = \frac{-cy + dxy}{ax - bxy} = \frac{y(-c + dx)}{x(a - by)}. \qquad (2)$$

This equation is separable, since we can write it in the form

$$\frac{a - by}{y} \frac{dy}{dx} = \frac{-c + dx}{x}.$$

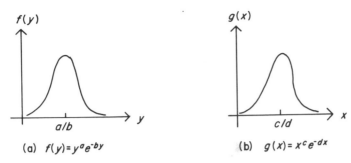

(a) $f(y) = y^a e^{-by}$ (b) $g(x) = x^c e^{-dx}$

Figure 15.1

Consequently, $a \ln y - by + c \ln x - dx = k_1$, for some constant k_1. Taking exponentials of both sides of this equation gives

$$\frac{y^a}{e^{by}} \frac{x^c}{e^{dx}} = K \tag{3}$$

for some constant K. Thus the orbits of (1) are the family of curves defined by (3), and these curves are *closed* as we now show.

Lemma 1. *Equation* (3) *defines a family of closed curves for* $x, y > 0$.

PROOF. Our first step is to determine the behavior of the functions $f(y) = y^a/e^{by}$ and $g(x) = x^c/e^{dx}$ for x and y positive. To this end, observe that $f(0) = 0, f(\infty) = 0$, and $f(y)$ is positive for $y > 0$. Computing

$$f'(y) = \frac{ay^{a-1} - by^a}{e^{by}} = \frac{y^{a-1}(a - by)}{e^{by}},$$

we see that $f(y)$ has a single critical point at $y = a/b$. Consequently, $f(y)$ achieves its maximum value $M_y = (a/b)^a/e^a$ at $y = a/b$, and the graph of $f(y)$ has the form described in Figure 15.1(a). Similarly, $g(x)$ achieves its maximum value $M_x = (c/d)^c/e^c$ at $x = c/d$, and the graph of $g(x)$ has the form described in Figure 15.1(b).

From the preceding analysis, we conclude that (3) has no solution $x, y > 0$ for $K > M_x M_y$, and the single solution $x = c/d$, $y = a/b$ for $K = M_x M_y$. Thus we need only consider the case $K = \lambda M_y$, where λ is a positive number less than M_x. Observe first that the equation $x^c/e^{dx} = \lambda$ has one solution $x = x_m < c/d$, and one solution $x = x_M > c/d$. Hence the equation

$$f(y) = y^a e^{-by} = \left[\frac{\lambda}{x^c e^{-dx}} \right] M_y$$

has no solution y when x is less than x_m or greater than x_M. It has the single solution $y = a/b$ when $x = x_m$ or x_M, and it has two solutions $y_1(x)$ and $y_2(x)$ for each x between x_m and x_M. The smaller solution $y_1(x)$ is always less than a/b, while the larger solution $y_2(x)$ is always greater than a/b. As x approaches either x_m or x_M, both $y_1(x)$ and $y_2(x)$ approach a/b. Consequently, the curves

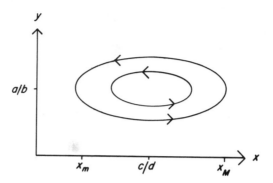

Figure 15.2. Orbits of (1) for x, y Positive

defined by (3) are closed for x and y positive and have the form described in Figure 15.2. Moreover, none of these closed curves (with the exception of $x = c/d$, $y = a/b$) contain any equilibrium points of (1). Therefore, all solutions $x(t)$, $y(t)$ of (1), with $x(0)$ and $y(0)$ positive, are *periodic* functions of time.[1] That is to say, each solution $x(t)$, $y(t)$ of (1), with $x(0)$ and $y(0)$ positive, has the property that $x(t + T) = x(t)$ and $y(t + T) = y(t)$ for some positive T. ☐

D'Ancona's data are really the *averages* over each one-year period of the proportion of predators. Thus, in order to compare these data with the predictions of (1), we must compute the "average values" of $x(t)$ and $y(t)$, for any solution $x(t)$, $y(t)$ of (1). Remarkably, we can find these average values even though we cannot compute $x(t)$ and $y(t)$ exactly. This is the content of Lemma 2.

Lemma 2. *Let $x(t)$, $y(t)$ be a periodic solution of (1), with period $T > 0$. Define the average values of x and y as*

$$\bar{x} = \frac{1}{T} \int_0^T x(t)\, dt, \qquad \bar{y} = \frac{1}{T} \int_0^T y(t)\, dt.$$

Then $\bar{x} = c/d$ and $\bar{y} = a/b$. In other words, the average values of $x(t)$ and $y(t)$ are the equilibrium values.

PROOF. Dividing both sides of the first equation of (1) by x gives $\dot{x}/x = a - by$, so that

$$\frac{1}{T} \int_0^T \frac{\dot{x}(t)}{x(t)}\, dt = \frac{1}{T} \int_0^T [a - by(t)]\, dt.$$

[1] For a detailed proof of this statement see [3, section 4.6].

Now, $\int_0^T \dot{x}(t)/x(t)\,dt = \ln x(T) - \ln x(0)$, and this equals zero since $x(T) = x(0)$. Consequently,

$$\frac{1}{T}\int_0^T by(t)\,dt = \frac{1}{T}\int_0^T a\,dt = a,$$

so that $\bar{y} = a/b$. Similarly, by dividing both sides of the second equation of (1) by $Ty(t)$ and integrating from 0 to T, we obtain that $\bar{x} = c/d$. □

We are now ready to include the effects of fishing in our model. Observe that fishing decreases the population of food fish at a rate $\varepsilon x(t)$ and decreases the population of selachians at a rate $\varepsilon y(t)$. The constant ε reflects the intensity of fishing; i.e., the number of boats at sea, and the number of nets in the water. Thus the true state of affairs is described by the modified system of differential equations.

$$\frac{dx}{dt} = ax - bxy - \varepsilon x = (a - \varepsilon)x - bxy$$

$$\frac{dy}{dt} = -cy + dxy - \varepsilon y = -(c + \varepsilon)y + dxy. \tag{4}$$

This system is exactly the same as (1) (for $a - \varepsilon > 0$), with a replaced by $a - \varepsilon$, and c replaced by $c + \varepsilon$. Hence the average values of $x(t)$ and $y(t)$ are now

$$\bar{x} = \frac{c + \varepsilon}{d}, \qquad \bar{y} = \frac{a - \varepsilon}{b}. \tag{5}$$

Consequently, a moderate amount of fishing ($\varepsilon < a$) actually increases the number of food fish, on the average, and decreases the number of selachians. Conversely, a reduced level of fishing increases the number of selachians, on the average, and *decreases* the number of food fish. This remarkable result, which is known as Volterra's principle, explains the data of D'Ancona and completely solves our problem.

Volterra's principle has spectacular applications to insecticide treatments which destroy both insect predators and their insect prey. It implies that the application of insecticides will actually increase the population of those insects which are kept in control by other predatory insects. A remarkable confirmation comes from the cottony cushion scale insect (*Icerya purchasi*), which, when accidentally introduced from Australia in 1868, threatened to destroy the American citrus industry. Thereupon, its natural Australian predator, a ladybird beetle (*Novius Cardinalis*) was introduced, and the beetles reduced the scale insects to a low level. When DDT was discovered to kill scale insects, it was applied by the orchardists in the hope of further reducing the scale insects. However, in agreement with Volterra's principle, the effect was an increase of the scale insect!

Oddly enough, many ecologists and biologists refused to accept Volterra's model as accurate. They pointed to the fact that the oscillatory behavior predicted by Volterra's model is not observed in most predator–prey systems. Rather, most such systems tend to equilibrium states as time evolves. Our answer to these critics is that the system of differential equations (1) is not intended as a model of the general predator–prey interaction. This is because the food fish and selachians do not compete intensively among themselves for their available resources. A more general model of predator–prey interactions is the system of differential equations

$$\dot{x} = ax - bxy - ex^2 \qquad \dot{y} = -cy + dxy - fy^2. \qquad (6)$$

Here, the term ex^2 reflects the internal competition of the prey x for their limited external resources, and the term fy^2 reflects the competition among the predators for the limited number of prey. The solutions of (6) are not, in general, periodic. Indeed, it is possible (but rather difficult) to show that all solutions of (6) approach equilibrium solutions if both e and f are positive.

Surprisingly, some ecologists and biologists even refuse to accept the more general model (6) as accurate. As a counterexample, they cite the experiments of the mathematical biologist E. F. Gause. In these experiments, the population was composed of two species of protozoa, one of which, *Didinium nasatum*, feeds on the other, *Paramecium caudatum*. In all of Gause's experiments, the Didinium quickly destroyed the Paramecium and then died of starvation. This situation cannot be modeled by (6), since no solution of (6) with $x(0)y(0) \neq 0$ can reach $x = 0$ or $y = 0$ in finite time.

Our answer to these critics is that the *Didinium* are a special and atypical type of predator. On the one hand, they are ferocious attackers and require a tremendous amount of food: a *Didinium* demands a fresh *Paramecium* every three hours. On the other hand, the *Didinium* do not perish from an insufficient supply of *Paramecium*. They continue to multiply, but give birth to smaller offspring. Thus the system of equations (6) does not accurately model the interaction of *Paramecium* and *Didinium*. A better model, in this case, is the system of differential equations

$$\frac{dx}{dt} = ax - b\sqrt{x}\,y \qquad \frac{dy}{dt} = \begin{cases} d\sqrt{x}\,y, & x \neq 0 \\ -cy, & x = 0 \end{cases}. \qquad (7)$$

It can be shown (see Exercise 6) that every solution $x(t)$, $y(t)$ of (7) with $x(0)$ and $y(0)$ positive reaches $x = 0$ in finite time. This does not contradict the existence uniqueness theorem, since the function

$$g(x, y) = \begin{cases} d\sqrt{x}\,y, & x \neq 0 \\ -cy, & x = 0 \end{cases}$$

does not have a partial derivative with respect to x or y at $x = 0$.

Finally, we mention that several predator–prey interactions exist in nature which cannot be modeled by any system of ordinary differential equations. These situations occur when the prey are provided with a refuge that is inaccessible to the predators. In these situations, it is impossible to make any definitive statements about the future number of predators and prey, since we cannot predict how many prey will be stupid enough to leave their refuge. In other words, this process is now *random*, rather than *deterministic*, and therefore cannot be modeled by a system of ordinary differential equations. This was verified directly in a famous experiment by Gause. He placed five *Paramecium* and three *Didinium* in each of 30 identical test tubes, and provided the *Paramecium* with a refuge from the Didinium. Two days later, he found the predators dead in four tubes, and a mixed population containing from two to 38 *Paramecium* in the remaining 26 tubes.

Exercises

1. Find all biologically realistic equilibrium points of (6) and determine their stability.

2. It can easily be shown that $y(t)$ ultimately approaches zero for all solutions $x(t)$, $y(t)$ of (6), if $c/d > a/e$. Show that solutions $x(t)$, $y(t)$ of (6) exist for which $y(t)$ increases at first to a maximum value and then decreases to zero. (To an observer who sees only the predators without noticing the prey, such a case of a population passing through a maximum to total extinction would be very difficult to explain.)

3. In many instances, it is the adult members of the prey who are chiefly attacked by the predators, while the young members are better protected, either by their smaller size or by living in a different station. Let x_1 be the number of adult prey, x_2 the number of young prey, and y the number of predators. Then

$$\dot{x}_1 = -a_1 x_1 + a_2 x_2 - bx_1 y$$

$$\dot{x}_2 = nx_1 - (a_1 + a_2)x_2$$

$$\dot{y} = -cy + dx_1 y$$

where $a_2 x_2$ represents the number of young (per unit time) growing into adults and n represents the birth rate proportional to the number of adults. Find all equilibrium solutions of this system.

4. There are several situations in nature where species 1 preys on species 2 which in turn preys on species 3. One case of this kind of population is the island of Komodo in Malaya which is inhabited by giant carnivorous reptiles and by mammals—their food—which feed on the rich vegetation of the island. We assume that the reptiles have no direct influence on the vegetation and that only the plants compete among themselves for their available resources. A system of differential equations governing this interaction is

$$\dot{x}_1 = -a_1 x_1 - b_{12} x_1 x_2 + c_{13} x_1 x_3$$

$$\dot{x}_2 = -a_2 x_2 + b_{21} x_1 x_2$$

$$\dot{x}_3 = a_3 x_3 - a_4 x_3^2 - c_{31} x_1 x_3.$$

Find all equilibrium solutions of this system.

5. Consider a predator–prey system where the predator has alternate means of support. This system can be modeled by the differential equations

$$\dot{x}_1 = \alpha_1 x_1(\beta_1 - x_1) + \gamma_1 x_1 x_2$$

$$\dot{x}_2 = \alpha_2 x_2(\beta_2 - x_2) - \gamma_2 x_1 x_2$$

where $x_1(t)$ and $x_2(t)$ are the predator and prey populations, respectively, at time t.

a) Show that the change of coordinates $\beta_i y_i(t) = x_i(t/\alpha_i \beta_i)$ reduces this system of equations to

$$\dot{y}_1 = y_1(1 - y_1) + a_1 y_1 y_2$$

$$\dot{y}_2 = y_2(1 - y_2) - a_2 y_1 y_2$$

where $a_1 = \gamma_1 \beta_2/\alpha_1 \beta_1$ and $a_2 = \gamma_2 \beta_1/\alpha_2 \beta_2$.

b) What are the stable equilibrium populations when
(i) $0 < a_2 < 1$, (ii) $a_2 > 1$?

c) It is observed that $a_1 = 3a_2$ (a_2 is a measure of the aggressiveness of the predator). What is the value of a_2 if the predator's instinct is to maximize its stable equilibrium population?

6. a) Let $x(t)$ be a solution of $\dot{x} = ax - M\sqrt{x}$, with $M > a\sqrt{x(t_0)}$. Show that

$$a\sqrt{x} = M - (M - a\sqrt{x(t_0)})e^{a(t-t_0)/2}$$

b) Conclude from (a) that $x(t)$ approaches zero in finite time.

c) Let $x(t)$, $y(t)$ be a solution of (7), with $by(t_0) > a\sqrt{x(t_0)}$. Show that $x(t)$ reaches zero in finite time. *Hint:* Observe that $y(t)$ is increasing for $t \geq t_0$.

d) It can be shown that $by(t)$ will eventually exceed $a\sqrt{x(t)}$ for every solution $x(t)$, $y(t)$ of (7) with $x(t_0)$ and $y(t_0)$ positive. Conclude, therefore, that all solutions $x(t)$, $y(t)$ of (7) achieve $x = 0$ in finite time.

References

[1] V. Volterra, *Lecons sur la theorie mathematique de la lutte pour la vie* Paris, 1931.
[2] G. F. Gause, *The Struggle for Existence*, New York Hafner, 1964.
[3] M. Braun, *Differential Equations and Their Applications*, 2nd ed. New York: Springer Verlag, 1978.

Notes for the Instructor

Objectives The module contains a discussion of the D'Ancona–Volterra (or Lotka–Volterra) model of the shark–food fish community. It is shown that the lack of fishing is more beneficial to sharks than to their prey, the food fish.

Prerequisites. Separable differential equations, orbits of autonomous systems.

Time. The module may be covered in one or two lectures.

CHAPTER 16
Quadratic Population Models: Almost Never Any Cycles

Courtney S. Coleman*

1. Quadratic Population Models

The populations of lynx and hare in the Canadian North Woods wax and wane together in a mysterious, 10-year cycle. The numbers of sharks and of food fish in the Adriatic jointly oscillate in a curious way.[1] Our aim is to construct simple mathematical models for the interaction of any pair of species and then to look for cyclical steady states in these models. Our results appear to be contradictory. As we shall see, most simple systems have no cycles at all, but the famous system of D'Ancona–Volterra (or Lotka–Volterra) has nothing but cycles [5]. The attempt to construct a synthesis going beyond this contradiction leads to an inclusive view of modeling. Several distinct models of a phenomenon give us more understanding of the phenomenon than one model can ever impart.

Let us begin, then, with two interacting species. Let $x(t)$ and $y(t)$ denote the respective *populations* (i.e., the number of individuals) at time t of the two species. (Sometimes *population densities*, numbers per unit area or volume of habitat, are used instead, but we shall not do this.) The values of $x(t)$ and $y(t)$ are integers and change by integer amounts as time goes on. However, for large populations an increase by one or two over a short time span is "infinitesimal" relative to the total, and we may think of the populations as changing continuously instead of by discrete jumps. Once we assume that $x(t)$ and $y(t)$ are continuous, we might as well go all the way, smooth off any corners on the graphs of $x(t)$ and of $y(t)$ and assume that both

* Department of Mathematics, Harvey Mudd College, Claremont, CA 91711.

[1] See [5], [6], [8]–[14] for studies of these and similar phenomena.

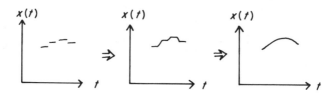

Figure 16.1. Smoothing Population Curve

functions are differentiable. Figure 16.1 shows this smoothing process for $x(t)$.

To say much about the changing populations of a pair of interacting species we must know something about the "laws" of birth and mortality for each species. Typically, these laws are expressed in terms of rates of change. Let us see what form a rate law might take, say, for the x species. Averaged over time and over all classes of age, sex, and fertility, a "typical" x individual makes a net contribution r to the x rate of change. The rate of the total population $x(t)$ is then,

$$\frac{dx(t)}{dt} = [\text{individual contribution to the rate}] \times [\text{number of individuals}] \tag{1}$$
$$= r\,x(t).$$

The *rate coefficient* r will differ, of course, from species to species, but it always has the meaning given above.

Finding $x(t)$ depends upon solving (1), which in turn depends upon the form of the rate coefficient. If r is a constant, then $x(t)$ grows or decays exponentially, $x(t) = x_0\,e^{rt}$, but no interaction takes place with the y species at all. Since we assume that the two species do interact, r should be a function of y, at least, and may also depend upon x. Not knowing exactly how r varies with x and y, let us adopt the engineer's motto, "When in doubt, linearize!", and assume that r is a linear function of x and y, $r = a_0 + a_1 x + a_2 y$, where the a's are constants.

We can treat the rate equation for the y species in the same way and obtain the *quadratic population model*,

$$\frac{dx}{dt} = (a_0 + a_1 x + a_2 y)x$$
$$\frac{dy}{dt} = (b_0 + b_1 x + b_2 y)y, \tag{2}$$

where the a's and the b's are constants. If $a_2 = b_1 = 0$, then (2) models a pair of noninteracting species, each of which follows a generalized logistic law (see, e.g., [4] or [16, example 3]). On the other hand, if $a_1 = b_2 = 0$, (2) becomes the D'Ancona–Volterra model of shark–food fish interactions treated by Braun [5]. Thus (2) is a simple model which is still complex enough to include the logistic, D'Ancona–Volterra, and, of course, the exponential growth ($a_1 = a_2 = b_1 = b_2 = 0$) models.

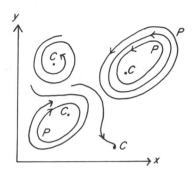

Figure 16.2. Population Orbits

Now let us give some biological meaning to the constants appearing in this model.

Natural Growth Coefficients: The constants a_0 and b_0 represent the *natural growth* of the respective species in the absence of any other interactions. The terms $a_0 x$ and $b_0 y$ in (7) are often called the *biotic potentials* of the species because they are the natural growth rates.

Single-Species Social Coefficients: The constant a_1 measures the effect of the population $x(t)$ on the growth rate of species x itself. If $a_1 < 0$, a_1 is the *overcrowding* (or *selflimiting*) *coefficient* and measures the negative effects of overpopulation; if $a_1 > 0$, then it is called the *mutualism* coefficient and indicates an increase in productivity with an increase in population. The first case indicates *competition* among the members of species x for limited resources, while the second case implies *cooperation* among the members. A similar analysis applies to b_2.

Two-Species Social Coefficients: The coefficients a_2 and b_1 measure the effect of each species on the growth rate of the other; they are called *two-species social coefficients* (or, simply, *interaction constants*). We have the following cases if $a_2 b_1 \neq 0$:

1. $a_2 > 0$, $b_1 > 0$ the case of *interspecies mutualism*,
2. $a_2 < 0$, $b_1 > 0$ (or $a_2 > 0$, $b_1 < 0$), in which case x is the *prey* of the *predator* y (or y is the prey of the predator x),
3. $a_2 < 0$, $b_1 < 0$, where *mutual predation* occurs or, perhaps, *competition* for common resources

The *social terms* in (2) (i.e., the terms involving the social coefficients) are all quadratic; this kind of a term is sometimes called *mass action* and represents interactions between individuals. One might say that "it takes two to tango, or to tangle." In any event, the quadratic model is plausible biologically even though its derivation has been mostly theoretical.

A solution, $x = x(t)$, $y = y(t)$, of (2) defines a parametric curve in the xy plane with t as the parameter. This curve is called a *population orbit* or *trajectory.*[2] Since negative populations do not make sense, we shall only be interested in population orbits in the *population quadrant*, $x \geq 0$, $y \geq 0$. We shall be particularly interested in *steady-state orbits*, which correspond to the constant and the periodic solutions of (2) and which are portrayed in the population quadrant by points and simple closed curves, respectively. See Figure 16.2 for various hypothetical orbits, C indicating constant and P periodic orbits. The arrowheads on the orbits show the direction of increasing time.

2. Equilibrium Populations: No Cycles

Now that we have set up system (2) as a model for interacting species, let us see what we can say about the possible functions $x(t)$ and $y(t)$ which satisfy (2); i.e., what are the solutions of (2) like? Without more information about the coefficients, it is impossible to actually solve for $x(t)$ and $y(t)$ as specific functions of t. Instead, let us proceed somewhat indirectly.

Two interacting species surviving over a long period of time in a common environment may well have reached some kind of balance, a so-called *equilibrium state*. It may be that each of the populations remains essentially constant in time, $x(t) \equiv x_0, y(t) \equiv y_0$. In this case, we have that $dx(t)/dt \equiv 0$ and $dy(t)/dt \equiv 0$. Conversely, if the growth rates of the x species and the y species both vanish identically, then the two populations are constant. These constant populations correspond to the *critical points* of (2) and can be determined by setting the right-hand sides of (2) equal to 0,

$$\begin{cases} x(a_0 + a_1 x + a_2 y) = 0 \\ y(b_0 + b_1 x + b_2 y) = 0, \end{cases} \tag{3}$$

and then solving this pair of equations simultaneously for x and y.

The first equation of (3) is satisfied whenever $x = 0$. If $x = 0$, then the second equation holds either if $y = 0$ or if $y = -b_0/b_2$ (assuming that $b_2 \neq 0$). The second equation of (3) is satisfied if $y = 0$, while if $y = 0$ then the first equation holds if $x = 0$ or if $x = -a_0/a_1$ (assuming that $a_1 \neq 0$). Thus we have determined so far three critical points (i.e., constant population points):

$$\begin{cases} x = 0, & y = 0, \\ x = 0, & y = -b_0/b_2, \\ x = -a_0/a_1, & y = 0. \end{cases}$$

---

[2] "Orbit" commonly denotes the elliptical path of a planet or a satellite and "trajectory" the path of a missile. In the current context, however, the terms refer to arbitrary solution curves of (2) in the xy plane.

Observe that in each case at least one of the species is extinct. This is hardly consistent with a nontrivial long-term balance between the two species, so we are not much interested in any of these three points.

We have, however, overlooked a possible constant population point. Equations (3) will be satisfied if each of the bracketed terms vanishes. Solving the equations

$$\begin{cases} a_0 + a_1 x + a_2 y = 0 \\ b_0 + b_1 x + b_2 y = 0 \end{cases}$$

simultaneously, we have that a fourth constant population point is given by

$$x_0 = \frac{a_2 b_0 - a_0 b_2}{a_1 b_2 - a_2 b_1} \qquad y_0 = \frac{a_0 b_1 - a_1 b_0}{a_1 b_2 - a_2 b_1}, \tag{4}$$

as long as $a_1 b_2 \neq a_2 b_1$.

Thus, if the two species do indeed interact in such a way that (2) is an accurate model and that both remain essentially constant but not vanishing over a long time span, then the two populations must be given by (4). Observe that the critical point given by (4) represents a true population state only if $x_0 > 0$ and $y_0 > 0$ (i.e., if (x_0, y_0) lies in the population quadrant).

Now, another kind of simple equilibrium state exists that could conceivably be attained by two interacting species. It could happen that the population numbers of neither species remain constant in time but fluctuate periodically. Thus we may have that $x(t)$ and $y(t)$ are nonconstant periodic functions of time with a common period. The corresponding population orbit is called a *cycle* [see Figure 16.2]. The possibility of such cyclical behavior has been the subject of a good deal of controversy in biological and ecological circles recently [13], [14]. We shall not say much about the biological or ecological reasons for believing that specific systems of interacting species do or do not exhibit a cyclical equilibrium. *We shall show, however, that "almost no" system of the form of (2) can possess cycles.* (We shall explain later what "almost no" means.)

Let us begin by imposing two conditions on the coefficients of system (2).

Condition (A): The number $A = a_1 b_2 - a_2 b_1$ does not vanish.
Condition (B): The number $B = a_1 b_0 (a_2 - b_2) - a_0 b_2 (a_1 - b_1)$ does not vanish.

Condition (A) asserts that the two lines

$$\begin{cases} a_0 + a_1 x + a_2 y = \text{const.} \\ b_0 + b_1 x + b_2 y = \text{const.} \end{cases}$$

are not parallel. Parallelism would indicate a highly unlikely exact relationship between the rate coefficients of the two species being modeled.

Condition (B) does not have a simple interpretation, but its violation under certain conditions has a simple consequence: if any cycle at all exists, then a continuous band of cycles exists (see the D'Ancona–Volterra model later in this chapter). However, for the number B to vanish is not reasonable if (2) is to model any real system. For, in such a case the constants $a_0, a_1, a_2, b_0, b_1,$ and b_2 are only approximate and to require that $B = 0$ would be virtually impossible. Thus conditions (A) and (B) are likely to be met by any system for which we have reason to believe (2) is an exact model. We have the following central result.

No Cycles Theorem. *If the coefficients of the quadratic population model (2) satisfy conditions (A) and (B), then (2) has no cycles in the population quadrant.*

Remark. Thus whatever may be the fate of the two species being modeled, they can never establish a nonconstant periodic equilibrium (i.e., a cycle) if conditions (A) and (B) are satisfied.

PROOF. For $x > 0$ and $y > 0$ define the function

$$K(x, y) = x^\alpha y^\beta,$$

where

$$\alpha = \frac{b_2(b_1 - a_1)}{A} - 1 \quad \text{and} \quad \beta = \frac{a_1(a_2 - b_2)}{A} - 1.$$

Now let f and g denote the respective right-hand sides of the two rate equations of (2) and consider the function

$$\frac{\partial}{\partial x}(Kf) + \frac{\partial}{\partial y}(Kg).$$

A straightforward calculation (Exercise 4) shows that

$$\frac{\partial}{\partial x}(Kf) + \frac{\partial}{\partial y}(Kg) = \frac{B}{A}K, \tag{5}$$

whether or not $B = 0$, as long as $A \neq 0$.

Now suppose that (2) *does* have a cycle Γ inside the population quadrant. we shall derive a contradiction, thus showing that there are no cycles. Let R denote the union of Γ and its interior (see Figure 16.3). Since $K(x, y) > 0$ when $x > 0$ and $y > 0$ and since A and B are nonvanishing constants by conditions (A) and (B), we see that the quantity, $(B/A)K(x, y)$, has a fixed sign in the population quadrant and does not vanish. Thus according to (5) we have that

$$0 \neq \int_R \int \frac{B}{A} K(x, y)\, dx\, dy = \int_R \int \left[\frac{\partial}{\partial x}(Kf) + \frac{\partial}{\partial y}(Kg) \right] dx\, dy. \tag{6}$$

By Green's theorem in the plane, the last double integral can be replaced by a line integral around Γ:

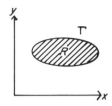

Figure 16.3. Cycle Γ and Region R

$$\int_R \int \left[\frac{\partial}{\partial x}(Kf) + \frac{\partial}{\partial y}(Kg) \right] dx\, dy = -\oint_\Gamma [Kg\, dx - Kf\, dy]. \qquad (7)$$

Since Γ is an orbit, we have that along Γ, $dx/dt = f$ and $dy/dt = g$. Thus the line integral in (7) vanishes, since

$$\oint_\Gamma [Kg\, dx - Kf\, dy] = \int_0^T \left[Kg\frac{dx}{dt} - Kf\frac{dy}{dt} \right] dt = \int_0^T [Kgf\, dt - Kfg\, dt] = 0,$$

$$(8)$$

where the period of the cycle Γ is T. From (6)–(8) we have that

$$0 \neq \int_R \int \left[\frac{\partial}{\partial x}(Kf) + \frac{\partial}{\partial y}(Kg) \right] dx\, dy = 0.$$

The assumption that (2) has a cycle is untenable and the No Cycles theorem is proved. □

Thus when conditions (A) and (B) are satisfied, the only possible equilibrium state for a pair of nonvanishing species modeled by (2) is the constant population state given by (4).

3. The D'Ancona–Volterra Model: All Cycles

The best-known model of interacting species is the D'Ancona–Volterra model of the shark and food fish communities in the Adriatic Sea. Braun has given a delightful and thorough account of the model in Chapter 15, and we shall only summarize the results. The shark population y feeds on another fish x. In the absence of x, the biotic potential of the shark population is negative and $dy/dt < 0$, while in the absence of y, the potential of the prey is positive and $dx/dt > 0$. The interaction is proportional to xy and is favorable to the sharks and unfavorable to the food species. Under these conditions, the D'Ancona–Volterra predator–prey model has the form

$$\begin{cases} \dfrac{dx}{dt} = x(a - by) \\[2mm] \dfrac{dy}{dt} = y(-c + dx), \end{cases} \qquad (9)$$

where all the constants are positive. We can find the orbits in the xy population quadrant by dividing the two equations of (9),

$$\frac{dy}{dx} = \frac{y(-c + dx)}{x(a - by)},$$

separating variables,

$$\frac{a - by}{y}\frac{dy}{dx} = \frac{-c + dx}{x},$$

and integrating,

$$a \ln y - by + c \ln x - dx = k_1, \tag{10}$$

where k_1 is a constant of integration. Equation (10) defines the family of orbits in the population quadrant, and, as Braun shows, every orbit is a cycle, except for the constant population point $(c/d, a/b)$. See Figure 16.4 for a sketch of these cycles.

Cycles? How could there be any cycles? Doesn't the No Cycles theorem assert that there are no cycles at all? The catch is in the hypotheses of that theorem. No cycles exist *if* each of the numbers A and B is nonzero, but a simple calculation shows that $B = 0$ (although $A \neq 0$) in the D'Ancona–Volterra model. So the No Cycles theorem does not apply to (9) since one of its hypotheses does not hold.

Not only do cycles abound in the model, but it is now known that if $A = 0$, or $B = 0$, or both, in any quadratic model, and if even one cycle exists, then infinitely many cycles exist, and they completely fill up some region in the population quadrant or the quadrant itself. It is cycle-feast or cycle-famine in quadratic population models. Isolated cycles, i.e., limit cycles, cannot exist. See van der Vaart [15] for proofs of all these results.

4. Stability

A biological cycle which perseveres over any substantial length of time must be somewhat insensitive to the inevitable shocks and disturbances of the "real world." That is, if a disturbance at time t_1 suddenly knocks a population point p off a cycle Γ onto a nearby point q and then the disturbance stops, the orbit starting at q at time t_1 should move back towards Γ as time goes on. Such a cycle is *ecologically stable* (see Figure 16.5). (The mathematical term for this is *asymptotic orbital stability*, but we shall not give the formal definition here.) It seems likely that the only exact population cycles that could exist in nature are those that are ecologically stable.

The D'Ancona–Volterra cycles of Section 3 are not ecologically stable, for a sudden shock may drive a population point p off a cycle Γ and onto a completely different cycle Γ_1. Once on Γ_1, of course, the population remains

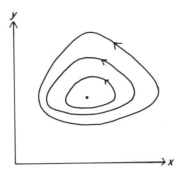

Figure 16.4. D'Ancona–Volterra Cycles

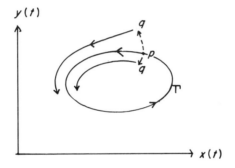

Figure 16.5. Ecologically Stable Cycle

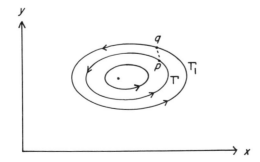

Figure 16.6. D'Ancona–Volterra Cycles Are Not Ecologically Stable

there unless another shock sends the population onto yet another cycle. In any event, a return to the original cycle is unlikely (see Figure 16.6). Thus an ecologically stable cycle must be *isolated*; i.e., no other cycles are "nearby." As we noted earlier, quadratic population models cannot have isolated cycles and hence cannot have ecologically stable cycles. Does this mean that none of the quadratic models are of any use in modeled oscillatory population behavior?

5. Which Model?

The cycles of the D'Ancona–Volterra model may not be ecologically stable, but that is no reason to throw out the model. Rather, we should think of (9) and its orbits as an oversimplification of the true state of affairs, but one which focuses our attention on the oscillatory nature of predator–prey interactions. A. Winfree, in modeling another kind of natural cycle, quotes Picasso,

> *Art is the lie that helps us to see the truth.*

Winfree goes on to say that

> *By art in this context I mean simple equations intended to emphasize relevant essentials of a more intricate reality.*

The D'Ancona–Volterra model is a broad-brush picture of the cycles of the shark–food fish community: it is an impression, not a portrait. Although some details are distorted and others are missing altogether, we not only recognize the subject in the picture, but we perceive it in a new and revealing way.

Within the bounds of quadratic population models, we can add more details to this picture. We can include overcrowding terms and obtain

$$\begin{cases} \dfrac{dx}{dt} = x(a - \alpha x - by), \\[2mm] \dfrac{dy}{dt} = y(-c + dx - \beta y), \end{cases} \tag{11}$$

where the overcrowding coefficients α and β are positive and small in comparison to a, b, c, and d. This is not a D'Ancona–Volterra model, and the constant B no longer vanishes, nor does A.[3] Even though the No Cycles theorem applies and (16) has no cycles, it is still "close" to a D'Ancona–Volterra model, at least for bounded x and y, since α and β are small. Orbits change but little when the rate equations are only slightly altered, and the

---

[3] $B = \alpha c(-b + \beta) + a\beta(-\alpha - d) = -[\alpha bc + \alpha\beta a + \beta ad] + \alpha\beta c$, which is less than 0 if $\alpha\beta$ is sufficiently small. $A = \alpha\beta + bd > 0$.

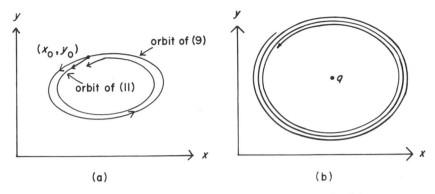

Figure 16.7. (a) Orbits of (9) and (11); (b) Spiral Orbit of (11)

orbits of (11) should be close to those of (9). Suppose we pick a population point (x_0, y_0) "near" the equilibrium point $(c/d, a/b)$ of (9) and track the orbit of (9) through (x_0, y_0) once around $(c/d, a/b)$ The orbit returns precisely to (x_0, y_0). Now let us begin again at (x_0, y_0), but this time we shall follow the orbit of (11) around one complete turn. When the orbit of (11) returns to the vicinity of (x_0, y_0), it just misses that point, slipping past on the inside (see Figure 16.7). This behavior is repeated in the next turn, and again in the next, and so on. The successive turns of the orbit form a tightly coiled spiral which slowly works its way toward the equilibrium point.[4]

When α and β are small and (x_0, y_0) is close to the equilibrium point, the inward movement of the spiraling orbit is so slight that to the observer the orbit appears to be a cycle. Thus (11) can still portray the predator–prey oscillations observed in the shark–fish community, but the system does not have the ecologically unstable cycles of the Lotka–Volterra model. Of course, a price must be paid for this extra detail: no longer do we have a formula for the orbits of (11) since the integral (10) only applies to the D'Ancona–Volterra model.

Which model is to be preferred? No overwhelming reason exists to choose one or the other. Each deepens our understanding of the natural phenomena being portrayed, but in different ways.

Exercises

1. Find the orbits of the system $dx/dt = a_2 xy$, $dy/dt = b_1 xy$, where $a_2 \neq 0$ and $b_1 \neq 0$. Show that no cycles exist even though $B = 0$. Interpret this system biologically. *Hint:* Look at dy/dx.

---

[4] The proof that the orbits of (11) have this spiraling character is omitted. The interested reader may read Braun's module [3] in this volume for an indication of the kind of analysis needed for such a proof.

2. Find the orbits of the system, $dx/dt = x(a_0 + a_2 y)$, $dy/dt = b_0 y$, where a_0, a_2, and b_0 are not zero. *Hint:* Find dy/dx and separate variables. Interpret biologically. Does the No Cycles theorem apply?

3. Show that the cubic system

$$\frac{dx}{dt} = (1 - x^2 - y^2)x - y, \qquad \frac{dy}{dt} = (1 - x^2 - y^2)y + x,$$

has a unique and ecologically stable isolated cycle. Note, however, that the system is not a model for interacting species. Why not? *Hint:* Use polar coordinates, $r = (x^2 + y^2)^{1/2}$, $\tan \theta = y/x$, and show that $dr/dt = r(1 - r^2)$, $d\theta/dt = 1$. Then find $r(t)$ by separation of variables. $r \equiv 1$ is the desired isolated cycle.

4. Verify (5). (You will have to treat the case $\alpha = -1$, $\beta = -1$ rather carefully.)

5. (See also [5]).
   a) Let (2) have a cycle Γ defined by the parametric equations, $x = x(t)$, $y = y(t)$, of period T in the population quadrant. Show that the average populations along Γ, $\bar{x} = (1/T)\int_0^T x(t)\, dt$ and $\bar{y} = (1/T)\int_0^T y(t)\, dt$, are given by $\bar{x} = x_0$, $\bar{y} = y_0$, where (x_0, y_0) is the constant equilibrium population given by (4). Assume that $A \neq 0$ and that $x_0 > 0$, $y_0 > 0$. *Hint:* $(1/x)(dx/dt) = a_0 + a_1 x + a_2 y$,

$$\frac{1}{T}\int_0^T \frac{1}{x}\frac{dx}{dt}\, dt = \frac{1}{T}\int_0^T (a_0 + a_1 x(t) + a_2 y(t))\, dt = a_0 + a_1 \bar{x} + a_2 \bar{y}.$$

   However, $\int_0^T (1/x)(dx/dt)\, dt = \ln x(T) - \ln x(0) = 0$ since $x(t)$ has period T. Similarly, show that $0 = b_0 + b_1 \bar{x} + b_2 \bar{y}$.
   b) Suppose the predator–prey model (11) with overcrowding has cycles and that $\beta < b$. Suppose that both species are "harvested" at a rate proportional to their respective populations: thus, replace the constants a and $-c$ in (11) by $a\text{-}h$ and $-c\text{-}h$, respectively. Suppose the harvested model still has cycles. Use a) to show that the average population of the prey species x along a cycle increases with harvesting but the average population of the predator species y decreases.
   c) Suppose that aphids are the prey and lady bugs are the predator. Should you use a broad-spectrum insecticide to kill the aphids? Explain.

6. (*Generalized Bendixson Criterion*) Let R be a region in the plane and suppose $K(x, y)$ is a continuous function on R. Suppose that $(\partial/\partial x)(Kf) + (\partial/\partial y)(Kg)$ has a fixed sign on R [and never vanishes on R]. Show that the system $dx/dt = f(x, y)$, $dy/dt = g(x, y)$, where f and g are continuous on R, has no cycles in R. *Hint:* Follow the proof of the No Cycles theorem.

References

[1] F. Albrecht, H. Gatzke, A. Haddad and N. Wax, "The Dynamics of two interacting populations," *J. Math. Anal. Appl.*, vol. 46, pp. 658–670, 1974. A research paper which gives a complete analysis of the general system, $\dot{x} = xf$, $\dot{y} = yg$, with various conditions appropriate to population dynamics imposed on f and g. Little discussion of the biological or ecological meaning.

[2]   M. Braun, *Differential Equations and Their Applications*, 2nd ed.   New York: Springer-Verlag, 1978. This is one of the best books on the sophomore level; the presentation of the applications is especially good. Several of the sections of the book are also included in this volume.

[3]   ——, "The principle of competitive exclusion in population biology," this volume, ch. 17. Braun presents a careful study from a mathematical point of view of a basic ecological principle, essentially the same as a section in [2].

[4]   ——, "Single species population model," this volume, ch. 5. This is a good presentation of exponential and of logistic growth and also appears as a section in [2].

[5]   ——, "Why the percentage of sharks caught in the Mediterranean Sea rose dramatically during World War I," this volume, ch. 15. This is a slight alteration of a section of [2] and contains a good exposition of the D'Ancona–Volterra model.

[6]   C. S. Coleman, "Biological cycles and the fivefold way," this volume, ch. 18. This introduces the lynx–hare cycle and the Poincaré–Bendixson theory of limit sets. The Kolmogorov theorem on biological cycles is also discussed.

[7]   W. A. Coppel, "A survey of quadratic systems," *J. Differential Equations*, vol. 2, pp. 293–304, 1966. An advanced monograph surveying what was known about such systems up to 1965, this contains a version of the No Cycle theorem and presents a variety of curiosa about quadratic systems; e.g., any cycle must bound a convex region.

[8]   D. L. De Angelis, "Estimates of predator–prey limit cycles," *Bull. Math. Bio.*, vol. 37, pp. 291–299, 1975. A predator–prey system with a unique cycle which is ecologically stable is discussed. The rate equation for the prey is cubic, for the predator linear.

[9]   L. B. Keith, *Wildlife's Ten-Year Cycle*, Madison: Univ. of Wisconsin Press, 1963. An intriguing, nonmathematical discussion of the evidence concerning the famous 10-year cycle of lynx, hare, grouse, and other populations in the Canadian coniferous forests.

[10] E. J. Kormondy, *Concepts of Ecology*. Englewood Cliffs, NJ: Prentice-Hall, 1969. This elementary textbook covers, among many other things, interacting species in a nonmathematical way.

[11] D. L. Lack, *The Natural Regulation of Animal Numbers*. Oxford: Clarendon Press, 1954. Other forms for mathematical models of interacting species are discussed.

[12] R. M. May, "Limit cycles in predator–prey communities," *Science* vol. 177, pp. 900–902, 1972. A readable discussion of the subject is given with extensive references.

[13] ——, *Stability and Complexity in Model Ecosystems*.   Princeton, NJ: Princeton Univ. Press, 1973. This excellent text is accessible to sophomores or juniors willing to think deeply. A moderate amount of differential equations is used. Particularly good discussions of the ecological implications are given, with a very good bibliography.

[14] J. Maynard Smith, *Models in Ecology*.   Cambridge, 1974. Comments similar to those for May's book apply. The analysis of differential equations is mostly heuristic and geometric, good bibliography.

[15] H. R. van der Vaart, "The phase portrait of differential systems of Volterra type," *Bull. Math. Bio.*, vol. 40, pp. 133–160, 1978. Corrects and extends the analysis in Coppel [7] on cycles of quadratic population models.

[16] B. H. West, "Setting up first-order differential equations from word problems," this volume, ch. 1. It is an excellent, student-oriented presentation.

Notes for the Instructor

Objectives. In the preceding chapter Braun shows that all of the population orbits in the D'Ancona–Volterra model of predator–prey interactions are closed curves, or cycles. In this module it is shown that "most" quadratic population models have no cycles at all. As it happens, a model with cycles and models without cycles can both be used to portray the oscillations in predator–prey communities. This apparent contradiction is resolved in the last section of the module. Although the article is essentially independent and selfcontained, it has a natural place as a supplement to a study of the D'Ancona–Volterra model.

Prerequisities. Some familiarity with nonlinear differential equations, Green's theorem in the plane.

Time. The material may be covered in one lecture if the students are already familiar with the D'Ancona–Volterra model.

CHAPTER 17

The Principle of Competitive Exclusion in Population Biology

Martin Braun*

It is often observed in nature that the struggle for existence between two similar species competing for the same limited food supply and living space nearly always ends in the complete extinction of one of the species. This phenomenon is known as the "principle of competitive exclusion," and was first enunciated, in a slightly different form, by Darwin in 1859. In his paper, "The origin of species by natural selection," he writes:

> "As the species of the same genus usually have, though by no means invariably, much similarity in habits and constitutions and always in structure, the struggle will generally be more severe between them, if they come into competition with each other, than between the species of distinct genera."

There is a very interesting biological explanation of the principle of competitive exclusion. The cornerstone of this theory is the idea of a "niche." A niche indicates what place a given species occupies in a community; i.e., what are its habits, food, and mode of life. It has been observed that as a result of competition two similar species rarely occupy the same niche. Rather, each species takes possession of those kinds of food and modes of life in which it has an advantage over its competitor. If the two species tend to occupy the same niche then the struggle for existence between them will be intense and result in the extinction of the weaker species.

An excellent illustration of this theory is the colony of terns inhabiting the island of Jorilgatch in the Black Sea. This colony consists of four different species of terns: sandwich tern, common tern, blackbeak tern, and little tern. These four species band together to chase away predators from the colony. However, sharp differences exist among them as regards the procuring of

* Department of Mathematics, Queens College, Flushing, NY 11367.

food. The sandwich tern flies far out into the open sea to hunt certain species, while the blackbeak tern feeds exclusively on land. On the other hand, the common tern and little tern catch fish close to the shore. They sight the fish while flying and dive into the water after them. The little tern seizes its fish in shallow swampy places, whereas the common tern hunts somewhat further from shore. In this manner, these four similar species of tern living side by side upon a single small island differ sharply in all their modes of feeding and procuring food. Each has a niche in which it has a distinct advantage over its competitors.

In this section we present a rigorous mathematical proof of the law of competitive exclusion. This will be accomplished by deriving a system of differential equations which govern the interaction between two similar species and then showing that every solution of the system approaches an equilibrium state in which one of the species is extinct.

In constructing a mathematical model of the struggle for existence between two competing species, it is instructive to look again at the logistic law of population growth

$$\frac{dN}{dt} = aN - bN^2. \tag{1}$$

This equation governs the growth of the population $N(t)$ of a single species whose members compete among themselves for a limited amount of food and living space. Recall (see Chapter 5) that $N(t)$ approaches the limiting population $K = a/b$, as t approaches infinity. This limiting population can be thought of as the maximum population of the species which the microcosm can support. In terms of K, the logistic law (1) can be rewritten in the form

$$\frac{dN}{dt} = aN\left(1 - \frac{b}{a}N\right) = aN\left(1 - \frac{N}{K}\right) = aN\left(\frac{K - N}{K}\right). \tag{2}$$

Equation (2) has the following interesting interpretation. When the population N is very low, it grows according to the Malthusian law $dN/dt = aN$. The term aN is called the "biotic potential" of the species. It is the potential rate of increase of the species under ideal conditions, and it is realized if no restrictions on food and living space exist and if the individual members of the species do not excrete any toxic waste products. As the population increases though, the biotic potential is reduced by the factor $(K - N)/K$, which is the relative number of still vacant places in the microcosm. Ecologists call this factor the environmental resistance to growth.

Now, let $N_1(t)$ and $N_2(t)$ be the population at time t of species 1 and 2, respectively. Further, let K_1 and K_2 be the maximum population of species 1 and 2 which the microcosm can support, and let $a_1 N_1$ and $a_2 N_2$ be the biotic potentials of species 1 and 2. Then, $N_1(t)$ and $N_2(t)$ satisfy the system of differential equations

$$\frac{dN_1}{dt} = a_1 N_1 \left(\frac{K_1 - N_1 - m_2}{K_1} \right), \qquad \frac{dN_2}{dt} = a_2 N_2 \left(\frac{K_2 - N_2 - m_1}{K_2} \right), \qquad (3)$$

where m_2 is the total number of places of the first species which are taken up by members of the second species, and m_1 is the total number of places of the second species which are taken up by members of the first species. At first glance, it would appear that $m_2 = N_2$ and $m_1 = N_1$. However, this is not generally the case, for it is highly unlikely that two species utilize the environment in identical ways. Equal numbers of individuals of species 1 and 2 do not, on the average, consume equal quantities of food, take up equal amounts of living space, and excrete equal amounts of waste products of the same chemical composition. In general, we must set $m_2 = \alpha N_2$ and $m_1 = \beta N_1$, for some constants α and β. The constants α and β indicate the degree of influence of one species upon the other. If the interests of the two species do not clash, and they occupy separate niches, then both α and β are zero. If the two species lay claim to the same niche and are very similar, then α and β are very close to one. On the other hand, if one of the species, say species 2, utilizes the environment unproductively, i.e., consumes a great deal of food or excretes poisonous waste products, then one individual of species 2 takes up the place of many individuals of species 1. In this case, then, the coefficient α is very large.

We restrict ourselves now to the case where the two species are nearly identical and lay claim to the same niche. Then, $\alpha = \beta = 1$ and $N_1(t)$ and $N_2(t)$ satisfy the system of differential equations.

$$\frac{dN_1}{dt} = a_1 N_1 \left(\frac{K_1 - N_1 - N_2}{K_1} \right) \qquad \frac{dN_2}{dt} = a_2 N_2 \left(\frac{K_2 - N_1 - N_2}{K_2} \right). \qquad (4)$$

In this instance, we expect the struggle for existence between species 1 and 2 to be intense and to result in the extinction of one of the species. This is indeed the case, as we now show.

Theorem. (Principle of Competitive Exclusion). *Suppose that K_1 is greater than K_2. Then, every solution $N_1(t)$, $N_2(t)$ of (4) approaches the equilibrium solution $N_1 = K_1$, $N_2 = 0$ as t approaches infinity. In other words, if species 1 and 2 are nearly identical and the microcosm can support more members of species 1 than of species 2, then species 2 will utimately become extinct.*

The first step in our proof is to show that $N_1(t)$ and $N_2(t)$ can never become negative. To this end, observe that

$$N_1(t) = \frac{K_1 N_1(0)}{N_1(0) + (K_1 - N_1(0)) e^{-a_1 t}} \qquad N_2(t) = 0$$

is a solution of (4) for any choice of $N_1(0)$. The orbit of this solution in the $N_1 - N_2$ plane is the point $(0, 0)$ for $N_1(0) = 0$; the line $0 < N_1 < K_1$, $N_2 = 0$ for $0 < N_1(0) < K_1$; the point $(K_1, 0)$ for $N_1(0) = K_1$; and the line $K_1 < N_1 < \infty$, $N_2 = 0$ for $N_1(0) > K_1$. Thus the N_1 axis, for $N_1 \geq 0$, is the

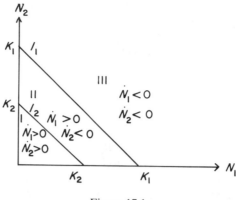

Figure 17.1

union of four distinct orbits of (4). Similarly, the N_2 axis, for $N_2 \geq 0$, is the union of four distinct orbits of (4). This implies that all solutions $N_1(t)$, $N_2(t)$ of (4) which start in the first quadrant ($N_1 > 0, N_2 > 0$) of the $N_1 - N_2$ plane must remain there for all future time.

The second step in our proof is to split the first quadrant into regions in which both dN_1/dt and dN_2/dt have fixed signs. This is accomplished in the following manner. Let l_1 and l_2 be the lines $K_1 - N_1 - N_2 = 0$ and $K_2 - N_1 - N_2 = 0$, respectively. Observe that dN_1/dt is negative if (N_1, N_2) lies above l_1 and positive if (N_1, N_2) lies below l_1. Similarly, dN_2/dt is negative if (N_1, N_2) lies above l_2 and positive if (N_1, N_2) lies below l_2. Thus the two parallel lines l_1 and l_2 split the first quadrant of the $N_1 - N_2$ plane into three regions (see Figure 17.1) in which both dN_1/dt and dN_2/dt have fixed signs. Both $N_1(t)$ and $N_2(t)$ increase with time (along any solution of (4)) in region I; $N_1(t)$ increases and $N_2(t)$ decreases with time in region II; and both $N_1(t)$ and $N_2(t)$ decrease with time in region III. Finally, we require the following three simple lemmas.

Lemma 1. *Any solution $N_1(t)$, $N_2(t)$ of (4) which starts in region I at $t = t_0$ must leave this region at some later time.*

PROOF. Suppose that a solution $N_1(t)$, $N_2(t)$ of (4) remains in region I for all time $t \geq t_0$. This implies that both $N_1(t)$ and $N_2(t)$ are monotonic increasing functions of time for $t \geq t_0$, with $N_1(t)$ and $N_2(t)$ less than K_2. Consequently, both $N_1(t)$ and $N_2(t)$ have limits ξ, η respectively, as t approaches infinity. This, in turn, implies that (ξ, η) is an equilibrium point of (4). Now, the only equilibrium points of (4) are $(0, 0)$, $(K_1, 0)$, and $(0, K_2)$, and (ξ, η) obviously cannot equal any of these three points. We conclude, therefore, that any solution $N_1(t)$, $N_2(t)$ of (4) which starts in region I must leave this region at a later time.   □

Lemma. 2. *Any solution $N_1(t)$, $N_2(t)$ of (4) which starts in region II at time $t = t_0$ will remain in this region for all future time $t \geq t_0$, and ultimately approach the equilibrium solution $N_1 = K_1$, $N_2 = 0$.*

PROOF. Suppose that a solution $N_1(t)$, $N_2(t)$ of (4) leaves region II at time $t = t^*$. Then either $\dot{N}_1(t^*)$ or $\dot{N}_2(t^*)$ is zero, since the only way a solution of (4) can leave region II is by crossing l_1 or l_2. Assume that $\dot{N}_1(t^*) = 0$. Differentiating both sides of the first equation of (4) with respect to t and setting $t = t^*$ gives

$$\frac{d^2 N_1(t^*)}{dt^2} = \frac{-a_1 N_1(t^*)}{K_1} \frac{dN_2(t^*)}{dt}.$$

This quantity is positive. Hence $N_1(t)$ has a minimum at $t = t^*$. However, this is impossible, since $N_1(t)$ is increasing whenever a solution $N_1(t)$, $N_2(t)$ of (4) is in region II. Similarly, if $\dot{N}_2(t^*) = 0$, then

$$\frac{d^2 N_2(t^*)}{dt^2} = \frac{-a_2 N_2(t^*)}{K_2} \frac{dN_1(t^*)}{dt}.$$

This quantity is negative, implying that $N_2(t)$ has a maximum at $t = t^*$—but this is impossible, since $N_2(t)$ is decreasing whenever a solution $N_1(t)$, $N_2(t)$ of (4) is in region II.

The previous argument shows that any solution $N_1(t)$, $N_2(t)$ of (4) which starts in region II at time $t = t_0$ will remain in region II for all future time $t \geq t_0$. This implies that $N_1(t)$ is monotonic increasing and $N_2(t)$ is monotonic decreasing for $t \geq t_0$, with $N_1(t) < K_1$ and $N_2(t) > K_2$. Consequently, both $N_1(t)$ and $N_2(t)$ have limits ξ, η, respectively, as t approaches infinity. This, in turn, implies that (ξ, η) is an equilibrium point of (4). Now, (ξ, η) obviously cannot equal $(0, 0)$ or $(0, K_2)$. Consequently, $(\xi, \eta) = (K_1, 0)$, and this proves Lemma 2. □

Lemma 3. *Any solution $N_1(t)$, $N_2(t)$ of (4) which starts in region III at time $t = t_0$ and remains there for all future time must approach the equilibrium solution $N_1(t) = K_1$, $N_2(t) = 0$ as t approaches infinity.*

PROOF. If a solution $N_1(t)$, $N_2(t)$ of (4) remains in region III for $t \geq t_0$, then both $N_1(t)$ and $N_2(t)$ are monotonic decreasing functions of time for $t \geq t_0$, with $N_1(t) > K_1$ and $N_2(t) > K_1$. Consequently, both $N_1(t)$ and $N_2(t)$ have limits ξ, η, respectively, as t approaches infinity. This, in turn, implies that (ξ, η) is an equilibrium point of (4). Now, (ξ, η) obviously cannot equal $(0, 0)$ or $(0, K_2)$. Consequently, $(\xi, \eta) = (K_1, 0)$. □

PROOF OF THEOREM. Lemmas 1 and 2 state that every solution $N_1(t)$, $N_2(t)$ of (4) which starts in regions I or II at time $t = t_0$ must approach the equilibrium solution $N_1 = K_1$, $N_2 = 0$ as t approaches infinity. Similarly, Lemma 3 shows that every solution $N_1(t)$, $N_2(t)$ of (4) which starts in region III at time $t = t_0$ and remains there for all future time must also approach the equilibrium solution $N_1 = K_1$, $N_2 = 0$. Next, observe that any solution $N_1(t)$, $N_2(t)$ of (4) which starts on l_1 or l_2 must immediately afterwards enter region II. Finally, if a solution $N_1(t)$, $N_2(t)$ of (4) leaves region III, then it must cross the line l_1 and immediately afterwards enter region II. Lemma 2 then forces this solution to approach the equilibrium solution $N_1 = K_1$, $N_2 = 0$. $\qquad\qquad\qquad\qquad\qquad\qquad\qquad\qquad\qquad\qquad\qquad\square$

The theorem deals with the case of identical species, i.e., $\alpha = \beta = 1$. By a similar analysis (see Exercises 4–6) we can predict the outcome of the struggle for existence for all values of α and β.

Exercises

1. Rewrite the system of equations (4) in the form

$$\frac{K_1}{a_1 N_1}\frac{dN_1}{dt} = K_1 - N_1 - N_2 \qquad \frac{K_2}{a_2 N_2}\frac{dN_2}{dt} = K_2 - N_1 - N_2,$$

   then subtract these two equations and integrate to obtain directly that $N_2(t)$ approaches zero for all solutions $N_1(t)$, $N_2(t)$ of (4) with $N_1(t_0) > 0$.

2. The system of differential equations

$$(*)\qquad \frac{dN_1}{dt} = N_1[-a_1 + c_1(1 - b_1 N_1 - b_2 N_2)]$$

$$\frac{dN_2}{dt} = N_2[-a_2 + c_2(1 - b_1 N_1 - b_2 N_2)]$$

   is a model of two species competing for the same limited resource. Suppose that $c_1 > a_1$ and $c_2 > a_2$. Deduce that $N_1(t)$ ultimately approaches zero if $a_1 c_2 > a_2 c_1$, and $N_2(t)$ ultimately approaches zero if $a_1 c_2 < a_2 c_1$.

3. In 1926, Volterra presented the following model of two species competing for the same limited food supply:

$$\frac{dN_1}{dt} = [b_1 - \lambda_1(h_1 N_1 + h_2 N_2)]N_1,$$

$$\frac{dN_2}{dt} = [b_2 - \lambda_2(h_1 N_1 + h_2 N_2)]N_2.$$

   Suppose that $b_1/\lambda_1 > b_2/\lambda_2$. (The coefficient b_i/λ_i is called the susceptibility of species i to food shortages.) Prove that species 2 will ultimately become extinct if $N_1(t_0) > 0$.

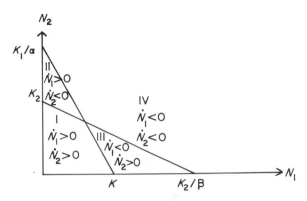

Figure 17.2

Exercises 4–6 are concerned with the system of equations

(*) $$\frac{dN_1}{dt} = \frac{a_1 N_1}{K_1}(K_1 - N_1 - \alpha N_2) \qquad \frac{dN_2}{dt} = \frac{a_2 N_2}{K_2}(K_2 - N_2 - \beta N_1).$$

4. a) Assume that $K_1/\alpha > K_2$ and $K_2/\beta < K_1$. Show that $N_2(t)$ approaches zero as t approaches infinity for every solution $N_1(t)$, $N_2(t)$ of (*) with $N_1(t_0) > 0$.
   b) Assume that $K_1/\alpha < K_2$ and $K_2/\beta > K_1$. Show that $N_1(t)$ approaches infinity for every solution $N_1(t)$, $N_2(t)$ of (*) with $N_2(t_0) > 0$. *Hint:* Draw the lines $l_1: N_1 + \alpha N_2 = K_1$ and $l_2: N_2 + \beta N_1 = K_2$ and follow the proof of our theorem.

5. Assume that $K_1/\alpha > K_2$ and $K_2/\beta > K_1$. Prove that all solutions $N_1(t)$, $N_2(t)$ of (*), with both $N_1(t_0)$ and $N_2(t_0)$ positive, ultimately approach the equilibrium solution

$$N_1 = N_1^0 = \frac{K_1 - \alpha K_2}{1 - \alpha\beta} \qquad N_2 = N_2^0 = \frac{K_2 - \beta K_1}{1 - \alpha\beta}.$$

Hint:
   a) Draw the lines $l_1: N_1 + \alpha N_2 = K_1$ and $l_2: N_2 + \beta N_1 = K_2$. The two lines divide the first quadrant into four regions (see Figure 17.2) in which both $\dot N_1$ and $\dot N_2$ have fixed signs.
   b) Show that all solutions $N_1(t)$, $N_2(t)$ of (*) which start in either region II or III must remain in these regions and ultimately approach the equilibrium solution $N_1 = N_1^0$, $N_2 = N_2^0$.
   c) Show that all solutions $N_1(t)$, $N_2(t)$ of (*) which remain exclusively in region I or region IV for all time $t \geq t_0$ must ultimately approach the equilibrium solution $N_1 = N_1^0$, $N_2 = N_2^0$.

6. Assume that $K_1/\alpha < K_2$ and $K_2/\beta < K_1$.
   a) Show that the equilibrium solution $N_1 = 0$, $N_2 = 0$ of (*) is unstable.
   b) Show that the equilibrium solutions $N_1 = K_1$, $N_2 = 0$ and $N_1 = 0$, $N_2 = K_2$ of (*) are asymptotically stable.

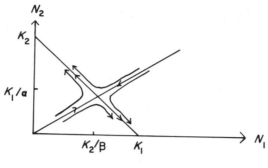

Figure 17.3

c) Show that the equilibrium solution $N_1 = N_1^0$, $N_2 = N_2^0$ (see Exercise 5) of (*) is a saddle point. (This calculation is very cumbersome.)

d) It is not too difficult to see that the phase portrait of (*) must have the form described in Figure 17.3.

References

[1] G. F. Gause, *The Struggle for Existence* New York: Hafner, 1964.

Notes for the Instructor

Objectives. The biological principle of competitive exclusion asserts that two similar species competing for a limited amount of resources cannot coexist in the same community. In this module the principle is derived mathematically from the coupled differential equations modeling the populations of the two species.

Prerequisites. Basic understanding of differential equations, equilibrium points, and the "phase" plane.

Time. The module may be covered in one or two lectures.

CHAPTER 18
Biological Cycles and the Fivefold Way

Courtney S. Coleman*

1. Prologue

One of the more curious natural phenomena is the regular periodic variation in the populations of certain interacting species, a variation which does not correlate with known periodic external forces such as the cycles of darkness and light, the seasons, or weather cycles. Most commonly, these species are involved in a predator–prey relationship. The phenomenon has been observed in the microcosm of a laboratory culture (*paramecium aurelia* (predator) and *saccharomyces exiguus* (prey)—see D'Ancona [11]) and in the macrocosm of the Canadian coniferous forests (Canadian lynx (predator) and snowshoe hare (prey)—see Keith [17]). Other examples include the budworm–larch tree cycle in the Swiss Alps [2] and a lemming–vegetation cycle in Scandinavia [19]. We shall discuss only the lynx–hare cycle, but the mathematical models we shall develop are in principle applicable to any predator–prey interaction. In the course of our analysis we shall study the equilibria of coupled differential systems and the possible limit behavior of bounded orbits of such systems. This will all then be applied to the question of the existence of cycles in a predator–prey system.

It is only fair to warn the reader at the outset that this whole area continues to be a source of much discussion and controversy among biologists, ecologists, and mathematicians. Thus our analysis is in no sense complete and, in fact, is intended to raise more questions than it resolves.

* Department of Mathematics, Harvey Mudd College, Claremont, CA 91711.

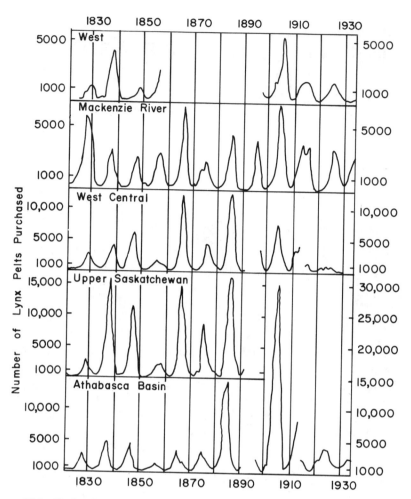

Figure 18.1. Hudson's Bay Company Purchases of Lynx Pelts, 1828–1935. (Adapted from Elton and Nicholson [12])

2. The Canadian Lynx Cycle[1]

The Canadian coniferous forests cover an area of several million square kilometers, extending from Quebec on the east into British Columbia and Alaska on the west. A number of years ago it was noticed that the populations of several species of birds, fish, and mammals living in this vast region

---

[1] The material of this section and the next is based largely on Keith [17] and on Elton and Nicholson [12].

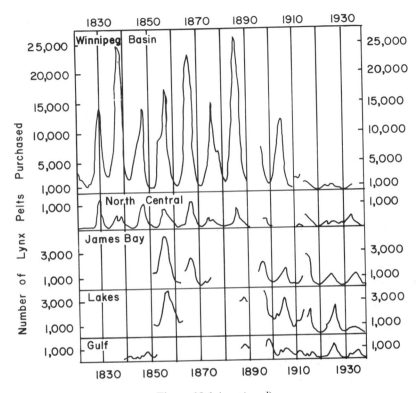

Figure 18.1 (*continued*).

vary periodically in a remarkably regular and predictable cycle of approximately 10 years. This cyclical behavior was first observed in the numbers of the Canadian lynx and for a curious reason. Lynx fur has long been valued for coats, caps, gloves, and other warm apparel. Since the middle of the 18th century, the lynx-trapping industry has thrived throughout the Canadian woods. The trading concession has been almost exclusively in the hands of the Hudson's Bay Company, and this company has kept more or less complete records of its purchases of pelts for over 200 years. Thus the trappers' experience with lynx abundance or scarcity has a quantitative expression in these records. The graphs of Figures 18.1 and 18.2 (adapted from Elton and Nicholson [12]) portray a 115-year segment of the records. The records from 1892–1896 and 1914–1915 are inadequate (except for the Mackenzie River region), and these years have been omitted from the graphs.

Although the maximal amplitudes vary, the regularity of the cycle and the great differences between the numbers in the peak and in the lean years are striking. What could be the cause? A number of explanations have been suggested. These suggestions can be classified into three categories:

(1) no cycle at all exists in the lynx population—the only cycle is in the numbers of pelts purchased;

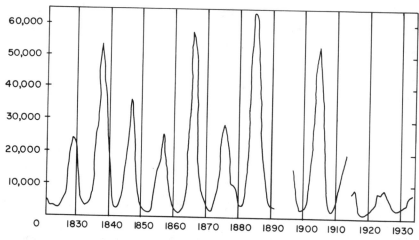

Figure 18.2. Totals from All Areas Indicated in Figure 1, except James Bay, Lakes, Gulf (Adapted from Elton and Nicholson [12])

(2) a lynx cycle does exist and is caused by an external force unaffected by the lynx (*exogenous* causes);
(3) the lynx and another species of fauna or flora form a closed interacting community whose internal interactions cause cycles in the numbers of both species (*endogenous* causes);
(4) none of the above.

Let us look at a sampling of the arguments and counterarguments in each category.

In category 1 the following explanations have been offered.

(1a) The Hudson's Bay Company is overstocked with lynx furs every 10 years and holds down its purchases until the excess inventory is sold. (The company's records do not support this argument).
(1b) The numbers of trappers or their efficiency follow a 10-year cycle.
(1c) The lynx migrate through the trapping regions on a 10-year cycle (but the cycle is virtually simultaneous over the entire lynx habitat and that habitat is almost entirely trapped).

These and other similar arguments have generally been discarded.[2] We shall take as given from this point on that:

(i) the number of lynx varies according to a 10-year cycle with marked highs and lows;[3]

---

[2] See Weinstein [28] for recent argument in support of 1b. See also the last section of this article.
[3] Sometimes the interval between successive highs has been 9 years. The average over the 200-year span of the records is about 9.6 years.

(ii) the number of lynx is directly proportional to the number of pelts purchased by the Hudson's Bay Company.

In other words, we shall assume that *a lynx cycle does exist, and the pelt records give an accurate indication* of its characteristics.

Explanations of type 2 assign the cause of the cycle to various external forces acting directly or indirectly upon the lynx community, the forces themselves remaining unaffected by that community. Listed below are some of these explanations.

(2a) The trapping itself causes the cycle (not likely, since the cycle has now been observed in regions with no trapping).

(2b) Ten-year cycles occur in the weather, the number or intensity of sunspots, or the amount of ozone in the atmosphere (evidence of such cycles is scanty and there appears to be no correlation with the lynx cycle).

(2c) A 10-year cycle occurs in the principal food source of the lynx, but lynx predation is only a minor factor in the mortality rate of that food source (a distinct possibility—see the Epilogue).

Although the last "explanation" is supported by some strong evidence, the category 3 argument of a two-species interacting community will occupy most of our attention.

3. The Canadian Lynx–Snowshoe Hare Cycle

The Hudson's Bay Company purchases the pelts of many species in addition to the lynx. One of these is the snowshoe hare. If the numbers of hare pelts purchased annually are plotted alongside the numbers of lynx pelts, a quite striking fact emerges: the hare have a 10-year cycle which appears to correlate with the lynx cycle (Figure 18.3). Irregularities in the purchasing policies for snowshoe hare pelts may account for some of the irregularities in the hare cycle. Observe also that the geographical area covered by these data is not as extensive as that used in Figure 18.2.

One is tempted to suggest a lynx–snowshoe hare cycle, but this requires an interaction between the two species. The first question is whether the hare is the principal food source of the lynx. Nellis *et al.* [25] and Brand *et al.* [4] have surveyed the lynx population in a 57-mi^2 region north of Edmonton, Alberta, Canada, over several winters beginning in 1964. By analyzing lynx scat (droppings) and the stomach and intestinal contents of dead lynx, and by trailing the animals, it was shown fairly conclusively that the hare is indeed the main food source of the lynx, at least during the winter. In fact, in the community studied the hare constituted about 75% of the lynx diet by weight during the height of the hare cycle, although this percentage dropped during times of hare scarcity.

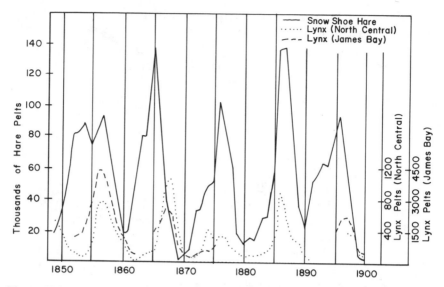

Figure 18.3. Hudson's Bay Company Pelt Collections for Snowshoe Hares at Posts along Hudson Bay and for Lynx Taken in Roughly the Same Region (Adapted from Elton and Nicholson [12] and MacLulich [21]).

The question is whether or not this predator–prey relationship between the lynx and the hare is the main cause of the changes in the birth and the mortality rates of each species. The evidence is strong in the case of the lynx [4], [25], but the question remains open concerning the hare. *If* the lynx are the dominant factor in the mortality rates of the hare, then we might reasonably consider the lynx–hare community to be a closed system, and the analysis that follows gives one explanation of the cycles.

On the other hand, if the hare cycles occur even in the absence of the lynx predation, we may have to fall back upon argument 2c, with a hare cycle of unknown cause "driving" the lynx cycle, or else resort to an argument of the type of "none of the above." As mentioned before, the main thrust of this presentation is based on the assumption of a closed community of interacting lynx and snowshoe hare in a predator–prey relationship. Let us now determine a general mathematical model for a two-species interaction.

4. A Two-Species Population Model

Let $x(t)$ and $y(t)$ denote the respective numbers of individuals, i.e., the *populations*, at time t of two species which form a closed interacting community.[4] Although we clearly have a lynx–hare community in mind, we shall

---

[4] Alternatively, $x(t)$ and $y(t)$ might denote the *population densities*, the numbers of individuals per unit area or volume of the habitat.

not at first make any assumption about the nature of the interaction. We shall assume that $x(t)$ and $y(t)$ vary continuously and even differentiably as functions of time. This is, of course, an idealization of the true state of affairs since populations must be integers and change only by integer amounts as time advances. However, one might argue that for a very large population, at least, an increase or decrease in the population by one or two is "infinitesimal" relative to the total, and we might as well allow the population to change by arbitrary small amounts, not just integers. For example, if the number of lynx at 8:00 a.m. one day is 50,000 and an hour later is 50,001, then it seems reasonable to set the "population" at 50,000.0833 $\cdots$ at 8:05 a.m. Of course, the argument is of dubious validity for small populations, and the analysis below has limited relevance in that case. Once we have made the idealization to continuous functions, we might as well go all the way and smooth away any "corners" on the graphs of $x(t)$ and $y(t)$ versus t; thus we assume differentiability as well as continuity.

Let us suppose that in the span of time from t to $t + \Delta t$ the two populations change by amounts Δx and Δy, respectively. The quantities $\Delta x/x(t)$ and $\Delta y/y(t)$ are called the *relative* (or *fractional*) *changes*, while the ratios,

$$\frac{\Delta x/\Delta t}{x(t)} \quad \text{and} \quad \frac{\Delta y/\Delta t}{y(t)} \tag{1}$$

are the average relative rates of change. Letting $\Delta t \to 0$ in (1), we obtain the *relative rate of change* of the population of each species,

$$\frac{1}{x(t)} \frac{dx(t)}{dt} \quad \text{and} \quad \frac{1}{y(t)} \frac{dy(t)}{dt}.$$

It is rare that we would know from the beginning simple formulas for $x(t)$ and $y(t)$, but it is common to have information about relative birth and mortality rates. Let us denote relative birth rates by B, relative mortality rates by M. One might expect each of these rates for each of the two species to depend upon the populations themselves; i.e., we might have $B_1(x, y)$ and $M_1(x, y)$ for the x species and $B_2(x, y)$ and $M_2(x, y)$ for the y species. If the community is closed, then no other forces play a significant role in affecting the relative rates of change, and we might have that

$$\begin{cases} \dfrac{1}{x} \dfrac{dx}{dt} = B_1(x, y) - M_1(x, y) \equiv f(x, y) \\[2mm] \dfrac{1}{y} \dfrac{dy}{dt} = B_2(x, y) - M_2(x, y) \equiv g(x, y). \end{cases} \tag{2}$$

We shall assume that *the relative rates f and g are continuously differentiable* in x and y for $x \geq 0$, $y \geq 0$. (Because of the interpretations of x and y as populations, we are only concerned with the population quadrant $x \geq 0$, $y \geq 0$ in the xy plane.)

Cross multiplying, we have the fundamental equations for a closed community of two interacting species:

$$\begin{cases} \dfrac{dx}{dt} = xf(x, y), \\[2mm] \dfrac{dy}{dt} = y\,g(x, y). \end{cases} \tag{3}$$

We shall take up the study of the solutions of this and more general systems in the next section.

5. Differential Systems—Solutions, Orbits, Equilibria

In this section and the next we shall depart from the specific model given by (3) and take up the general *planar autonomous differential system,*

$$\begin{cases} \dfrac{dx}{dt} = F(x, y) \\[2mm] \dfrac{dy}{dt} = G(x, y), \end{cases} \tag{4}$$

where F and G are assumed to be continuously differentiable functions for all values of x and y in some region R in the xy plane. For simplicity, we shall usually take R to be the entire plane. The term "autonomous" refers to the absence of any explicit dependence of the rate functions F and G in (4) upon time t. Observe that (3) is a special case of (4) with R being the first quadrant (see Exercise 2).

A pair of functions, $x(t)$ and $y(t)$, is a *solution* of (4) if

(a) $x(t)$ and $y(t)$ are defined on a common open interval I: $a < t < b$ (a may be $-\infty$, b may be $+\infty$);

(b) $x(t)$ and $y(t)$ satisfy (4) for every $t \in I$:

$$\begin{cases} \dfrac{dx(t)}{dt} = F(x(t), y(t)) \\[2mm] \dfrac{dy(t)}{dt} = G(x(t), y(t)); \end{cases}$$

(c) there are no functions, $x^*(t)$ and $y^*(t)$, defined on an interval $I^* \supset I$, which coincide with $x(t)$ and $y(t)$, respectively, on I and satisfy (4) for all $t \in I^*$.

Condition c means that we are only interested in solutions of (4) which are *maximally extended* in time. For example, the function $x_1(t) = e^t$, $-\infty < t < \infty$, is an acceptable solution of the differential equation $dx/dt = x$, while the function $x_2(t) = e^t$, $-1 < t < 10$, is not acceptable.

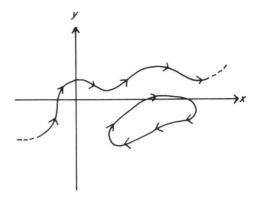

Figure 18.4. Three Types of Orbits of Planar System

Each solution $x = x(t)$, $y = y(t)$ of (4) defines a curve Γ in the xy plane called the *orbit* of the solution. We shall assume without proof that

through each point of R there passes an orbit,   (5a)

no two orbits intersect unless they coincide,   (5b)

each orbit is either a non-selfintersecting smooth curve,
or a simple smooth closed curve, or a single point.   (5c)

See Hirsch and Smale [16] or Braun [5] for proofs of similar "existence and uniqueness" theorems. See Figure 18.4 for sketches of the three types of orbits listed in (5c). The arrowheads indicate how the orbit is traced out with increasing t. Different solutions may generate the same orbit as the following example shows.

EXAMPLE 1. The system

$$\frac{dx}{dt} = y \qquad \frac{dy}{dt} = -x \qquad (6)$$

has a pair of distinct solutions

$$x_1(t) = \cos t, \qquad y_1(t) = -\sin t, \qquad -\infty < t < \infty,$$

$$x_2(t) = \sin t, \qquad y_2(t) = \cos t, \qquad -\infty < t < \infty,$$

as a direct calculation shows. However, the two solutions define the same orbit Γ, the unit circle with equation $x^2 + y^2 = 1$ and oriented clockwise by advancing time. See Exercise 3 for formulas for all solutions of (6). See Figure 18.5.

If the time parameterization of the orbits is not important, then we might divide the two equations of (4) to obtain

$$\frac{dy}{dx} = \frac{G(x, y)}{F(x, y)},$$

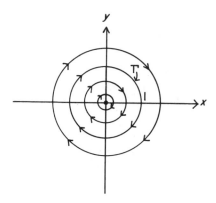

Figure 18.5. Orbits of (6)

which can possibly be "solved" by one of the techniques for handling first-order equations given in Braun [5] or Boyce and DiPrima [3]. For example, we would obtain $dy/dx = -x/y$ upon division in (6). Separating variables and integrating, we have that

$$\begin{cases} y\,dy = -x\,dx, \\ \tfrac{1}{2}(y^2 - y_0^2) = -\tfrac{1}{2}(x^2 - x_0^2), \\ x^2 + y^2 = A^2, \text{ where } A^2 = x_0^2 + y_0^2. \end{cases}$$

With $A = 1$ we obtain the unit circle illustrated in Figure 18.5, except that there is no longer any time orientation. Observe that all the orbits are circles centered at the origin, except for the point orbit at the origin itself (see Figure 18.5).

A nonconstant solution, $x = x(t)$, $y = y(t)$, of (4) is a *periodic solution* (or *cycle*) if a positive number T exists such that for all t

$$x(t + T) = x(t) \qquad y(t + T) = y(t).$$

The smallest such number T is called the *period* of the cycle. The corresponding orbit is a simple closed curve, which we shall also call a *cycle*. Observe that the cycle Γ of Example 1 has period $T = 2\pi$. Conversely, any orbit which is a simple closed curve corresponds to a periodic solution. This seems to be obvious, and we shall omit the somewhat tedious formal proof.

A cycle is one of the three possible orbits listed in (5c). Another type is the point orbit, or *critical point*. A point (x_0, y_0) is a critical point of (4) if and only if

$$F(x_0, y_0) = 0 \qquad G(x_0, y_0) = 0.$$

For, we have that $x(t) \equiv x_0, y(t) \equiv y_0$ defines a solution of (4) if and only if

$$\frac{dx_0}{dt} \equiv 0 = F(x_0, y_0) \qquad \frac{dy_0}{dt} \equiv 0 = G(x_0, y_0).$$

Critical points play an important role in the study of cycles since *inside every cycle is at least one critical point.* The proof of this fact is not easy—see Hirsch and Smale [16].

The system of Example 1 has exactly one critical point; all other orbits are cycles (Exercise 3). The system in the next example has exactly one critical point and exactly one cycle.

EXAMPLE 2. Let r and θ be polar coordinates,

$$\begin{cases} x = r \cos \theta \\ y = r \sin \theta \end{cases} \tag{7a}$$

or

$$\begin{cases} r = (x^2 + y^2)^{1/2} \\ \theta = \arctan\left(\dfrac{y}{x}\right) \end{cases} \tag{7b}$$

and consider the differential system in polar coordinates,

$$\frac{dr}{dt} = r(1 - r^2), \qquad r \geq 0, \tag{8a}$$

$$\frac{d\theta}{dt} = 1. \tag{8b}$$

First, let us express (8) in rectangular coordinates. From (7a) and (8) we have that

$$\begin{cases} \dfrac{dx}{dt} = \dfrac{dr}{dt} \cos \theta - r \sin \theta \dfrac{d\theta}{dt} = (1 - r^2) r \cos \theta - r \sin \theta \\ \qquad = (1 - x^2 - y^2) x - y \\ \dfrac{dy}{dt} = \dfrac{dr}{dt} \sin \theta + r \cos \theta \dfrac{d\theta}{dt} = (1 - r^2) r \sin \theta + r \cos \theta \\ \qquad = (1 - x^2 - y^2) y + x, \end{cases} \tag{9}$$

where we have used the chain rule (e.g., $(d/dt)(\cos \theta) = (d/d\theta)(\cos \theta)(d\theta/dt)$). Observe that any solution, $r = r(t)$, $\theta = \theta(t)$, of (8) gives via (7) a solution, $x = x(t)$, $y = y(t)$, of (9), and conversely.

Equation (8) can be solved more easily than (9). In fact, we have that

$$r(t) = \frac{r_0}{[r_0^2 + (1 - r_0^2) e^{-2t}]^{1/2}}, \qquad r_0 \geq 0, \tag{10}$$

$$\theta(t) = t + \theta_0, \qquad \theta_0 \text{ arbitrary.} \tag{11}$$

The solution for $r(t)$ is obtained by separating the variables in (8a), but the details are left to the reader (Exercise 4). Formula (11) follows from (8b) by integration. (We assume that $r = r_0$ and $\theta = \theta_0$ when $t = 0$.)

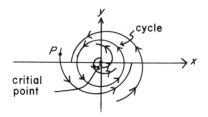

Figure 18.6. One Cycle, One Critical Point

We have that $r(t) \equiv 1$ if $r_0 = 1$, while $r(t) \equiv 0$ if $r_0 = 0$. From (8a) we see that $r(t)$ increases monotonically to 1 as t increases to ∞ if $0 < r_0 < 1$, while $r(t)$ decreases monotonically to 1 as t increases to ∞ if $r_0 > 1$. In rectangular coordinates, $r(t) \equiv 0$ corresponds to a critical point $(0, 0)$, while $r(t) \equiv 1$, $\theta(t) = t + \theta_0$ corresponds to the rectangular equations,

$$x(t) = \cos(t + \theta_0), \; y(t) = \sin(t + \theta_0),$$

the parametric equations of the unit circle.

Thus system (9) has a unique critical point at the origin and a unique cycle along the unit circle. See Figure 18.6 for a portrait of the orbits.

The most distinctive orbits in Examples 1 and 2 are the cycles and the critical points, orbits which we shall call *equilibria*. Observe that every orbit in Example 1 is an equilibrium orbit, while every orbit in Example 2 is an equilibrium orbit or else approaches such an orbit as time advances. Is there any easy way to determine the equilibrium orbits without having to solve system (4) and find formulas for all the solutions? We can find the critical points by solving simultaneously the equations,

$$F(x, y) = 0 \quad \text{and} \quad G(x, y) = 0.$$

Finding the cycles of a system (if there are any) is not usually so easy, however. In Examples 1 and 2 it was relatively easy because the corresponding systems could be solved, but this is often not possible. In the next section we shall develop a powerful method for detecting the presence of cycles among the orbits of a general planar autonomous system such as (4), a method which does not depend upon having solution formulas in hand.

The differential systems of Examples 1 and 2 are not of form (3) and thus cannot represent the dynamics of a pair of interacting species (Exercise 2). Nevertheless, each system possesses cycles, and we expect any model of the lynx–hare interaction to have a cycle. Thus we might imagine the orbit portraits in Figures 18.5 and 18.6 to be possible diagrams of lynx–hare population orbits, at least qualitatively. (Observe that the nonconstant orbits in Figures 18.5 and 18.6 traverse all four quadrants, while lynx–hare population orbits must, of course, remain in the first quadrant.) We leave it to the reader (Exercises 3–5) to analyze these two examples in more detail and to speculate about the "ecological meaning" of each system.

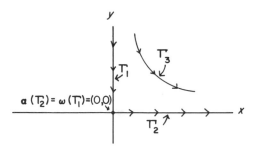

Figure 18.7. Some Orbits and Limit Sets of Example 3.

6. The Fivefold Way

The goal of this section is to develop a test for the existence of a cycle for system (4), a test that does not require that the system have already been solved. As before, a solution of (4) is a pair of functions, $x = x(t)$, $y = y(t)$, which satisfy the two equations of (4) for all time t in some maximal time interval. For convenience, we shall denote the pair $(x(t), y(t))$ by $z(t, P)$, where $P = (x(0), y(0))$, the "initial point" of the solution. Γ will denote the corresponding orbit.

The subset of Γ defined by $\{z(t,P): t \geq 0\}$ is the *positive semi-orbit from P* and is denoted by Γ_P^+. Γ_P^- can be similarly defined. If we have a positive semi-orbit of an orbit Γ but are indifferent about its "starting point" P, we will use the simpler notation Γ^+ (Γ^- has a like meaning). Likewise, we shall use $z(t)$ instead of $z(t, P)$. Look back at Figure 18.6 to see examples of positive semi-orbits, e.g., the semi-orbit beginning at P.

In connection with positive and negative semi-orbits, Γ_P^+ and Γ_P^-, we can define the *omega* and *alpha limit sets*, $\omega(\Gamma_P^+)$ and $\alpha(\Gamma_P^-)$ (also called the *positive* and *negative limit sets*, respectively) as follows.

Definition. Let Γ_P^+ be a positive semi-orbit. Then $\omega(\Gamma_P^+)$ is the set of all Q such that for some increasing sequence of times, $t_1, t_2, \cdots, t_n, \cdots$, diverging to ∞, we have that $z(t_n) \to Q$.

The alpha limit set is defined similarly but with t_n diverging to $-\infty$ and decreasing as n increases. The reader should verify that limit sets are independent of the choice of point P on Γ, hence we can speak of $\omega(\Gamma)$ and $\alpha(\Gamma)$ (Exercise 6). Observe that $\alpha(\Gamma) = \omega(\Gamma) = \Gamma$ if Γ is a critical point or a periodic orbit. The following examples illustrate some of the possibilities for planar limit sets.

EXAMPLE 3. The system $dx/dt = x$, $dy/dt = -y$ has an orbit Γ_1 defined by $x = 0$, $y = e^{-t}$; $\omega(\Gamma_1) = (0, 0)$, while $\alpha(\Gamma_1)$ is empty since $y \to \infty$ as $t \to -\infty$. For the orbit Γ_2, $x = e^t$, $y = 0$, we have $\omega(\Gamma_2)$ empty and $\alpha(\Gamma_2) = (0, 0)$. The orbit Γ_3, $x = e^t$, $y = e^{-t}$, has both limit sets empty. See Figure 18.7.

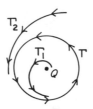

Figure 18.8. Orbits and Limit Sets for Example 4.

It is customary, if $\Gamma \neq \omega(\Gamma)$, to say that Γ is *positively asymptotic to* $\omega(\Gamma)$, as $t \to +\infty$. Thus in Example 3 Γ_1 is positively asymptotic to the origin. We could likewise define Γ as *negatively asymptotic to* $\alpha(\Gamma)$; in Example 3 Γ_2 is negatively asymptotic to the origin.

EXAMPLE 4. The orbits indicated in Figure 18.8 arise from the system of Example 2 and have the following limit sets (see Exercise 7):

$$\begin{cases} Q = \omega(Q) = \alpha(Q) \\ \Gamma = \omega(\Gamma) = \alpha(\Gamma) \\ \Gamma = \omega(\Gamma_1), \, Q = \alpha(\Gamma_1) \\ \Gamma = \omega(\Gamma_2), \, \alpha(\Gamma_2) \text{ is empty.} \end{cases}$$

From these examples it would appear that a limit set is either empty or else contains a single equilibrium orbit, i.e., critical point or cycle. This is not quite true, as the following example shows.

EXAMPLE 5. In polar coordinates, the system

$$\begin{cases} \dfrac{dx}{dt} = -y[y^2 + (x^2 + y^2 - 1)^2] + x - xy^2 - x^3 \\ \\ \dfrac{dy}{dt} = x[y^2 + (x^2 + y^2 - 1)^2] + y - x^2y - y^3 \end{cases} \tag{12}$$

becomes

$$\begin{cases} \dfrac{dr}{dt} = r(1 - r^2) \\ \\ \dfrac{d\theta}{dt} = r^2 \sin^2 \theta + (r^2 - 1)^2. \end{cases}$$

Observe that the critical points (in polar coordinates) are located at $r = 0$ for P_3, at $r = 1, \theta = 0$ for P_2, and at $r = 1, \theta = \pi$ for P_1. Observe that $d\theta/dt > 0$ except at these critical points, while $dr/dt > 0$ if $0 < r < 1$, $dr/dt < 0$ if $r > 1$, and $dr/dt = 0$ if $r = 1$. We have that $\omega(\Gamma_3) = \Gamma_1 \cup \Gamma_2 \cup \{P_1, P_2\}$, while $\alpha(\Gamma_3)$ is empty; $\omega(\Gamma_4) = \omega(\Gamma_3)$, and $\alpha(\Gamma_4) = \{P_3\}$; $\omega(\Gamma_2) = \{P_1\} =$

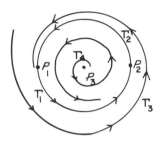

Figure 18.9. Orbits of (12)

$\alpha(\Gamma_1)$, $\alpha(\Gamma_2) = \{P_2\} = \omega(\Gamma_1)$. Thus the phase portrait is as sketched in Figure 18.9. Recall from Example 2 that $r(t)$ is given by (10). (Examples 2 and 5 have the same equation for dr/dt but not for $d\theta/dt$). We leave it to the reader to complete the analysis for this example, verifying the limit sets indicated below and the validity of Figure 18.9 (Exercise 8).

Example 5 contains an ω-limit set which is neither a single critical point nor a cycle, i.e., the set $\Gamma_1 \cup \Gamma_2 \cup \{P_1, P_2\}$. This is an example of a cycle graph. More generally, a *cycle graph* CG is a connected union of orbits such that:

(1) CG contains a finite nonzero number of critical points;
(2) if $P_1, \cdots, P_n$ are the critical points of CG, then there are noncritical orbits, $\Gamma_1, \cdots, \Gamma_n$, of CG such that

$$\alpha(\Gamma_i) = P_i, \qquad i = 1, \cdots, n$$

$$\omega(\Gamma_i) = P_{i+1}, \qquad i = 1, \cdots, n,$$

where $P_{n+1} \equiv P_1$;
(3) each limit set of each orbit in CG is a critical point of CG.

The orbits Γ_3 and Γ_4 in each part of Example 5 are positively asymptotic to the cycle graph $\Gamma_1 \cup \Gamma_2 \cup \{P_1, P_2\}$.

From Examples 2–5 it would appear that an orbit Γ which is bounded with increasing t [i.e., Γ^+ lies in a rectangle] may be positively asymptotic to a critical point, a cycle, or a cycle graph, if Γ is not a critical point or a cycle to begin with. Indeed, no other alternatives exist, as the following very deep result of Poincaré and of Bendixson shows.

The Fivefold Way. Let Γ^+ be a bounded positive semi-orbit of (4), and suppose (4) has only a finite number of critical points. Then exactly one of the following five alternatives is true:

(1) Γ is a critical point;
(2) Γ is positively asymptotic to a critical point;
(3) Γ is a cycle.
(4) Γ is positively asymptotic to a cycle.
(5) Γ is positively asymptotic to a cycle graph.

(1) $\bullet\,\Gamma$ (2) Γ (3) $\bigcirc\Gamma$ (4) Γ (5) Γ

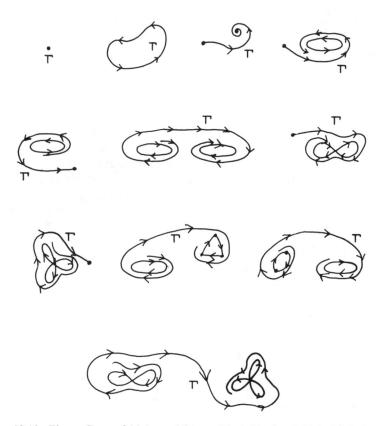

Figure 18.10. Eleven Cases of Alpha and Omega Limit Sets for Orbit Lying in Bounded Region Containing Only Finite Number of Critical Points.

See Coddington and Levinson [9] for a proof of this theorem. A similar result holds for $\alpha(\Gamma)$ if Γ^- is bounded. If we combine the possibilities for $\omega(\Gamma)$ with those for $\alpha(\Gamma)$, we obtain eleven alternatives for the limit sets of a bounded orbit (see Exercise 9). These are sketched in Figure 18.10.

According to the fivefold way we can conclude the existence of a cycle if there is a positive semi-orbit Γ^+ which is bounded and whose positive limit set $\omega(\Gamma)$ contains no critical points. In fact, we have in this case $\omega(\Gamma)$ itself is a cycle. In the next section we shall describe a predator model for which this situation is exactly what occurs.

For later use let us now introduce the notion of a repellent critical point.

A critical point P of (4) is *repellent* if there is a circular region D centered at P such that

(1) for all $Q \in D$, $z(t, Q) \to P$ as $t \to -\infty$,
(2) for all $Q \in D$ and all sequences $\{t_n\}$ such that $t_n \to \infty$ and $z(t_n, Q)$ is defined, we have that $z(t_n, Q) \nrightarrow P$.

There is a simple algebraic test for repellence, but we shall omit the proof.

Repellence Test. Let (x_0, y_0) be a critical point of (4), and let a, b, c, d be constants defined as follows

$$a = \frac{\partial F}{\partial x}\bigg|_{(x_0, y_0)} \qquad b = \frac{\partial F}{\partial y}\bigg|_{(x_0, y_0)} \qquad c = \frac{\partial G}{\partial x}\bigg|_{(x_0, y_0)} \qquad d = \frac{\partial G}{\partial y}\bigg|_{(x_0, y_0)}. \quad (13)$$

Then (x_0, y_0) is a repellent critical point if $a + d > 0$ and $ad - bc > 0$.[5] If $a + d < 0$ or if $ad - bc < 0$, then (x_0, y_0) is not repellent.

Observe that if P is repellent, then Γ is *not* in the positive limit set of any nonconstant orbit. The next example shows how the fivefold way and the repellence test might be used to deduce the existence of a cycle without having to know solution formulas.

EXAMPLE 6. We shall use the system of Example 2, but we shall not use any of the solution formulas. Let R be the circular region, $0 \leq r \leq 2$, centered at the origin. Inside R there is exactly one critical point $(0, 0)$. This point is repellent, for we have from (9) that

$$\begin{cases} a = \dfrac{\partial}{\partial x}\{(1 - x^2 - y^2)x - y\}_{(0,0)} = \{1 - x^2 - y^2 - 2x^2\}_{(0,0)} = 1 \\[2mm] b = \dfrac{\partial}{\partial y}\{(1 - x^2 - y^2)x - y\}_{(0,0)} = \{-2yx - 1\}_{(0,0)} = -1 \\[2mm] c = \dfrac{\partial}{\partial x}\{(1 - x^2 - y^2)y + x\}_{(0,0)} = \{-2xy + 1\}_{(0,0)} = 1 \\[2mm] d = \dfrac{\partial}{\partial y}\{(1 - x^2 - y^2)y + x\}_{(0,0)} = \{1 - x^2 - y^2 - 2x^2\}_{(0.0)} = 1. \end{cases} \quad (14)$$

Hence we have that $a + d = 2$, $ad - bc = 2$, and the hypotheses of the repellence test are satisfied.

Now at any point P on the perimeter of R we have that $dr/dt = r(1 - r^2) = 2(1 - 4) < 0$. Hence Γ_P^+ enters R through P and can never leave. Thus the fivefold way applies to Γ_P^+. Since the repellent critical point at the origin

---

[5] These conditions imply that the eigenvalues of the Jacobian matrix of F and G at (x_0, y_0) have positive real parts. See [3] or [9] for a discussion of the nature of orbits near a critical point.

cannot belong to $\omega(\Gamma_P^+)$ and since R contains no other critical points, we have that $\omega(\Gamma_P^+)$ is a cycle. (Γ_P cannot itself be a cycle if P is on the perimeter of R. Why not?) Of course, in this case $r \equiv 1$ is easily seen to be a cycle by inspection of (8).

7. Predator–Prey Models: Kolmogorov's Theorem

Let us return to system (3), that special form of (4) which we claim can be used to model two species interactions:

$$\begin{cases} \dfrac{dx}{dt} = xf(x, y) \\[2mm] \dfrac{dy}{dt} = yg(x, y). \end{cases}$$

We shall now do two things. First, we shall impose conditions on the functions f and g that reflect "predator–prey interactions" between the prey species x and the y predator. Next we shall show that the fivefold way can be used to deduce the presence of a cycle in certain cases. In order to keep track of the predator and the prey more easily, let us use H for hare instead of x and L for lynx instead of y, and consider the following system,

$$\frac{dH}{dt} = Hf(H, L)$$

$$\frac{dL}{dt} = Lg(H, L), \tag{15}$$

where f and g are continuously differentiable throughout the population quadrant, $H \geq 0, L \geq 0$.

The following conditions on f and g reflect possible predator–prey interactions. They are related to conditions originally given by Kolmogorov [18] and then modified by Bulmer [8].

(1) $\partial f / \partial L < 0$ an increase in the lynx population adversely affects the hare;

(2) $\partial g / \partial H > 0$ an increase in the hare population favorably affects the lynx;

(3)[6] $\partial f / \partial H < 0$ an increase in the hare population causes overcrowding, etc., and affects the hare rate adversely;

(4) $\partial g / \partial L < 0$ the lynx is limited by overpopulation effects of its own numbers;

(5) $f(0, L_1) = 0$ for some $L_1 > 0$; beyond the *threshhold population* L_1

---

[6] This assumption will be changed later.

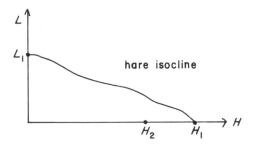

Figure 18.11. Hare Isocline

of lynx, the hare population decreases (by conditions 1 and 2) even when the hare population is small;

(6) $f(H_1, 0) = 0$ for some $H_1 > 0$; H_1 is the *carrying capacity* of the system for the hare in the absence of lynx; beyond H_1 decline due to overpopulation sets in (by condition 3);

(7) $g(H_2, 0) = 0$ for some $H_2 > 0$; H_2 is a lower bound on the numbers of hare needed to support the lynx community;

(8) $H_1 > H_2$ otherwise, the lynx population dies out;

(9)[7] the *hare isocline*, the smooth curve defined by $f(H, L) = 0$, has the shape sketched in Figure 18.11;

(10) the *lynx isocline*, the smooth curve defined by $g(H, L) = 0$, has the shape sketched in Figure 18.12.

Conditions 9 and 10 may seem both imprecise and nonecological. More precise versions are as follows.

(9) The equation $f(H, L) = 0$ of the hare isocline can be solved uniquely, with $H \geq 0$, $L \geq 0$, for L in terms of H, giving $L = h(H)$, where h is defined on the interval $[0, H_1]$, h is continuously differentiable, monotone decreasing, and $h(0) = L_1$, $h(H_1) = 0$.

(10) The equation $g(H, L) = 0$ of the lynx isocline can be solved uniquely for $H \geq 0$, $L \geq 0$, for H in terms of L, giving $H = k(L)$, where k is defined on the interval $[0, \infty)$, k is monotone increasing and continuously differentiable, and $k(0) = H_2$.

In fact, these precise statements can be deduced via the implicit function theorem from conditions 1–7 (see [1] or [16] for not-too-easy proofs of similar statements). We shall not carry out these derivations here.

The reader may easily show (Exercise 10) that (15) has exactly one critical point Q inside the population quadrant and that Q is the intersection point (x_0, y_0) of the lynx and the hare isoclines.

Recall that we are after cycles. Thus one might expect that the next step would be to proceed as in Example 6. That we shall do, but the region R we

---

[7] Condition 9 will be modified.

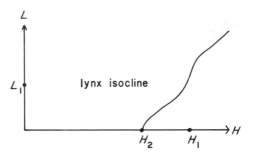

Figure 18.12. Lynx Isocline

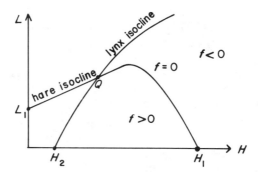

Figure 18.13. Isoclines According to Rosenzweig

shall construct which contains a bounded positive semi-orbit Γ_P^+ also contains the critical point Q. *The difficulty is that Q is not repellent* (Exercise 11). Thus it is conceivable that $\omega(\Gamma_P^+)$ could contain Q and there may be no cycle at all. This may indeed be the case in a predator–prey system. Nevertheless, we want cycles and will have to alter some of the conditions to make Q repellent.

This is a real difficulty and has caused much controversy among researchers in this area in recent years. Rosenzweig [26] has suggested on the basis of laboratory experiments with living predator–prey systems containing cycles (but not lynx and hare) that conditions 3, 9, and 10 should be replaced by the following.

(3a) $\partial f/\partial H\big|_{L=0,\,H\text{small}} > 0$ No lynx and low hare density imply social facilitation of hare reproduction.

(9a–10a) The hare and lynx isoclines have the shapes sketched in Figure 18.13, the critical point Q lying on the rising part of the hare isocline.

We need a final condition to ensure that Q is repellent. Let a, b, c, d be defined at the critical point $Q = (x_0, y_0)$ as in (13). Since $F = xf$ and $G = yg$, we have (recalling that $f(x_0, y_0) = 0 = g(x_0, y_0)$):

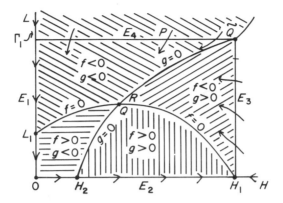

Figure 18.14. Region R for Kolmogorov's Cycle Theorem

$$\begin{cases} a = F_x(x_0, y_0) = f(x_0, y_0) + x_0 f_x(x_0, y_0) = x_0 f_x(x_0, y_0) \\ b = F_y(x_0, y_0) = x_0 f_y(x_0, y_0) \\ c = G_x(x_0, y_0) = y_0 g_x(x_0, y_0) \\ d = G_y(x_0, y_0) = g(x_0, y_0) + y_0 g_y(x_0, y_0) = y_0 g_y(x_0, y_0). \end{cases} \qquad (16)$$

Our last condition is

(11) $\qquad\qquad a + d > 0 \text{ and } ad - bc > 0.$

We shall not attempt to justify these new conditions on ecological grounds (see Rosenzweig [26]). They are controversial, but they do give us the following theorem.

Kolmogorov's Cycle Theorem. *Let the functions f and g of* (15) *satisfy conditions* 1, 2, 3a, 4–8, 9a, 10a, *and* 11 *in the population quadrant. Then a cycle exists inside the population quadrant.*

PROOF. For simplicity, we shall assume that the lynx isocline sketched in Figure 13 intersects the vertical line $H = H_1$ (see Bulmer [8] for the analysis when this is *not* the case.) Now construct the rectangular region R illustrated in Figure 18.14.

Let us show that a positive semi-orbit Γ_P^+, where P is, say, on the interior arc of the top edge E_4 of R, lies entirely in R. We shall do this by showing that Γ_P^+ cannot "escape" from R across the perimeter $E_1 \cup E_2 \cup E_3 \cup E_4$. E_1 lies along an orbit, the orbit Γ_1 defined by $H \equiv 0, L > 0$, while 0 is the critical point at the origin. Thus no orbit can escape from R across E_1 with increasing time, since in order to do so it would have to intersect Γ_1 or the critical orbit 0; but orbits cannot intersect. Edge E_2 consists of two critical points, at 0 and H_1 (we identify the number H_1 with the point $(H_1, 0)$, and the orbit defined by $0 < H < H_1, L = 0$, and moving away from 0 towards the point H_1. (The reader should verify that the arrowheads on E_1 and E_2 are as indicated (Exercise 12).) As before, no orbit can exit R across E_2.

Orbits cross the interior arc of E_3 moving from right to left with increasing t since the points of this arc are above the hare isocline where $f(H, L) = 0$ and we know by condition 1 that $\partial f/\partial L < 0$. Thus $dH/dt < 0$ along the arc. As before, no orbit can escape R through the critical point H_1. The orbit through $\tilde{Q}$ is horizontal and is moving to the left at that point since $\tilde{Q}$ is on the lynx isocline, i.e., $g(H, L) = 0$, and hence $dL/dt = 0$. Along the interior arc of E_4 orbits move downwards and to the left since points of the arc are above both isoclines and $\partial f/\partial L < 0$ and $\partial g/\partial L < 0$ by conditions 1 and 4. Thus no orbit can escape R across its boundary, and orbits enter R through the points of the top and right edges which lie inside the population quadrant.

Our aim now is to show that $\omega(\Gamma_P^+)$ is a cycle. By the fivefold way this will follow if we can prove that $\omega(\Gamma_P^+)$ contains no critical point. Now the critical points of (15) in the population quadrant are at the origin 0, at the point H_1 on the H axis, and at Q. Q cannot be in $\omega(\Gamma_P^+)$ because Q is repellent. We cannot have that $\omega(\Gamma_P^+) = \{0\}$. In order to "reach" 0, Γ_P^+ must intersect the interior of the horizontally shaded sector in Figure 18.14 and remain within the sector for all t large enough (if $\omega(\Gamma_P^+) = \{0\}$). However, if $(x(t), y(t))$ is in the interior of the sector, then $x(t) > 0$ and $dx(t)/dt > 0$, which means that we *cannot* have both $(x(t), y(t))$ remaining in the sector for all large enough t and $(x(t), y(t)) \to 0$ as $t \to \infty$. A similar argument (Exercise 13) shows that $\omega(\Gamma_P^+) \neq \{H_1\}$.

Thus if $\omega(\Gamma_P^+)$ contains any critical points at all, $\omega(\Gamma_P^+)$ must be a cycle graph CG. Q cannot belong to CG since Q is not the ω-limit set of any nonconstant orbit (cf., the definition of a cycle graph). If H_1 belongs to CG, then so must the orbit along E_2 with ω-limit set $\{H_1\}$ and α-limit set $\{0\}$, since that is the only orbit in R with ω-limit set $\{H_1\}$. By condition 3 of the definition of a cycle graph we must then have that $0 \in$ CG, but if $0 \in$ CG, the unbounded orbit Γ_1 also belongs to CG since it is the only orbit in the population quadrant whose ω-limit set is $\{0\}$. This is impossible since $\omega(\Gamma_P^+)$ is bounded because Γ_P^+ is bounded. Thus $\omega(\Gamma_P^+)$ contains no critical points. By the fivefold way we have that $\omega(\Gamma_P^+)$ is a cycle and Kolmogorov's cycle theorem is proved.   □

Remark. Observe that the above argument concerning Γ_P^+ holds for any point P on the edge of R and inside the population quadrant. Observe, also, that the proof that Γ_P^+ remains within R did *not* use the repellence of Q and thus would follow equally well from conditions 1–10 as originally formulated.

We have finally come up with a test for a cycle in a predator–prey system. See the Epilogue and Exercises 14 and 18 for systems which meet all the hypotheses of Kolmogorov's Theorem. It should be emphasized again that there may exist predator–prey systems without cycles (see [10]), but, of course, they are of little interest to us here.

We have confined our attention throughout to models using differential equations. Such equations instantaneously convert variations in the populations into changes in the rates of growth. It may be more realistic, however, to insert a time delay and use *differential-delay equations* such as

$$
\begin{cases}
\dfrac{dH(t)}{dt} = H(t)f(H(t - \alpha), L(t - \beta)) \\[2mm]
\dfrac{dL(t)}{dt} = L(t)g(H(t - \gamma), L(t - \delta)),
\end{cases}
$$

where the delays α, β, γ, and δ are positive constants. Essentially, this system asserts that the current rates are affected not only by the current populations, but also by what happened at some specific times in the past. Cycles can indeed be induced by using such equations [22].

8. Epilogue: Do Hares Eat the Lynx?

If the constants are suitably chosen, then the model

$$
\begin{cases}
\dfrac{dH}{dt} = H[\alpha^2 - \beta^2(H - \gamma^2)^2 - \delta^2 L] \\[2mm]
\dfrac{dL}{dt} = L[-a^2 + b^2 H - c^2 L]
\end{cases}
\tag{17}
$$

meets the conditions of the Kolmogorov cycle theorem (Exercise 14), and there is a cycle. (See Exercise 18 for another model.) We see from the form of dH/dt,

$$
\frac{dH}{dt} = (\alpha^2 - \beta^2\gamma^4)H + 2\beta^2\gamma^2 H^2 - \beta^2 H^3 - \delta^2 HL,
$$

that if $\alpha^2 > \beta^2\gamma^4 > 0$ and if $\delta > 0$, whenever L and H are sufficiently small, the hare population "explodes." What happens is that the quadratic term $2\beta^2\gamma^2 H^2$ dominates in this situation, and we have a very rapid growth in the hare population. Of course, once H or L is sufficiently large, then the negative rate terms predominate, the growth rate slackens and may even become negative.

Does (17) model the true state of affairs in the lynx–hare communities of the Canadian forests? Gilpin [15] attempted to fit a similar model to the actual data from the pelt records of Hudson's Bay Company. The "best-fit" choice of constants in his rate equations led to his replacing $-\delta^2$ and b^2 in (17) by δ^2 and $-b^2$, respectively, but this means that an encounter between a hare and a lynx increases the hare population and lowers the lynx population. Somehow the hare has eaten the lynx!

Gilpin suggests that this biological miracle might be accomplished by means of a disease carried by the hares which is fatal to lynx but not to the hares. Weinstein [28] says, though, that there is no evidence of such a disease. Weinstein thinks that the lynx cycle is imaginary and that the cycle indicated in the records of the Hudson's Bay Company reflects only changes in the hunting strategy of the trappers. In particular, he thinks that the trappers, (Canadian indians), are mainly interested in the hare as food. In years of hare plenty the trappers easily meet their own food needs and use their spare time trapping lynx for pocket money, but if the hares are scarce, the trappers have no time for the lynx, and consequently, the number of lynx pelts purchased by the Hudson's Bay Company plunges. This explanation seems somewhat far-fetched.

There are other difficulties. For instance, Nellis *et al*, [25] and Brand *et al*, [4] indicate that the lynx predation is probably not intensive enough to be the determinant factor in snowshoe hare mortality. Thus it is likely that something else is causing the hare cycle, but not the lynx. This would mean that the lynx-hare community is not closed.

Keith makes an even more telling argument against the existence of a closed lynx–hare community in his book [17]. He points out that other denizens of those northern woods also have a 10-year cycle: ruffed grouse, prairie grouse, Hungarian partridge, ptarmigan, spruce grouse, blue grouse, colored fox, mink, and fisher. Bulmer [7] adds to that list yet more species: wolverine, horned owl, coyote, wolf, skunk, marten, muskrat, and even the Atlantic salmon. The evidence of a cycle is scanty for some of these species, and the period is not always 10 years. Still, enough of a correlation exists both in the lengths of the periods and in the phases of the cycles (i.e., nearly simultaneous peaks and low points) of enough different species over the huge forest area to suggest that some external periodic driving force sets this whole ecosystem moving in its cyclic orbits. See [27] for a recent treatment of many of these questions.

Keith [17] feels that the evidence is not sufficient to single out any specific cause. Bulmer [7], [8], however, makes a convincing case that the snowshoe hare itself directly or indirectly causes all the other species to cycle. What is it that induces the 10-year cycle in the hare population? No one knows.

Exercises

1. Discuss the various explanations given in Sections 1, 2, 3, 4, and 8 for the cycles. Can you think of other possibilities?

2. One might be tempted to say that (4) can be written in the form of (3) by setting

$$F = x(F/x) \equiv xf \qquad G = y(G/y) \equiv yg.$$

   Why is this not acceptable? *Hint:* What might happen when $x = 0$ or $y = 0$?

3. a) Find solution formulas for (6). *Hint:* $d^2x/dt^2 = dy/dt = -x$. Solve the constant

coefficient equation, $d^2x/dt^2 + x = 0$ to get $x(t) = c_1 \cos t + c_2 \sin t$, where c_1 and c_2 are arbitrary constants. Then find y from $dx/dt = y$.

b) Use the results of part a to show that (6) has exactly one critical point, all other orbits being cycles. Show that 2π is the period of each cycle. (This is *not* typical for systems with a family of cycles. Usually the periods of distinct cycles differ.)

4. Solve $dr/dt = r(1 - r^2)$, $r \geq 0$. Hint: Find the general solution by writing $r^{-1}(1 - r^2)^{-1} dr = dt$ and using partial fractions or a table of integrals; assume $0 < r < 1$, or $r > 1$, and treat $r = 0$, 1 separately. Show that the unique cycle of (9) has period 2π.

5. Although the systems of Examples 1 and 2 do not represent interacting species, pretend that they do and discuss the orbits from an ecological point of view. For each discuss the effect of a sudden shock of short duration.

6. Let Γ_p^+ be a bounded positive semi-orbit, $\Gamma_p^+ \subset \Gamma$. Let $Q \in \Gamma$. Show that $\omega(\Gamma_Q^+) = \omega(\Gamma_p^+)$. Hint: Show that if $(x(t), y(t))$ is a solution of (4) for all t, $a < t < b$, then $(x_c(t), y_c(t))$ is also a solution for all t, $a + c < t < b + c$, if $x_c(t) = x(t - c)$, $y_c(t) = y(t - c)$. Show this by calculating $dx_c(t)/dt = dx(t - c)/dt = dx(t - c)/d(t - c) = F(x(t - c), y(t - c)) = F(x_c(t), y_c(t))$. Then show that $(x(t), y(t))$ and $(x_c(t), y_c(t))$ define the same orbit Γ.

7. Verify the limit sets of Example 4.

8. Verify the limit sets of Example 5.

9. Show that the 11 alternatives sketched in Figure 18.10 are the only ones possible for a bounded orbit.

10. a) Show that under conditions 1–10 exactly one internal critical point occurs in the population quadrant.
    b) Repeat a) if (15) satisfies the conditions 1, 2, 3a, 4–8, 9a, 10a.

11. Show that under conditions 1–10, the internal critical point Q is not repellent. Hint: Use (13) to show that $a < 0$ and $d < 0$, then use the repellence test.

12. Use the hypotheses of the Kolmogorov cycle theorem to verify the directions of the arrowheads on E_1 and E_2 in Figure 18.14.

13. In the proof of the Kolmogorov cycle theorem $\omega(\Gamma_p^+) \neq \{H_1\}$ is asserted. Prove it.

14. Let $f(x, y) \equiv \alpha^2 - \beta^2(x - \gamma^2)^2 - \delta^2 y$, $g(x, y) \equiv a^2 x - b^2 y - c^2$, where α, β, γ, δ, a, b, c are nonzero constants. Find values of these constants such that conditions 1, 2, 3a, 4–8, 9a, 10a, and 11 are satisfied.

15. Show that if R is a bounded region such that every orbit of (4) crossing the boundary of R enters R with increasing t and if R has no critical points of (4), then R contains at least one cycle.

16. a) The *D'Ancona–Volterra system* is $dx/dt = x(a - by)$, $dy/dt = y(-c + dx)$, where a, b, c, d are positive constants. Show that a unique critical point exists inside the population quadrant and that all the other orbits in the interior of that quadrant are cycles. Hint: Show that an unparameterized equation for the orbits is $y^a e^{-by} x^c e^{-dx} = \text{const}$. Obtain this by separating variables in $dy/dx = y(-c + dx)x^{-1}(a - by)^{-1}$, integrating, and then taking an exponential. To show that the orbits are cycles is harder [5], [6].

b) Show that the system of a) can be interpreted as a predator–prey system with x as the prey species and y as the predator, but with no overcrowding effect. Show that condition 4 fails and the Kolmogorov cycle theorem does not apply.

17. The fivefold way may not apply if a bounded region contains infinitely many critical points. Show that the system $dx/dt = (x + y)(1 - x^2 - y^2)$, $dy/dt = (-x + y)(1 - x^2 - y^2)$ has a circle C of critical points and that every other solution, except for the critical point at the origin, has C as its omega limit set.

18. The model $dH/dt = r(1 - (H/K))H - c(1 - e^{-aH/c})L$, $dL/dt = bc(1 - e^{-aH/c})L - dL$ represents a predator–prey interaction in which the prey H has a natural logistic growth rate, $r(1 - H/K)H$, with carrying capacity K. The predator–prey interaction is characterized by a *predator satiation* term, $c(1 - e^{-aH/c})L$, where c is the maximum rate of prey capture per predator. The parameter a is a measure of how easily a predator is satiated, while the factor b in the equation for dL/dt is an "efficiency factor" for the conversion of prey into improved predator growth.

a) Let $K = 3,500$, $r = 0.5$, $a = 0.01$, $b = 0.02$, $d = 0.1$, and $c = 10$. Show that the hypotheses of the Kolmogorov Cycle Theorem are satisfied and that there is a limit cycle.

b) Discuss the meaning of the predator satiation term. Why would you expect a limit cycle in this case?

References

[1] F. Albrecht, H. Gatzke, A. Haddad and N. Wax, "The dynamics of two interacting populations," *J. Math. Anal. Applic.*, vol. 46, pp. 658–670, 1974. This is the definitive mathematical treatment of systems of the form of (15) under various pertinent hypotheses of f and g. Contains almost none of the biological or ecological interpretations. A brief bibliography is included.

[2] W. Baltensweiler, "The cyclic population dynamics of the Grey Larch Tortix, *Zeiraphera Griseana* Hübner," in *Insect Abundance*, edited by T.R.E. Southwood, Ed. Oxford: Blackwell, pp. 88–97.

[3] W. E. Boyce and R. C. DiPrima, *Elementary Differential Equations and Boundary Value Problems*, 2nd ed. New York: Wiley, 1969. A very good book at the sophomore–junior level, with applications.

[4] C. J. Brand, L. B. Keith and C. A. Fischer, "Lynx responses to changing snowshoe hare densities in central Alberta," *J. Wildlife Management*, 40 (3), vol. 40, no. 3, pp. 416–428, 1976. Continuation over the winters of 1971–1975 of field studies begun earlier [25]. Field data shows the complexities of the lynx–hare interaction.

[5] M. Braun, *Differential Equations and Their Applications*, 2nd ed. New York: Springer-Verlag, 1978. This is one of the best books on the sophomore level; the presentation of the applications is especially good.

[6] ——, "Why the percentage of sharks caught in the Mediterranean Sea rose dramatically during World War I," this volume, ch. 15. This is a slight alteration of [5, Section 4.9] and contains a good exposition of the Lotka–Volterra model.

[7] M. G. Bulmer, "A statistical analysis of the 10-year cycle in Canada," *J. Anim. Ecol.*, vol. 43, pp. 701–718, 1971. A very interesting statistical analysis of the lynx–hare and several other 10-year cycles in the Canadian forests. Discusses 8-year cycles in the Siberian taiga. A very good bibliography.

[8] ——, "The theory of prey–predator oscillations," *Theoret. Pop. Bio.*, vol. 9, pp. 137–150, 1976. Further arguments that the hare cycle [cause unknown] drives the other cycles in the Canadian woods. Discussion of Kolmogorov's Theorem.

[9] E. A. Coddington and N. Levinson, *Theory of Ordinary Differential Equations.* New York: McGraw-Hill, 1955. Remains one of the standard graduate level texts.

[10] C. S. Coleman, "Quadratic population models: Almost never any cycles," this volume, ch. 16. Shows that if f and g in (15) are linear, then only "ecologically rare" cases such as the Lotka–Volterra system of Exercise 16 will contain any cycles at all.

[11] U. D'Ancona, *The Struggle for Existence* Leiden, Brill, 1954. An interesting and readable account by the man who brought the strange increase in the numbers of selachia (sharks, skates, etc.) in the fishing areas of the Meditterranean to the attention of Volterra, who then formulated the first mathematical models of predator–prey species (see Exercise 16).

[12] C. Elton and M. Nicholson, "The ten-year cycle in numbers of lynx in Canada," *J. Anim. Ecol.*, vol. 11 pp. 215–244, 1942. Contains the data from the records of the Hudson's Bay Company. A good bibliography of the earlier sources.

[13] J. C. Frauenthal, *Introduction to Population Modeling.* Birkhäuser, 1980. An outstanding treatment of population modeling. Treats stochastic and time-delay models as well as the deterministic and instantaneous. (In the UMAP Expository Monograph Series.)

[14] G. F. Gause, *The Struggle for Existence.* Williams and Wilkins, 1934 (also Dover, 1971, paperback). A fascinating account of predator–prey and competing species relationships.

[15] M. E. Gilpin, "Do hares eat lynx?," *Amer. Naturalist*, vol. 107, pp. 727–730, 1973. Gilpin uses the actual data from Elton and Nicholson to find the best-fitting coefficients in his polynomial model and "proves" that the hare must be eating the lynx.

[16] M. Hirsch and S. Smale, *Differential Equations, Dynamical Systems, and Linear Algebra.* New York: Academic, 1974. A very good treatment from a modern point of view, junior–senior level. Contains a good treatment of population models.

[17] L. B. Keith, *Wildlife's Ten-Year Cycle.* Madison: Univ. of Wisconsin, 1963. A thought-provoking account on a nonmathematical level of the various natural cycles in the vast Canadian forests. Should be read by anyone interested in natural cycles.

[18] A. Kolmogorov, "Sulla Teoria di Volterra della Lotta per l'Esistenza," *G. Ist. Ital. Attuari*, vol. 7, pp. 74–80, 1936. Certainly one of the more inaccessible sources. See May [23], Bulmer [8], or Albrecht *et al.* [1] for discussions of Kolmogorov's Theorem.

[19] D. L. Lack, *The Natural Regulation of Animal Numbers.* Oxford: Clarendon, 1954, pp. 212–217.

[20] E. G. Leigh, "The ecological role of Volterra's equations," in *Lectures on Mathematics in the Life Sciences*, M. Gerstenhaber, Ed. Amer. Math. Soc., 1968, pp. 1–61. Gives a brief history of the equations and then presents Kerner's statistical approach.

[21] D. A. MacLulich, "Fluctuations in numbers of the varying hare (*Lepus Americanus*)," *Univ of Toronto Studies, Biol. Ser.*, no. 43, pp. 1–136, 1937. A basic source for the snowshoe hare cycle.

[22] R. M. May, *Stability and Complexity in Model Ecosystems.* Princeton, NJ: Princeton Univ. Press, 1973. A fascinating and readable account. Excellent bibliography.

[23]   ——, "Limit cycles in predator–prey communities," *Science*, vol. 177, pp. 900–902, 1972. A discussion of Kolmogorov's theorem. A good bibliography.

[24]   J. Maynard Smith, *Models in Ecology*.   Cambridge, 1974. Comparable to May [22]. Very good discussion of models. Good bibliography.

[25]   C. H. Nellis, S. P. Wetmore and L. B. Keith, "Lynx–prey interactions in central Alberta," *J. Wildlife Management*, vol. 36, pp. 320–329, 1972. An analysis of the lynx population in a region in Alberta, Canada. The study was quite thorough and was carried out over the winters of 1964–1968. Peripheral studies of the local hare and grouse populations were also made. See also Brand *et al.* [4].

[26]   M. L. Rosenzweig, "Why the prey curve has a hump," *Amer. Natur.*, vol. 103, pp. 81–87, 1969. An indication that the prey isocline may rise before it falls; argument is based on the data from an actual predator-prey system.

[27]   J. Roughgarden, *Theory of Population Genetics and Evolutionary Ecology: An Introduction*.   New York: Macmillan, 1979. A fascinating introduction to biological models. The first third of the book requires only a calculus background. The later chapters take the reader to the edge of current research. Limit cycles, Kolmogorov's theorem, stochastic models, and much more are all discussed in informal terms.

[28]   M.S. Weinstein, "Hares, lynx, and trappers," *Amer. Naturalist*, vol. 111, pp. 806–808, 1977. Weinstein claims that the cycles are not real but only reflect the trapping strategies of the hunters.

Notes for the Instructor

Objectives. The aims of the model are to summarize current thinking on the lynx–hare cycle, to develop mathematical models of a general sort for predator–prey interactions, and to convey some idea of the incompleteness of any mathematical model. Mathematically, the module introduces limit sets of orbits and the five alternatives for the positive limit of a bounded orbit. The examples and problems provide some practice in separating variables in addition to introducing the reader to limit sets. If the students have had advanced calculus, the instructor may want to prove that the limit set of a bounded orbit is bounded, closed, connected and invariant [16].

Prerequisites. Separation of variables, some mathematical maturity, experience with phase plane methods. It would be helpful if the reader has seen an existence and uniqueness theorem and the D'Ancona–Volterra (Lotka–Volterra) model [6].

Time. The material can be covered in three or four lectures with some material left for the students to read.

CHAPTER 19

Hilbert's 16th Problem: How Many Cycles?

Courtney S. Coleman*

> "... the aeolian harp, a pneumatic hammer, the scratching noise of a knife on a plate, the waving of a flag in the wind, the humming noise sometimes made by a water-tap, the squeaking of a door, the tetrode multivibrator, ..., the intermittent discharge of a condensor through a neon tube, the periodic re-occurrence of epidemics and of economical crises, the periodic density of an even number of species of animals living together and the one species serving as food for the other, the sleeping of flowers, the periodic recurrence of showers behind a depression, the shivering from cold, menstruation, and finally, the beating of the heart."[1]

1. Introduction

In the above quotation Balthazar van der Pol has grouped the most varied phenomena into a single category of systems which exhibit periodic oscillations, or cycles, even though no "external periodic forces" cause such oscillations. (One wonders, however, about some of van der Pol's systems; e.g., the sleeping of flowers is surely due to the diurnal cycle of darkness and light.) According to van der Pol, most of these systems can be modeled more or less accurately by a pair of differential equations of the form,

$$\begin{cases} \dfrac{dx}{dt} = X(x, y) \\[2mm] \dfrac{dy}{dt} = Y(x, y). \end{cases} \qquad (1)$$

* Department of Mathematics, Harvey Mudd College, Claremont, CA 91711.

[1] B. van der Pol, *Phil. Mag.*, vol. 6, ser. 7, pp. 763–765, 1928.

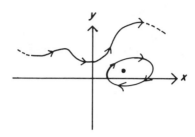

Figure 19.1. Three Kinds of Orbits of (1).

The symbols x and y denote the magnitude of the principal quantities being modeled. For example, x might denote the voltage across the condensor (capacitor) of a neon tube and y the rate at which that voltage changes. Or, as we shall see in Section 4, x and y might denote the population densities of a predator and its prey. For the present, however, let us simply assume that the differential system (1) is given and that we are interested in any cycles which may be present in its solution set. (See Section 2 for the definition of a cycle. See Exercise 1 for an example of a system with an "external periodic force.")

Specifically, let us suppose that $X(x, y)$ and $Y(x, y)$ are continuously differentiable in both variables x and y for all x and all y. A pair of functions $x = x(t)$ and $y = y(t)$ is a *solution* of (1) if the following conditions hold.

(1) The functions $x(t)$ and $y(t)$ are defined on a common open interval $I: a < t < b$; a may be $-\infty$ and b may be $+\infty$.
(2) The functions $x(t)$ and $y(t)$ are continuously differentiable on I and satisfy the system (1) for every $t \in I$:

$$\begin{cases} \dfrac{dx(t)}{dt} = X(x(t), y(t)) \\[2mm] \dfrac{dy(t)}{dt} = Y(x(t), y(t)). \end{cases}$$

(3) No functions $x^*(t)$ and $y^*(t)$ are defined on an interval I^*, where $I^* \supset I$ but $I^* \neq I$, which coincide with $x(t)$ and $y(t)$, respectively, on I and satisfy system (1) for all t in I^*.

The last condition means that we only look at solutions of (1) which are *maximally extended* in time. For example, the solution $x(t) = e^t$, $-\infty < t < \infty$, of $dx/dt = x$ is acceptable, while the "solution" $\tilde{x}(t) = e^t$, $1 < t < 2$, is not since we can clearly extend the time interval.

Each solution $x = x(t)$, $y = y(t)$, of (1) defines a curve Γ in the xy plane called the *orbit* of the solution. We shall assume without proof that

(i) *no two orbits intersect unless they coincide*,
(ii) *an orbit is either a non-selfintersecting curve, a simple closed curve, or a single point (a critical point).*

See Hirsch and Smale [13] for a proof of a similar "existence and unique-ness" theorem. See Figure 19.1 for examples of the three types of orbits listed in (ii). The arrow heads indicate how the orbit is traced out with increasing t.

Distinct solutions may generate the same orbit. For example, the system,

$$\frac{dx}{dt} = y \qquad \frac{dy}{dt} = -x$$

has the pair of solutions,

$$x_1(t) = \cos t, \qquad y_1(t) = -\sin t,$$
$$x_2(t) = \cos(t - \pi/3), y_2(t) = -\sin(t - \pi/3),$$

which define the same orbit Γ, the unit circle $x^2 + y^2 = 1$.

2. Cycles

A nonconstant solution, $x = x(t)$, $y = y(t)$, of (1) is a *cycle* or *periodic solution* if a positive number T (a *period* of the cycle) exists such that

$$x(t + T) = x(t) \qquad \text{and} \qquad y(t + T) = y(t)$$

for all t. We shall also call the corresponding orbit in the xy plane a *cycle*. It is clearly a simple closed curve. Conversely, any orbit which is a simple closed curve corresponds to a periodic solution. (This seems obvious, and we shall omit the formal proof. See Hirsch and Smale [13].) We shall now give several examples of systems of the form of (1) which possess cycles. It should be emphasized once more that any cycles which do appear are internally generated by the interactions of the "x substance" and the "y substance." They cannot be due to some external, periodically varying source since the right-hand sides of (1) have no explicit dependence upon time at all, much less a periodic dependence (see Exercise 1).

EXAMPLE 1 (Infinitely Many Cycles). The system

$$\frac{dx}{dt} = y \qquad \frac{dy}{dt} = -x \qquad (2)$$

is not hard to solve. For example, if we differentiate the first equation and use the second, we have that

$$\frac{d^2x}{dt^2} = \frac{dy}{dt} = -x,$$

or

$$\frac{d^2x}{dt^2} + x = 0. \qquad (3)$$

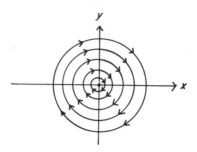

Figure 19.2. Cycles of $dx/dt = y$, $dy/dt = -x$

Equation (3) is a second-order constant coefficient ordinary differential equation whose general solution is given by

$$x(t) = c_1 \cos t + c_2 \sin t, \tag{4}$$

where c_1 and c_2 are arbitrary constants. Using the trigonometric identity,

$$a \cos \theta + b \sin \theta = A \cos(\theta - \delta),$$

where

$$A = (a^2 + b^2)^{1/2}, \qquad \cos \delta = a/A, \qquad \sin \delta = b/A,$$

we can rewrite (4) in the simpler form

$$x(t) = A \cos(t - \delta), \tag{5}$$

where A and δ are arbitrary constants. Using the second equation of (2) we then have that every solution of (2) has the form

$$\begin{cases} x = x(t) = A \cos(t - \delta) \\ y = y(t) = -A \sin(t - \delta) \end{cases} \tag{6}$$

for some constants A and δ. If $A \neq 0$, the corresponding orbit is a circle of radius $|A|$ centered at the origin (observe that $x^2(t) + y^2(t) = A^2$). See Figure 19.2.

If $A = 0$, we have the *critical point* $(0, 0)$, which is also a solution of (2). Observe that all orbits of (2), except for the single critical point, are cycles. The arrowheads on the orbits in Figure 19.2 indicate as before how the orbit is traced out with advancing time.

A *critical point* of (1) is any point (x_0, y_0) for which $X(x_0, y_0) = 0$ and $Y(x_0, y_0) = 0$. Such a point defines a solution of (1) since

$$\frac{dx_0}{dt} \equiv 0 = X(x_0, y_0) \qquad \frac{dy_0}{dt} \equiv 0 = Y(x_0, y_0).$$

Critical points play an important role in the search for cycles of (1) since inside *every cycle is at least one critical point*. (The proof of this fact is rather difficult. See Hirsch and Smale [13].) The system of Example 1 has a critical point—indeed it must have one, since it has cycles—as do most of the following examples.

EXAMPLE 2 (A Single Cycle). Let r and θ be the usual polar coordinates,

$$x = r\cos\theta \qquad y = r\sin\theta, \tag{7}$$

and consider the polar coordinate system

$$\begin{cases} \dfrac{dr}{dt} = r(1 - r^2), & r \geq 0 \tag{8a} \\[2ex] \dfrac{d\theta}{dt} = 1. \tag{8b} \end{cases}$$

First, let us express (7) in rectangular xy coordinates. We have that from (7)

$$\begin{cases} \dfrac{dx}{dt} = \dfrac{dr}{dt}\cos\theta - r\sin\theta\dfrac{d\theta}{dt} = (1 - r^2)r\cos\theta - r\sin\theta \\[2ex] \qquad = (1 - x^2 - y^2)x - y \tag{9} \\[2ex] \dfrac{dy}{dt} = \dfrac{dr}{dt}\sin\theta + r\cos\theta\dfrac{d\theta}{dt} = (1 - r^2)r\sin\theta + r\cos\theta \\[2ex] \qquad = (1 - x^2 - y^2)y + x, \end{cases}$$

where we have also used the chain rule, e.g., $(d/dt)(\cos\theta) = (d/d\theta)(\cos\theta)$ $(d\theta/dt)$. Observe that any solution $(r(t),\ \theta(t))$ of (8) gives a solution $(x(t),\ y(t))$ of (9). Conversely, a solution of (9) defines a solution of (8) if we use

$$r = (x^2 + y^2)^{1/2} \qquad \tan\theta = y/x. \tag{10}$$

Now (8) can be solved more easily than (9). In fact, we have

$$r(t) = \frac{r_0}{[r_0^2 + (1 - r_0^2)e^{-2t}]^{1/2}}, \qquad r_0 \geq 0, \tag{11a}$$

$$\theta(t) = t + \theta_0, \qquad \theta_0 \text{ arbitrary.} \tag{11b}$$

The solution for $r(t)$ is left to the reader (Exercise 2) and is obtained by separating variables in (8a). The equation for $\theta(t)$ follows immediately from (8b). Observe that we have assumed that at $t = 0$, $r = r_0$ and $\theta = \theta_0$.

We have that if $r_0 = 1$, then $r(t) = 1$, while if $r_0 = 0$, then $r(t) = 0$. If $0 < r_0 < 1$, then from (11a) we see that $r(t)$ increases monotonically to 1 as $t \to \infty$, while if $1 < r_0$, then $r(t)$ decreases monotonically to 1 as $t \to \infty$. In rectangular coordinates, $r(t) \equiv 0$ corresponds to a critical point at $x = 0$, $y = 0$, while $r(t) \equiv 1$, $\theta(t) = t + \theta_0$, corresponds to

$$x(t) = \cos(t + \theta_0) \qquad y(t) = \sin(t + \theta_0),$$

which are parametric equations for the unit circle.

Thus system (9) has a unique critical point at the origin and a unique

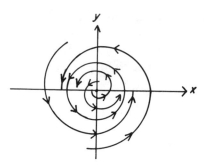

Figure 19.3. Single Cycle

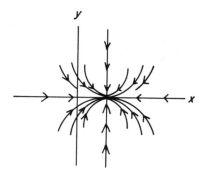

Figure 19.4. No Cycles at All

cycle along the unit circle. See Figure 19.3 for a portrait of the orbits. System (9) is said to be a *cubic system* since the right-hand sides are cubic polynomials in x and y. Observe also that the cycle is *asymptotically stable* in that all "nearby" orbits approach the cycle as time increases. In fact, all orbits, except for the critical point, approach the cycle in this example.

EXAMPLE 3 (No Cycles). The system

$$\frac{dx}{dt} = -x + 1 \qquad \frac{dy}{dt} = -2y \qquad (12)$$

is easy to solve by separating the variables (Exercise 3):

$$x(t) = 1 - (1 - x_0)e^{-t} \qquad y(t) = y_0 e^{-2t}.$$

No cycles at all occur, and a single critical point exists at $(1, 0)$. We leave it to the reader to show (Exercise 3) that the portrait of the orbits is as given in Figure 19.4.

Each cycle of Example 1 differs from the cycle of Example 2 in that the latter cycle is "isolated" from other cycles of the same system—indeed, no other cycles exist. More precisely, a cycle Γ of (1) is *isolated* if there is a positive number a such that the *annulus A centered on* Γ defined by

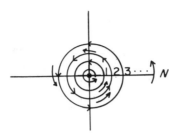

Figure 19.5. N Isolated Cycles

$$A = \{(x, y): \text{distance from } (x, y) \text{ to } \Gamma \text{ is less than } a\}$$

contains no other cycles. In this definition, by the distance from (x, y) to Γ we mean

$$\min_{(u, v) \in \Gamma} ((x - u)^2 + (y - v)^2)^{1/2}.$$

It can be shown that this minimum always exists and is positive unless (x, y) lies on Γ. A cycle Γ is *nonisolated* if every annulus centered on Γ contains at least one other cycle. From these definitions it is clear that each cycle of Example 1 is nonisolated, while the cycle of Example 2 is isolated.

EXAMPLE 4 (N Isolated Cycles). The polar coordinate system

$$\begin{cases} \dfrac{dr}{dt} = r(1 - r^2)(4 - r^2) \cdots (N^2 - r^2), r \geq 0, N \text{ a positive integer,} \\ \dfrac{d\theta}{dt} - 1, \end{cases} \tag{13}$$

corresponds to a rectangular coordinate system of the form,

$$\begin{cases} \dfrac{dx}{dt} = X(x, y) \\ \dfrac{dy}{dt} = Y(x, y), \end{cases} \tag{14}$$

where $X(x, y)$ and $Y(x, y)$ are polynomials in x and y of degree $2N + 1$ (Exercise 4). The system has a unique critical point at the origin, N isolated cycles coinciding with the circles,

$$r \equiv 1, r \equiv 2, \cdots, r \equiv N,$$

and no nonisolated cycles. The reader should prove these facts (Exercise 4) and should verify the validity of the corresponding portrait orbits in Figure 19.5.

These examples show that cycles do exist or at least that it is not too

difficult to construct systems of the form of (1) which possess cycles.[2] The
question is, given a system of the form of (1), how many isolated cycles does
it have? (A system with a nonisolated cycle must have infinitely many
cycles—see Exercise 5.) This leads us to Hilbert's 16th problem.

3. Hilbert's 16th Problem

The following problem was posed by David Hilbert in 1900 and is still
unsolved.

> "... *I wish to bring forward a question which, it seems to me, may be attacked
> by the ... method of continuous variation of coefficients, and whose answer is of
> ... value for the topology of families of curves defined by differential equations.
> This is the question as to the maximum number and position of Poincaré's boundary
> cycles (limit cycles) for a differential equation of the first order and degree of the
> form*
>
> $$\frac{dy}{dx} = \frac{Y}{X},$$
>
> *where X and Y are rational integral functions of the nth degree in x and y ...*"

"Rational integral functions" are polynomials,

$$\sum_{i+j=0}^{n} a_{ij}x^{i}y^{j} \equiv a_{00} + a_{10}x + a_{01}y + a_{20}x^{2} + a_{11}xy + \cdots + a_{0n}y^{n}.$$

(15)

"Poincare's boundary cycles (limit cycles)" are our isolated cycles.[3] Finally,
we treat (1) rather than $dy/dx = Y/X$, which is obtained from (1) by division.

The problem is the second part of the 16th of 23 problems posed by
Hilbert at the Second International Congress of Mathematicians, Paris,
1900. In his opening remarks at the Congress, Hilbert said:

> "... *History teaches the continuity of the development of science. We know that
> every age has its own problems, which the following age either solves or casts
> aside as profitless and replaces by new ones. If we would obtain an idea of the
> probable development of mathematical knowledge in the immediate future, we
> must let the unsettled questions pass before our minds and look over the problems
> which the science of today sets and whose solution we expect from the future. To
> such a review of problems the present day, lying at the meeting of the centuries,
> seems to me well adapted. For the close of a great epoch not only invites us to look
> back into the past but also directs our thoughts to the unknown future.*"

---

[2] The cycles in the examples are circles, but this need not be the case. Indeed, the shape of a
cycle can be very complicated—unless X and Y are quadratic polynomials in x and y, in which
case the cycle must bound a convex region. Also, inside each cycle in the examples is only one
critical point. Again, this is not generally true except in the quadratic case.

[3] Poincare (and probably Hilbert) included cycle graphs among his limit cycles, but we shall
not. A cycle graph (singular or separatrix cycle) is a closed curve which is a coherently oriented
union of critical points and orbits. See the previous chapter.

Some of Hilbert's problems were solved within a few years, some just recently (no. 10 by the 20-year old Soviet mathematician Matiyaśević in 1970), and a few remain unsolved. See [12a]–[12c] for a list of sources on Hilbert's problems and their current status.

Returning to the question of the maximal number of isolated cycles for polynomial systems as posed by Hilbert in his 16th problem, let us first define $\mathscr{P}_n$ to be the set of all real polynomials of degree no larger than n. Thus $\mathscr{P}_{n-1} \subset \mathscr{P}_{n'}$ for $n = 1, 2, \cdots$, since the coefficients of all terms of exact degree n of a polynomial of form (15) may vanish. In 1923 Dulac [11] proved that, given a specific n and X and Y in $\mathscr{P}_{n'}$, system (1) has a finite number K of isolated cycles, K depending upon the choice of X, Y, and n. The question of just how large K can be for each fixed n leads us to define the Hilbert numbers $H(n)$, $n = 0, 1, 2, \cdots$ as follows:

$$H(N) = \underset{X,\,Y \in \mathscr{P}_n}{\text{supremum}} \{\text{number of isolated cycles of system (1)}\}.$$

Thus a finite $H(n)$ for a particular n means that polynomials X and Y exist in $\mathscr{P}_n$ for which system (1) has exactly $H(n)$ isolated cycles. On the other hand, $H(n) = \infty$ for a particular n means that, for every integer K, polynomials X and Y exist in $\mathscr{P}_n$ for which (1) has at least K isolated cycles. This does not contradict Dulac's result. (Why not?)

Since $\mathscr{P}_{n-1} \subset \mathscr{P}_n$, it follows that $H(n-1) \leq H(n)$. From this and the result given in Example 4 for odd integers n, we conclude that for all integers $n = 0, 1, 2, \cdots$

$$H(n) \geq \left[\frac{n-1}{2}\right],$$

where $[a]$ denotes the largest integer not larger than a. Hence $H(n) \to \infty$ as $n \to \infty$, and $H(n) = \infty$ for $n \geq n_0$ if $H(n_0) = \infty$.

It is not hard to show that $H(0) = H(1) = 0$ (Exercise 7). The first possibility for a nontrivial Hilbert number [i.e., $H \neq 0, \infty$] is $H(2)$. The following example of Chin [6] gives a quadratic system, i.e., $n = 2$, with an isolated cycle, which implies that $H(2) \geq 1$:

$$\begin{cases} \dfrac{dx}{dt} = -1 - 2y + x^2 + xy + y^2 \\[2mm] \dfrac{dy}{dt} = 2x - x^2. \end{cases}$$

The reader may verify that $x^2 + y^2 = 1$ defines a cycle for this system (Exercise 8), but the proof that this cycle is isolated is omitted. In 1939 Bautin announced that $H(2) \geq 3$, but the publication of the proof was delayed by the rigours of World War II until 1952[3]. Bautin's constructions led to the feeling that, in fact, $H(2)$ is exactly three and hence that quadratic systems have at most three isolated cycles.

In a pair of deep papers in the mid-1950's Petrovskii and Landis [21]–

[23] "proved" that $H(2) = 3$ and that for $n > 2$,

$$H(n) \leq \begin{cases} \frac{1}{2}(6n - 7n^2 - 11n + 6), & \text{if } n \text{ is odd}; \\ \frac{1}{2}(6n^3 - 7n^2 + n + 4), & \text{if } n \text{ is even}. \end{cases} \tag{16}$$

This appeared to be a major contribution to the solution of the problem, the only thing remaining being to see if these upper bounds could be lowered to determine $H(n)$ itself, $n > 2$. Unfortunately, the whole question was reopened in 1967 when S. P. Novikov found a mistake in the proof of a crucial lemma and both papers had to be withdrawn [24].

It was still hoped that $H(2)$ would turn out to be three even though after 1967 it was apparent that the problem was far from simple. In 1979 that hope was shattered by Shi [25] and by Chen and Wang [5]. Using results and techniques of Chin [6], [7], Tung [28], and Yeh [31], [32], they found quadratic systems with at least four isolated cycles; hence we have that

$$H(2) \geq 4.$$

These systems are as follows:

$$\text{(Chen, Wang)} \quad \begin{cases} \dfrac{dx}{dy} = -\delta_2 x - y - 3x^2 + (1 - \delta_1)xy + y^2 \\ \dfrac{dy}{dt} = x + \dfrac{2}{9}x^2 - 3xy, \end{cases}$$

where δ_1 and δ_2 are small, positive numbers;

$$\text{(Shi)} \quad \begin{cases} \dfrac{dx}{dt} = \lambda x - y - 10x^2 + (5 + \delta)xy + y^2 \\ \dfrac{dy}{dt} = x + x^2 + (-25 + 8\varepsilon - 9\delta)xy, \end{cases}$$

where $\lambda = -10^{-200}$, $\delta = -10^{-13}$, $\varepsilon = -10^{-52}$. The relative location of the four known isolated cycles is the same for each system, three cycles enclosing the critical point $(0, 0)$ and a fourth cycle enclosing the critical point $(0, 1)$. See Figure 19.6. Whether these systems have any other isolated cycles is not yet known. None of these results will be proved here. The reader who wishes to go more deeply into all this should first become familiar with the Poincaré–Bendixson theorem (see Hirsch and Smale [13]). Detailed information on quadratic systems can be obtained from the papers by Chin, Tung, and Yeh given in the references. Coppel [10] summarizes most of the known results, including the following:

(1) Each cycle of a quadratic system bounds a convex region containing a single critical point.
(2) In a quadratic system cycles enclosing a common critical point have a common orientation (as t increases), while those enclosing distinct critical points are oppositely oriented.

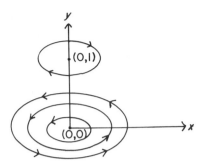

Figure 19.6. Four Isolated Cycles for Quadratic Systems of [5], [25]

Thus the first solves the problem of the possible shape of a cycle of a quadratic system, and the second implies that a quadratic system has at most two eyes (i.e., at most two critical points, each of which is enclosed by one or more isolated cycles). The examples of Chen and Wang and of Shi, in fact, do have two eyes. Thus for quadratic systems the possible shapes and relative locations of cycles are now known, but we still do not know $H(2)$, not even whether or not it is finite.

For $n > 2$, the situation is no clearer. Sibirskii [27] and Shi [25] have shown that $H(3) \geq 5$. For $n \geq 6$ Otrokov [20] has obtained a significantly better lower bound for $H(n)$ than the $[(n-1)/2]$ derived from Example 4. Otrokov's bound is

$$H(n) \geq \begin{cases} \frac{1}{2}(n^2 + 5n - 14), & n \text{ even,} \\ \frac{1}{2}(n^2 + 5n - 26), & n \text{ odd.} \end{cases}$$

In his derivation Otrokov constructs (implicitly) a polynomial system with its known isolated cycles forming a single eye. On the other hand, Il'jašenko [14] has shown that, for any n, X and Y exist in $\mathscr{P}_n$ for which system (1) has at least $(1/2)(n-1)(n+2)$ eyes with disjoint interiors. For $n > 2$ the maximal number of eyes is not known, nor is it known just which complex patterns of eyes within eyes, or eyes enclosing more than a single critical point, can exist.

4. An Ecological Interpretation

Let us return to the physical phenomena mentioned by van der Pol at the beginning. In particular, let $x(t)$ and $y(t)$ denote the respective population densities (i.e., numbers per unit area, or volume, of habitat) of two interacting species. For fixed time t, $x(t)$ and $y(t)$ are assumed to be the same for every subregion of the habitat; i.e., the distribution is assumed uniform. One species may be the predator feeding on the other, or both may be competing for a common food source in limited supply, or some other more exotic form of interaction between the two species may be going on.

The ratios,

$$\frac{dx(t)/dt}{x(t)} \quad \text{and} \quad \frac{dy(t)/dt}{y(t)},$$

are the *relative rates of change* of the two species. Each of these relative rates is the difference between the *relative birth rate B* and the *relative mortality rate M* of the species:

$$\frac{1}{x}\frac{dx}{dt} = B_1(x, y) - M_1(x, y) \equiv f(x, y)$$

(17)

$$\frac{1}{y}\frac{dy}{dt} = B_2(x, y) - M_2(x, y) \equiv g(x, y).$$

Observe that we have assumed that the relative rates depend upon the population densities but not explicitly upon time. This assumption appears approximately correct in a number of real systems of interactions. As before, any cycles of (17) must be internally produced and not due to external periodic forces. We made the same assumption, of course, in Section 1. We assume, as before, that $x(t)$, $y(t)$, $f(x, y)$, and $g(x, y)$ are continuously differentiable functions of their variables.

A good deal of controversy exists over the precise nature of the functions $f(x, y)$ and $g(x, y)$ and, of course, the form depends upon the nature of the interaction. For example, suppose that the y species is the principal predator of the food species x. One might expect that $f(x, 0) > 0$ in this case since in the absence of the predator the food species could be expected to have a positive relative growth rate. On the other hand, we might have that $f(x, 0) < 0$ if x is sufficiently large; this would account for the deleterious effects of overcrowding in the habitat even when no predator is about. It also seems reasonable to suppose in this case that, for each $x, f(x, y)$ is a decreasing function of y since the y species preys on the x species. Similar arguments apply to the relative rate function g.

In any case, the basic rate equations for a pair of interacting species are taken to be

$$\begin{cases} \dfrac{dx}{dt} = xf(x, y) \\[2mm] \dfrac{dy}{dt} = yg(x, y). \end{cases}$$

(18)

Given the interpretation of x and y as population densities, we are only interested in solutions of (18) in the quadrant $x \geq 0$, $y \geq 0$. Given the topic of this module, *the basic question is whether or not systems of two interacting*

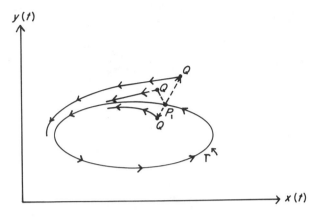

Figure 19.7. Ecologically Stable Cycle

species exist which can be modeled by a differential system of the form of (18) *and which possess one or more isolated cycles.*[4]

Let us outline the argument that the only cycles at all likely in a "natural" setting must be isolated and, in fact, *ecologically stable*, by which is meant the following. A "natural" cycle persevering over a long time span must be somewhat insensitive to the inevitable shocks and disturbances of the "real world." If a disturbance at time t_1 suddenly knocks a population density point $P_1 = (x_1, y_1)$ off such a cycle Γ and onto a nearby point Q and if the disturbance promptly ceases, then the orbit through Q should move back towards Γ as time goes on. Such a cycle is *ecologically stable* (see Figure 19.7). The mathematical term is asymptotic orbital stability, but we shall not give the formal definition here. It seems certain that only ecologically stable cycles can persist in nature.

An ecologically stable cycle Γ must be isolated. If it were not isolated, a sudden shock might drive a population density point P on Γ into a point Q^* on a different cycle, with a different set of amplitudes and, possibly, a different period. Once on the new cycle, the population density remains there unless some other shock sends the density point onto yet another cycle. In any event, we do not have the gradual return to the original cycle Γ. Hence Γ is not ecologically stable if it is nonisolated. See Figure 19.8. (Warning: An isolated cycle need not be ecologically stable (see Exercise 9).)

Several natural predator–prey communities have been studied, each of which possesses an ecologically stable cycle. The Canadian lynx–snowshoe hare system has maintained its 10-year cycle with astonishing regularity for at least the last two centuries [9], [15], [17], [19]. The curious bud moth–

[4] If (18) has isolated cycles and if f and g are both polynomials, the degree of f or g must be at least two [29], [8].

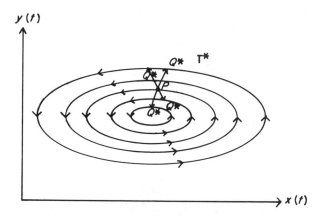

Figure 19.8. Nonisolated Cycle Is Not Ecologically Stable.

larch tree cycle in the Alps [2] and the lemming–vegetation cycle [16] both seem to be well-established. Other cycles have been established in the laboratory. In each of these cases the cycle appears to be internally generated by the interactions of the species themselves and is not the result of an external periodic source such as a weather cycle. See Keith and Bulmer [10], [4] for detailed arguments concerning the lynx–hare cycle; see also [9]. Most of the systems mentioned by van der Pol also possess isolated and ecologically stable cycles.

Each of these systems is also widely believed to be suitable for modeling by equations of the form of (18).[5] We shall not discuss possible forms for f and g here. See [17], [19], [9]. We can now formulate an ecological question suggested by Hilbert's 16th problem. *In there any natural (or laboratory) predator–prey system with at least two ecologically stable cycles?* The answer is not yet known.

Let us suppose that a natural system exists with two ecologically stable cycles. Then the orbits might be disposed an indicated in Figure 19.9. Now let Γ be an ecologically stable cycle. Its *domain of attraction* is the set of all points P such that the orbit through P tends to Γ as $t \to \infty$. The shaded region of Figure 19.9 is the domain of attraction of Γ_1. In the same figure the domain of attraction of Γ_3 consists of all points exterior to Γ_2.

Now, how could the population density orbit of such a natural system move from the domain of attraction of one cycle to the domain of attraction of another cycle? This could occur by means of a massive shock to the system which drives the system from point P, say, at time t_1 to point Q. Of course, this is all hypothetical, but it is at least conceivable that a natural system could act in this fashion.

---

[5] Recent evidence indicates that *differential-delay* equations of the form $dx(t)/dt = x(t - \alpha)$ $f(x(t - \alpha), y(t - \beta))$, and a like equation for dy/dt, may be more accurate. See May [17].

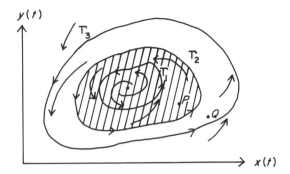

Figure 19.9. Two Ecologically Stable Cycles and Unstable Cycle in between.

5. Three Almost Impossible Projects

A. Find $H(2)$.

B. By using the computer to find numerical solutions of the quadratic system,

$$\begin{cases} \dfrac{dx}{dt} = a_0 + a_1x + a_2y + a_3x^2 + a_4xy + a_5y^2 \\[2mm] \dfrac{dy}{dt} = b_0 + b_1x + b_2y + b_3x^2 + b_4xy + b_5y^2, \end{cases} \tag{19}$$

where the coefficients are those used by Chen and Wang or by Shi, locate the four known cycles. Are there other cycles?

C. Find a predator–prey or other interacting system in nature, or construct one in the laboratory, with at least two ecologically stable cycles.

Clearly, A is very deep, very hard, and not likely to be answered soon. B is slightly easier but requires some computer experience. Alternatively, one might deal with the general quadratic system (19); however, the number of coefficients can be reduced from 12 to 6. Read [10] before doing anything. Problem C may actually be feasible in the engineering laboratory by using a suitable combination of diodes, in the chemistry laboratory using interacting chemicals, or in the biology lab with appropriate little creatures.

Exercises

1. The system, $dx/dt = y$, $dy/dt = -4x + \cos t$, is driven by the external periodic force, $x \equiv 0$, $y \equiv \cos t$. Show that the system has a cycle with the same period 2π as the external force. *Hint:* Observe that $d^2x/dt^2 = dy/dt = -4x + \cos t$. Thus find a solution $x(t)$ of $(d^2x/dt^2) + 4x = \cos t$ of period 2π (try a solution of the

form $x(t) = a \cos t$. Then the desired cycle will be defined by $x(t)$ and $y(t) = dx(t)/dt$.

2. Show that (11a) gives all solutions of (8a). *Hint:* $r \equiv r_0 \equiv 0$ and $r \equiv r_0 \equiv 1$ are seen to be solutions. Then assume $r_0 \neq 0, 1$ (hence $r(t) \neq 0, 1$) and write the differential equation as $dr/r(1 - r^2) = dt$. Use partial fractions or a table of integrals, being careful to distinguish between the cases $0 < r < 1$ and $r > 1$, to integrate and obtain an expression of the form $f(r, r_0) = t$ (the r-integration is from r_0 to r, the t-integration from 0 to t). Then solve for r as a function of t.

3. Solve (12). *Hint:* Write as $dx/(-x + 1) = dt$, $dy/y = -2dt$, and integrate. Verify Figure 19.4.

4. Show that $X(x, y) \equiv x(1 - x^2 - y^2)(4 - x^2 - y^2) \cdots (N^2 - x^2 - y^2) - y$ and that $Y(x, y) \equiv y(1 - x^2 - y^2)(4 - x^2 - y^2) \cdots (N^2 - x^2 - y^2) + x$ in (14). Hence show that X and Y are polynomials of degree $2N + 1$ in x and y *Hint:* Follow the technique used in Example 2. Verify the assertions made in Example 4.

5. Show directly from the definition that a system with a nonisolated cycle must have infinitely many cycles.

6. Show that the system $dx/dt = x \sin(x^2 + y^2) - y$, $dy/dt = y \sin(x^2 + y^2) + x$ has infinitely many isolated cycles. *Hint:* Use polar coordinates and find dr/dt and $d\theta/dt$. Why doesn't Hilbert's 16th problem apply?

7. a) Show that $H(0) = 0$ by solving $dx/dt = a$, $dy/dt = b$, where a and b are constants.
   b) Show that $H(1) = 0$. *Hint:* Consider the linear system, $dx/dt = a_1 x + a_2 y$, $dy/dt = b_1 x + b_2 y$. Suppose the system has the periodic solution $x_1 = x(t)$, $y_1 = y(t)$ with orbit Γ_1. Show that $x_2 = cx(t)$, $y_2 = cy(t)$, is also a periodic solution for *any* constant c and hence that Γ_1 is not an isolated cycle. Then extend to the general linear system $dx/dt = a_0 + a_1 x + a_2 y$, $dy/dt = b_0 + b_1 x + b_2 y$.

8. Show that $x^2 + y^2 - 1 = 0$ is an orbit of the system $dx/dt = -1 - 2y + x^2 + xy + y^2$, $dy/dt = 2x - x^2$ by showing that $x(dx/dt) + y(dy/dt) = 0$ if $x^2 + y^2 - 1 = 0$.

9. Show that the system (in polar coordinates) $dr/dt = r(r^2 - 1)$, $d\theta/dt = 1$ has an isolated cycle, which is ecologically unstable. *Hint:* Compare with Example 2.

10. Show that the system (in polar coordinates) $dr/dt = r(1 - r^2)(4 - r^2)(9 - r^2)$, $d\theta/dt = 1$, has two ecologically stable isolated cycles separated by an isolated cycle which is not ecologically stable.

11. Show that by proper choice of the function $f(r)$ in the polar coordinate system $dr/dt = rf(r)$, $d\theta/dt = 1$, one can construct almost any desired arrangement of cycles, isolated, not isolated, or some of each, all of which are circles centered at the origin.

12. a) (*Bendixson's Negative Criterion*). Show that if $\partial f/\partial x + \partial g/\partial y$ has a fixed sign and does not vanish in the planar region R, then the system $dx/dt = f(x, y)$, $dy/dt = g(x, y)$ cannot have a cycle lying with its interior entirely in R (assume f and g to be continuously differentiable everywhere). *Hint:* Apply Green's

Theorem in the plane to $\iint_A (\partial f/\partial x + \partial g/\partial y)\,dx\,dy$ if the system *does* have a cycle $\Gamma \subset R$ and A is the union of Γ and its interior. Show that this integral is not zero but that the integral $= \oint g\,dx - f\,dy = \int_0^T (g(dx/dt) - f(dy/dt))\,dt = 0$ if Γ has period T.

b) Find all half-planes which cannot contain any cycles of the quadratic system (19).

References

[1] F. Albrecht, H. Gatzke, A. Haddad and N. Wax, "The dynamics of two interacting populations," *J. Math. Anal. Applic.*, vol. 46, pp. 658–670, 1974. This is the definitive mathematical treatment of systems of the form of (18) under various pertinent hypotheses on f and g. Contains almost none of the biological or ecological interpretations. There is a brief bibliography.

[2] W. Baltensweiler, "The relevance of changes in the composition of larch budworm populations for the dynamics of its numbers," in *Dynamics of Populations*, P. J. den Boer and G. R. Gradwell, Eds. Wageningen: Centre for Agricultural Publishing and Documentation, 1971, pp. 208–219.

[3] N. N. Bautin, "On the number of limit cycles which appear with the variation of coefficients from an equilibrium position of focus or center type," *Mat. Sb.*, vol. 30 (72), pp. 181–196, (1952). English translation in *Amer. Math. Soc. Transl.*, no. 100 (1954). A research level paper containing very detailed analysis.

[4] M. G. Bulmer, "A statistical analysis of the 10-year cycle in Canada," *J. Anim. Ecol.*, vol. 43, pp. 701–718, 1971. A very interesting statistical analysis of the lynx–hare and several other 10-year cycles in the Canadian forests. Discusses 8-year cycles in the Siberian taiga. A very good bibliography.

[5] Lan Sun Chen and Ming-Shu Wang, "The relative position and number of limit cycles of a quadratic differential system," *Acta Math. Sinica*, vol. 22, pp. 751–758, 1979. A straightforward but somewhat advanced account of the most recent work on quadratic systems with an example showing that $H(2) \geqq 4$.

[6] Yuan-Shun Chin, "On the algebraic limit cycles of second degree of the differential equation . . . ," *Acta Math. Sinca*, vol. 8, pp. 23–35, 1958. An English translation appears in *Chinese Math.*, vol. 8, pp. 608–619, 1966. An illuminating study of quadratic systems with elliptical isolated cycles.

[7] ——, "Concrete examples of the existence of three limit cycles for the system . . . ," *Acta Math. Sinica*, vol. 9, pp. 213–336, 1959. In English translation in *Chinese Math.*, vol. 9, pp. 521–534, 1967. A detailed presentation of a specific quadratic system with three isolated cycles enclosing a single critical point. Pu Fu-quan assisted with the preparation of this paper.

[8] C. S. Coleman, "Quadratic population models: Almost never any cycles," this volume, ch. 16. Shows that if f and g in system (18) are linear, then cycles are present only in very special cases.

[9] ——, "Biological cycles and the fivefold way," this volume, ch. 18. Discusses in some detail the lynx–hare cycle, the Poincaré–Bendixson theory of limit sets (no proofs), and a version of Kolmogorov's theorem for predator–prey interactions.

[10] W. A. Coppel, "A survey of quadratic systems," *J. Diff. Eq.*, vol. 2, pp. 293–304, 1966. Should be read by anyone pursuing Hilbert's problem in the quadratic case. Readable by anyone with advanced calculus and a good course in differential equations.

[11]   M. H. Dulac, "Sur les cycle limites," *Bull. Soc. Math. de France*, vol. 51, pp. 45–188, 1923. Proves that the number of limit cycles of a polynomial system is finite.

[12a] David Hilbert, "Mathematical problems," presented at the 2nd Internat. Congr. of Mathematicians, Paris, 1900. English translation in *Bull. Amer. Math. Soc.*, vol. 8, pp. 437–479, 1901–1902. This is readable by a senior math major with a good background.

[12b] P. S. Aleksandrov, Ed., *Problemy Gil'berta*, Izdat. Nauka, Moscow, 1969. German transl.: *Die Hilbertische Probleme*, Akad. Verlag., Leipzig, 1971. Contains Russian and German versions of Hilbert's problems and a description of their current status. Nothing is said about the results concerning the second half of Problem 16.

[12c] F. E. Browder, Ed., *Mathematical Developments Arising from Hilbert Problems*, in *Proc. Symposia in Pure Math.*, Amer. Math. Soc., 1976. Goes considerably beyond Aleksandrov's book, lists other problems of current interest, but devotes only a few sentences to the second half of Hilbert's 16th problem.

[13]   M. Hirsch and S. Smale, *Differential Equations, Dynamical Systems, and Linear Algebra*. New York: Academic, 1974. A very good treatment from a modern point of view—junior, senior level. Contains a good treatment of population models.

[14]   Ju. S. Il'jašenko, "The origin of limit cycles under perturbation of the equation $dw/dx = -R_z/R_w$, where $R(z, w)$ is a polynomial," *Mat. Sbornik*, vol. 78, (120), pp. 360–373, 1969. An English translation appears in *Math. USSR Sbornik*, vol. 7, pp. 363–364, 1969. An advanced treatment of complex polynomial systems, but with applications to the real case, and the problem of estimating the number of disjoint eyes.

[15]   L. B. Keith, *Wildlife's Ten-Year Cycle*. Madison: Univ. of Wisconsin, 1963. A thought-provoking account on a nonmathematical level of the various natural cycles in the vast Canadian forests. Should be read by anyone interested in natural cycles.

[16]   D. L. Lack, *The Natural Regulation of Animal Numbers*. Oxford, Clarendon, 1954. Still a good source.

[17]   R. M. May, *Stability and Complexity in Model Ecosystems*. Princeton, NJ: Princeton Univ. Press, 1973. A fascinating and readable account. Excellent bibliography.

[18]   ——, "Limit cycles in predator–prey communities," *Science*, vol. 177, pp. 900–902, 1972. A discussion of Kolmogorov's theorem. A good bibliography.

[19]   J. Maynard Smith, *Models in Ecology*. Cambridge, 1974. An excellent source of differential equations models. Good bibliography.

[20]   N. F. Otrokov, "On the number of limit cycles of a differential equation in the neighborhood of a critical point," *Mat. Sbornik*, vol. 34 (76), pp. 127–144, 1954. A detailed construction of a polynomial system with a multitude of isolated cycles forming a single eye.

[21]   I. G. Petrovskii and E. M. Landis, "On the number of limit cycles of the equation $dy/dx = P(x, y)/Q(x, y)$, where P and Q are polynomials of the second degree," in Russian, *Mat. Sb.*, vol. 37 (79), pp. 209–250, 1955. English transl. in *Amer. Math. Soc. Transl.*, ser. 2, vol. 10, pp. 177–221, 1958. This paper and the following two require a background in topology and complex analysis.

[22]   E. M. Landis and I. G. Petrovskii, "On the number of limit cycles of the equation $dy/dx = P(x, y)/Q(x, y)$, where P and Q are polynomials," in Russian, *Mat. Sb.*, vol. 43 (85), pp. 149–168, 1957. English transl., *Amer. Math. Soc. Transl.* ser. 2, vol. 14, pp. 181–199, 1960.

[23]   I. G. Petrovskii and E. M. Landis, "Corrections" (to the above two papers), *Mat. Sb.*, vol. 48 (90), pp. 253–255, 1959.

[24] E. M. Landis and I. G. Petrovskii, "Letter to the Editor," *Mat. Sb.*, vol. 73 (115), p. 160, 1967. English transl., *Math of the USSR*, vol. 2, p. 144, 1967.

[25] Song-ling Shi, "A concrete example of the existence of four limit cycles for plane quadratic systems," *Scientific Sinica*, vol. 11, pp. 1051–1056, 1979 (Chinese edition) and vol. 23, pp. 154–158, 1980 (English edition). Just what the title says!

[26] ——, "Examples of five cycles for the system $(E'_3)\ \cdots$," *Acta Math Sinica*, vol. 4 (18), pp. 300–304, 1975. By means of examples shows that $H(3) \geqq 5$.

[27] K. S. Sibirskii, "On the number of limit cycles in the neighbourhood of a singular point," *Diff. Urav.* vol. 1, pp. 53–66, 1965. In English translation in *Diff. Eq.*, vol. 1, pp. 36–47, 1965. Shows that a cubic system may have five isolated cycles enclosing a single critical point.

[28] Chin-Chu Tung, "Positions of limit cycles of the system $\cdots$," *Scientia Sinica*, vol. 8, pp. 151–171, 1959. Reprinted in *Chinese Math.*, vol. 8, pp. 854–874, 1966. An excellent introduction to the properties of isolated cycles of quadratic systems together with several specific examples worked out in detail.

[29] H. R. van der Vaart, "Conditions for periodic solutions of Volterra differential systems," *Bull of Math. Bio.*, vol. 40, pp. 133–160, 1978. An exhaustive analysis of system (18), where f and g are affine functions of x and y.

[30] E. O. Wilson and W. H. Bossert, *A Primer of Population Biology*. Stamford, CT: Sinauer Assoc., 1971. A good, not very technical account of the subject.

[31] Yenchien Yeh, "Periodic solutions and limit cycles of certain nonlinear differential systems," *Sci. Record* (N.S.), vol. 1, pp. 391–394, 1957. Statements of properties of isolated cycles of quadratic systems.

[32] ——, "A qualitative study of the integral curves of the differential equation . . . , II: Uniqueness of limit cycles," *Acta Math. Sinica*, vol. 12, pp. 60–67, 1962. In English translation in *Chinese Math.*, vol. 3, pp. 62–70, 1963. An interesting treatment of the behaviour of isolated cycles of system (1).

[33] (added in proof) C. Chicone and Tian Jinghuang, "On general properties of quadratic systems," *Amer. Math. Monthly*, vol. 89, pp. 167–178, 1982. Brings [10] and [28] up to date; mentions [25]; lists a number of open problems.

Notes for the Instructor

Objectives. The basic aim is to introduce the student to a significant unsolved problem in mathematical analysis and to relate that problem to some ecological questions. The problem has to do with the number of cycles (periodic solutions) of a planar autonomous differential system.

Prerequisites. Elementary differential equations, polar coordinates, some experience with drawing the orbits of a differential equation, and for section 4, some prior acquaintance with predator–prey models.

Time. The material may be covered in two lectures.

PART VI

MODELS LEADING TO PARTIAL DIFFERENTIAL EQUATIONS

CHAPTER 20
Surge Tank Analysis

Donald A. Drew*

1. Surge Tanks

Hydrodynamic generation of electrical power requires the efficient transfer of water under pressure from an elevated storage area, or reservoir, to the generating plant, often several hundred meters away. The obvious solution of connecting a large pipe to both ends has a serious shortcoming. A large surge in the power supplied to the electrical customers requires an abrupt increase in the amount of water flowing into the turbine. A falling-off in power usage causes the water suddenly to flow more slowly. In both cases, a large mass of water in the pipe must suddenly change its velocity. The slight compressibility of the water, combined with the slight elasticity of the pipe, causes a wave of high pressure to propagate up and down the pipe, resulting in "water hammer" and often rupturing the pipes.

To overcome this difficulty, a large tank is connected to the pipe just ahead of the turbine. This large tank, called a surge tank, is supposed to fill with water when the turbine needs less water and supply more water quickly when a larger demand occurs (see Figure 20.1). We would like to analyze the operation of the pipe-surge tank system.

2. Equations of Flow

We shall divide the analysis into two sections. We shall first consider the flow of an incompressible fluid in a pipe. Then we shall consider the surge tank as a large pipe and connect it to the relevant pipe flow. Those interested

* Department of Mathematical Sciences, Rensselaer Polytechnic Institute, Troy, NY 12181.

Figure 20.1. Reservoir-Pipe-Surge Tank-Power Plant System

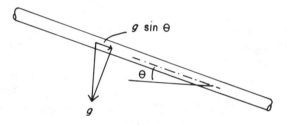

Figure 20.2

in a more complete set of equations governing the compressible flow of
fluid in a pipe should read the Appendix.

Consider a section of pipe of length L and cross sectional area A. If the
density of the fluid is ρ, the mass of fluid in the pipe is ρAL. If $v(t)$ is the
(longitudinal) velocity of the fluid in the pipe, the rate of change of linear
momentum of the fluid in the pipe is

$$\rho AL \frac{dv}{dt}. \tag{1}$$

By Newton's law, this rate of change of momentum is equal to the sum of
the applied forces.

Three types of applied forces are found in pipe flow: 1) pressure, 2)
gravitational, and 3) viscous (drag). We first consider the pressure force. If
p_i is the pressure at the inlet and p_0 is the pressure at the outlet, the total
pressure force (toward the outlet end) is $(p_i - p_0)A$. Now let us consider
the gravitational force. If the pipe makes an angle θ with the horizontal
(Figure 20.2), the component of the gravitational force in the direction of
motion is $\rho ALg \sin \theta$, where g is the gravitational acceleration.

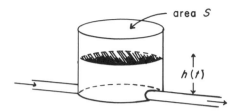

Figure 20.3. Surge Tank Geometry

Finally, we consider the drag force. Fluid flowing in a pipe experiences a resistance to flow which increases with increasing flow. Empirical evidence indicates that the drag force is proportional to the square of the velocity v. Thus the viscous force is $-cLv^2$, where c is an empirical constant.

Thus the equation of motion for the fluid in the pipe is

$$\rho AL \frac{dv}{dt} = A(p_i - p_0) + \rho ALg \sin \theta - cLv^2. \tag{2}$$

Now let us consider the dynamics of the surge tank. Essentially, the surge tank can be considered as a large pipe. In a large pipe, the flow velocities are small. Thus the momentum equation becomes

$$(p_a - p_0)S = -\rho Shg, \tag{3}$$

where p_a is atmospheric pressure, S is the cross sectional area of the surge tank, and $h(t)$ is the height of the water level above the surge tank inlet. Note that we have taken the pressure at the bottom of the tank equal to the pressure at the outlet.

We shall assume that the outlet velocity from the surge tank, which is also equal to the inlet velocity to the power plant, is a prescribed function of t given by $v_T(t)$, and that the cross sectional area of this pipe is A (see Exercise 1). Since the rate of change of the mass of fluid in the surge tank is equal to the rate of flow in, minus the rate of flow out, we see that

$$d(\rho Sh)/dt = \rho A[v(t) - v_T(t)]. \tag{4}$$

Equation (4) is often called the balance of mass equation.

3. Stability of Steady State

If we solve (3) for p_0 as a function of h and use (4) to eliminate the velocity from (2), we have

$$\rho SL \frac{d^2h}{dt^2} = A(p_i - p_a) - \rho g Ah + \rho ALg \sin \theta$$

$$- cL \left(\frac{S}{A} \frac{dh}{dt} + v_T \right)^2 + \rho AL \frac{dv_T}{dt} \tag{5}$$

(see Exercise 2).

If we then set $dh/dt = d^2h/dt^2 = 0$, and $dv_T/dt = 0$, we obtain the *steady-state* solution of (5), given by

$$h_s = L\sin\theta - \frac{(p_i - p_a)}{\rho g} - \frac{cLv_0^2}{\rho g A} \tag{6}$$

with $v = v_T = v_0$, a constant.

Let us now examine how small deviations (or *perturbations*) from this steady state can change in time. A small change in the turbine operating conditions, like more demand for power, may well result in a change in v_T. This change will induce changes in the level of the water in the surge tank. We would like to know what happens to the water in the surge tank.

The mathematical analysis of the evolution of $h(t)$ as governed by (5) is beyond the scope of this module. However, if we look at *small* deviations in the height of the water, we shall be able to complete a meaningful analysis. Moreover, we claim that the information gained from the approximate analysis, assuming small deviations, will be quite useful and, therefore, well worth the effort. On the other hand, we may not be able to obtain meaningful solutions at all if we do not make the assumption of small deviations.

We let $h = h_s + \varepsilon h_1$ and $v_T = v_0 + \varepsilon v_1$, where ε is a measure of the amplitude of the perturbation h_1. We shall assume that ε is very small. By substituting $h = h_s + \varepsilon h_1$ into (5), we have

$$\rho SL\frac{d^2 h_1}{dt^2} = -\rho g A h_1 - \frac{2cL}{A}v_0\frac{dh_1}{dt} - \varepsilon\frac{cLS^2}{A^2}\frac{dh_1^2}{dt} + \rho AL\frac{dv_1}{dt}$$

$$- 2\varepsilon cLv_0 v_1 - 2\varepsilon^2 cLv_1\frac{dh_1}{dt}. \tag{7}$$

We ignore terms multiplied by ε and ε^2. Thus the approximate equation for h_1 is

$$\frac{d^2 h_1}{dt^2} + \beta\frac{dh_1}{dt} + kh_1 = \alpha\frac{dv_1}{dt} \tag{8}$$

where $\beta = 2cv_0/\rho A$, $k = Ag/S$, and $\alpha = A/S$.

Exercise 3 asks you to solve this equation when the velocity of fluid into the generating plant steps up, corresponding to a sudden increase in power demand.

The general homogeneous solution of (8) can be written as

$$h_{1H} = h_0 e^{-\beta t/2}\cos(\omega t + \phi_0), \tag{9}$$

where h_0 and ϕ_0 are constants, and the frequency ω is given as

$$\omega = \sqrt{k - \frac{\beta^2}{4}}.$$

Physically, β is usually quite small, so that the solution (9) represents an oscillation which is decaying according to $e^{-\beta t/2}$. If β is small, any oscillation will persist for a long period of time. This is an undesirable feature of the surge tank. This problem has been solved to a high degree of satisfaction by introducing a riser, a thin vertical pipe with holes in it, in the surge tank. The inlet water from the pipe flows into the riser instead of directly into the tank. The analysis of this situation is somewhat more complicated; the interested reader should see Noble's book [1].

Appendix

Let us consider the flow of fluid in a pipe. Let x be the coordinate along the pipe measured from an arbitrary reference point. In this problem, as in almost all problems involving motions and faces, the appropriate quantities to balance are mass and momentum.

We consider the balance of mass first. Suppose $\rho(x, t)$ is the density of the fluid at x at time t. The mass of fluid in the section from x to $x + \Delta x$ is

$$A \int_x^{x+\Delta x} \rho(\tilde{x}, t)\, d\tilde{x}, \tag{A1}$$

where A is the cross sectional area of the pipe.

The flux of mass across a cross section at x is $A\rho(x, t)v(x, t)$ where v is the fluid velocity in the x direction. No mass is created or destroyed in any cross section, so that balance of mass becomes

$$A \frac{d}{dt} \int_x^{x+\Delta x} \rho(\tilde{x}, t)\, d\tilde{x} = A\rho(x, t)v(x, t) - A\rho(x + \Delta x, t)v(x + \Delta x, t). \tag{A2}$$

Dividing by $A \cdot \Delta x$ and letting $\Delta x \to 0$, we obtain

$$\frac{\partial \rho(x, t)}{\partial t} = -\frac{\partial}{\partial x}[\rho(x, t)v(x, t)]. \tag{A3}$$

It is instructive to write this equation as

$$\frac{\partial \rho}{\partial t} + v\frac{\partial \rho}{\partial x} + \rho\frac{\partial v}{\partial x} = 0. \tag{A4}$$

The quantity $\partial\rho/\partial t + v(\partial\rho/\partial x)$ is called the *material derivative* of ρ and is sometimes denoted by $D\rho/Dt$ or $d\rho/dt$. Let us discuss its significance.

Suppose $\phi(x, t)$ is a function of two variables, and $t = f(\tau)$ and $x = g(\tau)$ are functions of one variable τ. Then

$$\frac{d}{d\tau}\phi(g(\tau), f(\tau)) = \frac{\partial\phi}{\partial x}\frac{dx}{d\tau} + \frac{\partial\phi}{\partial t}\frac{dt}{d\tau}.$$

Now suppose that $f(\tau) = \tau$, so that τ and t are the same variable, and that $g(\tau) = g(t)$ is the position of a slice of fluid as a function of t. Then $dg/dt = dx/dt = $ velocity of that slice, so that $dx/dt = v$. Then

$$\frac{d\phi}{dt} = \frac{\partial\phi}{\partial t} + v\frac{\partial\phi}{\partial x} \tag{A5}$$

is the rate of change of ϕ with $x = x(t)$, where $x(t)$ is the position of the fluid slide at time t. Thus $d\phi/dt$ is the derivative of ϕ following the fluid, or material derivative. (It makes no difference that v depends on x.)

In terms of fluid density, we have

$$\frac{d\rho}{dt} = -\rho\frac{\partial v}{\partial x}. \tag{A6}$$

The right-hand side represents a squeezing of the fluid. If $\partial v/\partial x < 0$, then the velocity decreases to the right, and thus we expect that the fluid in that vicinity will increase in density.

Let us now consider the balance of momentum in the x direction. Momentum, as we shall define it, is simply mass times velocity. More precisely, the momentum in the section between x and $x + \Delta x$ is

$$A\int_x^{x+\Delta x} \rho(\tilde{x}, t)v(\tilde{x}, t)\, d\tilde{x}. \tag{A7}$$

The statement of balance of momentum is

Rate of change of linear momentum
= Net flux of momentum + Applied forces.

This differs slightly from the classical statement of Newton's second law in the presence of the flux term.

Let us consider the flux term. The velocity of fluid motion across a section at x is $v(x, t)$, and the momentum carried by the fluid is $\rho(x, t)v(x, t)$. Thus the flux at any x is

$$A \cdot \rho(x, t)v^2(x, t). \tag{A8}$$

The applied forces which we consider are of three types: pressure, drag, and gravity. Let us consider gravity first. We assume that the component of the gravitational acceleration in the x direction is $g\sin\theta$, where θ is the

angle made by the pipe from horizontal. The total gravitational force on
the section between x and $x + \Delta x$ is

$$A \int_x^{x+\Delta x} \rho(\tilde{x}, t)\, d\tilde{x}\, g \sin \theta. \tag{A9}$$

We next consider the pressure terms. To understand the concept of
pressure, let us isolate the section of fluid between x and $x + \Delta x$ by removing
the fluid to the left of x and to the right of $x + \Delta x$. In order to make the
fluid between x and $x + \Delta x$ behave as though the fluid removed is still
there, we must replace its effect by forces. If those forces are $Ap_L(x)$ and
$Ap_R(x + \Delta x)$, we note that by action–reaction $Ap_L(x) = Ap_R(x)$, and
$Ap_L(x + \Delta x) = Ap_R(x + \Delta x)$. Thus $p_L(x) = p_R(x) = p(x)$.

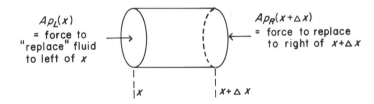

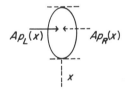

By the way we have defined $p_L(x)$ and $p_R(x + \Delta x)$, the net pressure force
on the section between x and $x + \Delta x$ is

$$A(p_L(x) - p_R(x + \Delta x)) = A(p(x) - p(x + \Delta x)). \tag{A10}$$

The effect of pressure appears in the same form as the flux terms, that is,
in terms of some function at x minus some function at $x + \Delta x$.

Finally, let us consider the drag force. Empirically, the drag force is
proportional to the *square* of the fluid velocity. The appropriate generalization of this force is

$$-\int_x^{x+\Delta x} cv^2(\tilde{x}, t)\, d\tilde{x}. \tag{A11}$$

Thus balance of linear momentum becomes

$$A\frac{d}{dt}\int_x^{x+\Delta x}\rho(\tilde{x},t)v(\tilde{x},t)\,d\tilde{x}=A\rho(x,t)v^2(x,t)$$

$$-A\rho(x+\Delta x,t)v^2(x+\Delta x,t)$$
$$+Ap(x,t)-Ap(x+\Delta x,t)$$
$$+A\int_x^{x+\Delta x}\rho(\tilde{x},t)\,d\tilde{x}\cdot g\sin\theta \qquad \text{(A12)}$$
$$-\int_x^{x+\Delta x}cv^2(\tilde{x},t)\,d\tilde{x}.$$

Dividing by $A\Delta x$ and letting $\Delta x\to 0$ gives

$$\frac{\partial\rho v}{\partial t}+\frac{\partial\rho v^2}{\partial x}=-\frac{\partial p}{\partial x}+\rho g\sin\theta-\frac{c}{A}v^2. \qquad \text{(A13)}$$

If we differentiate the left-hand side of (A13), we have

$$\left(\frac{\partial\rho}{\partial t}+v\frac{\partial\rho}{\partial x}+\rho\frac{\partial v}{\partial x}\right)v+\rho\left(\frac{\partial v}{\partial t}+v\frac{\partial v}{\partial x}\right)=\rho\left(\frac{\partial v}{\partial t}+v\frac{\partial v}{\partial x}\right) \qquad \text{(A14)}$$

after using (A4).

Exercises

1. Derive the balance of mass equation corresponding to (4) when the area of the pipe outlet from the surge tank is $B\neq A$.

2. Derive (5).

3. Find the solution of (8) corresponding to $v_1(t)=H_0(t)$, where H_0 is the Heaviside function, defined by

$$H_0(t)=\begin{cases}1, & \text{for } t\geq 0\\ 0, & \text{for } t<0.\end{cases}$$

For "initial" conditions, you may assume that $h_1=dh_1/dt=0$ for $t<0$. *Hint:* If $v_1(t)=H_0(t)$, $dv_1/dt=\delta(t)$, where δ is the Dirac delta function. This corresponds to the problem for h_1 as given by

$$\frac{d^2h_1}{dt^2}+\beta\frac{dh_1}{dt}+kh_1=0, \qquad t>0$$

$$h_1(0+)=0$$

$$\frac{dh_1}{dt}(0+)=\alpha.$$

References

[1] B. Noble, *Applications of Undergraduate Mathematics in Engineering*. MAA and Macmillan, 1967.

Texts in differential equations which consider the Dirac delta function are:

[2] W. E. Boyce and R. C. DiPrima, *Elementary Differential Equations and Boundary Value Problems*, 3rd ed.   New York: Wiley, 1977.
[3] M. Braun, *Differential Equations and Applications*, 2nd ed.   New York: Springer-Verlag, 1978.

Notes for the Instructor

Objectives. This module considers the stability of a pipe-surge tank system. The equations governing the amount of water in the surge tank are derived, and the stability of the steady state is studied. The module can be used as an introduction to the concepts of fluid dynamics. The basis for the module comes from Noble's book [1].

Prerequisites. Elementary mechanics, multivariable calculus, differential equations.

Time. If the Appendix is omitted, the material may be covered in one lecture.

CHAPTER 21
Shaking a Piece of String to Rest

Robert L. Borrelli*

1. Introduction

Periodic disturbances play an important role in many diverse areas within science and engineering. Examples of periodic disturbances are the motion of a pendulum in a gravitational field, the motion of celestial bodies, and oscillations in an electrical circuit. When periodic disturbances travel in space, they are described collectively as "wave motion." Some familiar examples of wave motion are the motion of ripples on the surface of a pond, sound waves, and transverse vibrations on a taut, flexible string.

In this module we shall analyze in some detail what is perhaps the simplest and most intuitive system involving wave motion: the transverse vibrations of a taut, flexible string. After constructing the linearized model for the motion of a vibrating string, we shall give a characterization of solutions of the vibrating string equation from which many properties of wave motion can be deduced. The objective of this module, however, is the solution of the following optimal control problem for the vibrating string: letting a taut, flexible string be fastened at one end, show the existence of a minimal time T within which the string, regardless of its initial deflection and initial velocity, can be brought to rest by applying an appropriate shaking action to the other end. Surprisingly, this problem was not treated until 1971 when P. C. Parks [1] presented a simple but elegant solution to this optimal control problem.

* Department of Mathematics, Harvey Mudd College, Claremont, CA 91711.

2. Linear Model for the Vibrating String

Suppose that a string is stretched between the points $x = 0$ and $x = L > 0$ of the x axis. We would like to construct a model for the motion of the string under the action of its tension when no external applied forces are acting. To simplify the problem, let us assume that the string is *flexible*; i.e., that the force of tension in the string acts only tangentially to the string and hence offers no resistance to bending. Since the equilibrium position of the string under the action of tension alone is the interval $0 \leq x \leq L$, we may describe the motion of the string by specifying three functions $u_1(x, t)$, $u_2(x, t)$, and $u_3(x, t)$ which give the coordinates at time t of the point x on the string in equilibrium position. On physical grounds it is clear that the motion of the string depends not only on the initial deflection and velocity of the string but also on how the string is restrained at the boundary points $x = 0$ and $x = L$. We shall discuss boundary conditions later.

We shall make some further simplifying assumptions in order to obtain a simple linear partial differential equation describing the motion of the string. See H. Weinberger [3] for a more general model for the vibrating string. Specifically, we assume the following.

(1) The motion of the string takes place in a fixed plane, and points on the string are constrained to move only in a direction transverse to the string. Thus the motion of the string may be described by a single function $u(x, t)$ defined in the region $R = \{(x, t): 0 < x < L, t > 0\}$.
(2) The motion[1] $u \in C^2(R) \cap C^1(\bar{R})$, and only "small" deflections from equilibrium are allowed in the sense that second-order terms in u_x may be ignored when compared to terms of lower order.
(3) The tension in the string $\vec{T}(x, t) \in C^1(\bar{R})$. The density of the string ρ is constant.
(4) No external transverse forces act on the string. In particular, the weight of the string is neglected.

Now we are ready to derive the equation of motion of the string. Let us compute the forces acting on the segment $[x, x']$ of the string. The forces acting on this segment are due to tension alone. We shall find the parallel and transverse components of the tension forces acting on $[x, x']$ at each of these endpoints. Referring to Figure 21.1, we have the following computation (the labeling in Figure 21.1 defines the symbols used):

$$|\vec{T}_u(x', t)| = |\vec{T}(x', t)||\sin \alpha| \qquad |\vec{T}_x(x', t)| = |\vec{T}(x', t)||\cos \alpha|$$
$$|\vec{T}_u(x, t)| = |\vec{T}(x, t)||\sin \beta| \qquad |\vec{T}_x(x, t)| = |\vec{T}(x, t)||\cos \beta|.$$

However, since we have

---

[1] We use $\bar{R}$ to denote the closure of the region R; i.e., $\bar{R} = \{(x, t): 0 \leq x \leq L, t \geq 0\}$.

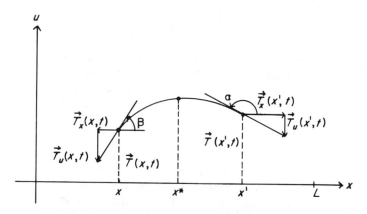

Figure 21.1. Profile on $[x, x']$ of String at Time t

$$|\sin \alpha| = \frac{|u_x(x', t)|}{\sqrt{1 + |u_x(x', t)|^2}} \qquad |\cos \alpha| = \frac{1}{\sqrt{1 + |u_x(x', t)|^2}}$$

and similar formulas for $\sin \beta$, $\cos \beta$, we observe from assumption 2 that

$$|\sin \alpha| = |u_x(x', t)| \qquad |\cos \alpha| = 1$$
$$|\sin \beta| = |u_x(x, t)| \qquad |\cos \beta| = 1,$$

where second- and higher order terms in u_x have been ignored. Thus supplying the proper orientations to the components, we see that the transverse and parallel components of the net tension forces acting on the segment $[x, x']$ are given, respectively, by

$$|\vec{T}(x', t)|u_x(x', t) - |\vec{T}(x, t)|u_x(x, t)$$

and by

$$|\vec{T}(x', t)| - |\vec{T}(x, t)|,$$

but since we assumed that only transverse motions of the string were possible, we conclude that $|\vec{T}(x', t)| = |\vec{T}(x, t)|$. Since x and x' were arbitrary we must have that $T = |\vec{T}|$ is independent of $x \in [0, L]$. Now applying Newton's second law to the segment $[x, x']$ we have

$$\rho(x' - x)u_{tt}(x^*, t) = T\{u_x(x', t) - u_x(x, t)\} \tag{1}$$

where x^* is the x coordinate of the center-of-mass of the string segment between x and x'. Dividing (1) through by $\rho \cdot (x' - x)$, denoting T/ρ by c^2, and taking limits as $x' \to x$ we obtain the *wave equation in one space dimension*,

$$u_{tt} = c^2 u_{xx}, \tag{PDE}$$

which must be satisfied for all points $(x, t) \in R$. We shall assume that c^2 is constant.

3. Conditions at the End of the String

Since only transverse motions of the string have been considered here, we must assume that if the endpoints are allowed to move, we can physically picture this situation by thinking of the string as being looped around copies of the u axis at the endpoints but free to move otherwise (of course, other physical interpretations are possible). Then three types of conditions exist at the boundary which we will find of interest.

First, let the endpoints of the string be driven according to the given actions $\alpha(t)$ and $\beta(t)$ at $x = 0$ and $x = L$, respectively. Then the solution $u(x, t)$ of (PDE) must satisfy the boundary conditions

$$u(0, t) = \alpha(t), \qquad u(L, t) = \beta(t), \qquad t \geq 0.$$

If $\alpha(t) \equiv \beta(t) \equiv 0$ then the string is fastened securely to the x axis at $x = 0$ and $x = L$. This special case is sometimes referred to as a *fixed* boundary condition at each endpoint.

Second, let the endpoints of the string be acted upon by given transverse forces $\gamma(t)$ and $\delta(t)$ at $x = 0$ and $x = L$, respectively. Now let us determine what condition the solution $u(x, t)$ of (PDE) must satisfy at $x = 0$. The transverse force acting on the segment $[0, h]$, $0 < h < L$, is given by

$$Tu_x(h, t) + \gamma(t).$$

Thus by Newton's second law we must have that

$$phu_{tt}(x^*, t) = Tu_x(h, t) + \gamma(t),$$

where x^* is the center of mass of segment of the string over $[0, h]$. Now if we let $h \to 0^+$ we arrive at the condition

$$Tu_x(0, t) + \gamma(t) = 0$$

for each $t \geq 0$ because $u_{tt}(x, t)$ is bounded in x for each $t \geq 0$. Similarly, we arrive at the condition

$$Tu_x(L, t) - \delta(t) = 0$$

for each $t \geq 0$ at the other endpoint. When $\gamma(t) \equiv \delta(t) \equiv 0$, then these conditions are sometimes referred to as *free* boundary conditions.

Third, suppose that the left-hand endpoint of the string is connected to the point $x = 0$ on the x axis by a spring with spring constant k. Then the reasoning in the second case above would imply that the solution of (PDE), $u(x, t)$, would satisfy the condition

$$Tu_x(0, t) = ku(0, t), \qquad t \geq 0.$$

If the right-hand endpoint were connected to the point $x = L$ with a spring, then a similar condition would be obtained. For this reason, such a boundary condition is sometimes referred to as an *elastic* boundary condition.

4. The Mixed Initial/Boundary Value Problem for the Vibrating String

As we have seen in Section 2, the deflection of a vibrating string, $u(x, t)$, must satisfy the so-called *wave equation* (PDE) in the region $R = \{(x, t): 0 < x < L, t > 0\}$, provided that $u(x, t) \in C^2(R)$ satisfies some simplifying conditions. In Section 3 we observed that if $u \in C^1(\bar{R})$ it is appropriate to demand that u and its first partial derivative with respect to x satisfy certain conditions at the boundaries $x = 0$ and $x = L$ for all $t \geq 0$ (depending on the character of the restraint of the string at the endpoints). Of course the initial deflection $u(x, 0)$ and the initial velocity $u_t(x, 0)$ of the string may also be specified. Thus in mathematical terms the motion of a vibrating string with given initial and endpoint conditions will be a function $u \in C^2(R) \cap C^1(\bar{R})$ which satisfies all of the relations (see Figure 21.2)

$$\text{(PDE)} \qquad u_{tt} = c^2 u_{xx}, \qquad \text{for } 0 < x < L, t > 0$$

$$\text{(IC)} \qquad \begin{cases} u(x, 0) = f(x), & 0 \leq x \leq L \\ u_t(x, 0) = g(x), & 0 \leq x \leq L \end{cases}$$

$$\text{(BC)} \qquad \begin{cases} B_0[u] = \phi(t), & t \geq 0 \\ B_L[u] = \mu(t), & t \geq 0 \end{cases} \qquad \text{(BP)}$$

where $f(x)$, $g(x)$, $\phi(t)$, $\mu(t)$ are given functions and the boundary operators B_0 and B_L may be (independently) any of the three types of boundary conditions mentioned in Section 3. Observe that all of these boundary operators are linear. The *boundary problem* (BP) with *initial conditions* (IC) and *boundary conditions* (BC) will be our object of attention for the remainder of this module. We shall see that the *initial data* $[f(x)$ and $g(x)]$ and the *boundary data* $[\phi(t)$ and $\mu(t)]$ can be chosen in a completely arbitrary way, subject only to some appropriate smoothness and consistency conditions.

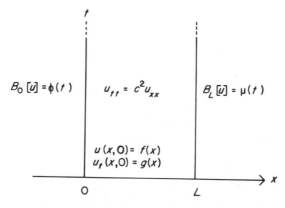

Figure 21.2. Geometry for Problem (BP)

5. Characteristics

The following elementary observation will be useful for finding solutions of (PDE) and ultimately the boundary problem (BP).

Theorem 1. *Let Ω be an open set in the xt plane, and let $u \in C^2(\Omega)$ be any solution of (PDE) in Ω. Then*

(i) $\partial_t u - c \partial_x u$ *is constant on any line segment in Ω of the form $x - ct = $ constant, and*

(ii) $\partial_t u + c \partial_x u$ *is constant on any line segment in Ω of the form $x + ct = $ constant.*

PROOF. We prove only assertion (i), the proof of (ii) is very similar. Observe that (PDE) can be written as

$$(\partial_t + c\partial_x)[\partial_t u - c\partial_x u] = 0$$

where the outer derivative is proportional to a directional derivative in the xt plane in the direction defined by the vector with x coordinate c and t coordinate unity. Thus on the line $x - ct = $ constant in Ω the function $\partial_t u - c\partial_x u$ has a vanishing directional derivative, and hence the desired result follows (see Exercise 1 for a proof of this theorem via the chain rule). □

Thus we see that the two families of lines

$$x - ct = \text{constant} \tag{2}$$

$$x + ct = \text{constant} \tag{3}$$

have special significance for the wave equation (PDE). Each line of the form (2) or (3) is called a *characteristic* for the wave equation (PDE).

6. The Cauchy Problem for the Wave Equation

In order to exploit the special property of characteristics for deriving a solution for (BP) we will first consider a pure initial value problem for (PDE). For convenience, let us assume that our initial data f and g are defined on the entire x axis. The boundaries of our string have therefore receded to $+\infty$ and $-\infty$, and we shall impose no boundary conditions in this case. Letting $S = \{(x, t): x \in \mathbf{R}, t > 0\}$, we seek solutions $u \in C^2(S) \cap C^1(\overline{S})$ for the initial value problem

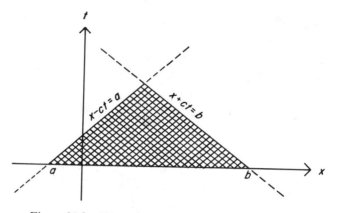

Figure 21.3. Characteristic Triangle $\Delta(a, b)$ for (PDE).

$$
\text{(CP)} \quad
\begin{cases}
u_{tt} = c^2 u_{xx}, & x \in \mathbf{R}, \quad t > 0 \\
u(x, 0) = f(x), & x \in \mathbf{R} \\
u_t(x, 0) = g(x), & x \in \mathbf{R}.
\end{cases}
$$

The problem (CP) is known as the *Cauchy problem* for the wave equation (PDE). In spite of the fact that one does not often encounter infinitely long (or even very long) strings, we shall find this physically idealistic case useful for obtaining a solution of the more realistic problem (BP).

First, we shall show that (CP) cannot have more than one solution: let $a < b$ be any two points in $\mathbf{R}$ and denote by $\Delta(a, b)$ the interior of the triangle whose base is the interval (a, b) and whose sides are segments of the characteristic lines $x - ct = a$, $x + ct = b$. The set $\Delta(a, b)$ is called a *characteristic triangle*; see Figure 21.3. Now the fact that (CP) cannot have more than one solution is a consequence of the following.

Theorem 2 (Uniqueness). *Let* $u, v \in C^2(\Delta(a, b))$ *be any two solutions of the PDE:* $u_{tt} = c^2 u_{xx}$ *which are continuously differentiable to the base of* $\Delta(a, b)$ *and satisfy the initial conditions of (CP) there. Then* $u = v$ *in* $\Delta(a, b)$.

PROOF. The function $w \equiv u - v \in C^2(\Delta(a, b))$ and satisfies (PDE). Moreover, we see that

$$
\lim_{\substack{(x,t)\to(x_0,0) \\ (x,t)\in\Delta(a,b)}} w(x, t) = 0, \qquad \text{for each } x_0 \in (a, b)
$$

$$
\lim_{\substack{(x,t)\to(x_0,0) \\ (x,t)\in\Delta(a,b)}} w_t(x, t) = 0, \qquad \text{for each } x_0 \in (a, b).
$$

Thus by Theorem 1 we see that for all $(x, t) \in \Delta(a, b)$, $w_t + w_x = 0$ and

$w_t - w_x = 0$. Hence $w_t \equiv w_x \equiv 0$ in $\Delta(a, b)$, and therefore[2] $w \equiv$ constant in $\Delta(a, b)$. Clearly, the constant must be zero, and so $w \equiv 0$ in $\Delta(a, b)$ and we are done.   □

We shall show that (CP) has a solution by actually constructing one from the initial data. Towards that end assume that the initial data f and g are defined over the interval (a, b) on the x axis, then we have the next theorem.

Theorem 3 (D'Alembert's Formula). *Let $f \in C^2(a, b)$, $g \in C^1(a, b)$ be given functions on the open interval (a, b). Then the function*

$$u(x, t) = \frac{f(x + ct) + f(x - ct)}{2} + \frac{1}{2c} \int_{x-ct}^{x+ct} g(s)\,ds, \qquad (4)$$

defined for $(x, t) \in \Delta(a, b)$ is the unique solution of the restricted Cauchy problem:

$$(RCP) \quad \begin{cases} u_{tt} = c^2 u_{xx}, & (x, t) \in \Delta(a, b) \\[2mm] \lim_{\substack{(x,t)\to(x_0,0) \\ (x,t)\in\Delta(a,b)}} u(x, t) = f(x_0), & \text{for each } x_0 \in (a, b) \\[2mm] \lim_{\substack{(x,t)\to(x_0,0) \\ (x,t)\in\Delta(a,b)}} u_t(x, t) = g(x_0), & \text{for each } x_0 \in (a, b). \end{cases}$$

Remark. Theorems 2 and 3 easily extend to the case where the initial data f and g are defined for all $x \in \mathbf{R}$ and satisfies appropriate smoothness conditions. Indeed, when this is the case, then the function $u(x, t)$ defined by (4) for $x \in \mathbf{R}$, $t \geq 0$ is the *unique* solution of the Cauchy problem (CP).

PROOF. For any $(x, t) \in \Delta(a, b)$ denote by X_1 and X_2, $X_1 < X_2$, the intersection with the x axis of the two characteristic lines through (x, t). Clearly, $a < X_1 < X_2 < b$, and $X_1 = x - ct$, $X_2 = x + ct$. From Theorem 1 we conclude that

$$\begin{aligned} u_t(x, t) + cu_x(x, t) &= u_t(X_2, 0) + cu_x(X_2, 0) \\ u_t(x, t) - cu_x(x, t) &= u_t(X_1, 0) - cu_x(X_1, 0), \end{aligned} \qquad (5)$$

but since

$$\begin{aligned} u_t(X_2, 0) &= g(X_2), & u_t(X_1, 0) &= g(X_1) \\ u_x(X_2, 0) &= f'(X_2), & u_x(X_1, 0) &= f'(X_1) \end{aligned}$$

we see that (5) becomes

---

[2] Recall the theorem: let $R \subset \mathbf{R}^n$ be a region and let $f \in C^1(R)$ be such that $\partial f/\partial x_i = 0$ in R, for each index $i = 1, 2, \cdots, n$. Then $f \equiv$ constant in R.

$$u_t(x, t) + cu_x(x, t) = g(x + ct) + cf'(x + ct)$$

$$u_t(x, t) - cu_x(x, t) = g(x - ct) - cf'(x - ct). \tag{6}$$

Solving (6) we obtain

$$2u_t(x, t) = [g(x + ct) + g(x - ct)] + c[f'(x + ct) - f'(x - ct)] \tag{7}$$

$$2u_x(x, t) = \frac{1}{c}[g(x + ct) - g(x - ct)] + [f'(x + ct) + f'(x - ct)] \tag{8}$$

for all $(x, t) \in \Delta(a, b)$. Letting

$$U(x, t) \equiv u(x, t) - \frac{f(x + ct) + f(x - ct)}{2} - \frac{1}{2c} \int_{x-ct}^{x+ct} g(s)\, ds \tag{9}$$

we see via (7) and the Leibniz Rule[3] that $U_t(x, t) \equiv 0$ for $(x, t) \in \Delta(a, b)$, and hence we conclude that $U(x, t)$ is a function of x alone. Using (8) we see that $U_x(x, t) \equiv 0$, and so $U(x, t) \equiv$ constant. Finally, taking limits we see via (8) that $U(x, t) \equiv 0$ for all $(x, t) \in \Delta(a, b)$, and we are done. ☐

7. Properties of Solutions of the Wave Equation

Having the explicit solution of (CP), we can now examine not only its mathematical properties but also its significance to the problem of the vibrating string.

1) *Superposition of Traveling Waves:* First, observe that the function $u(x, t)$ in (4) can be written as

$$u(x, t) = P(x + ct) + Q(x - ct) \tag{10}$$

where

$$\begin{cases} P(s) = \frac{1}{2}f(s) + \frac{1}{2c}\int_0^s g(s')\, ds' \\[2mm] Q(s) = \frac{1}{2}f(s) + \frac{1}{2c}\int_s^0 g(s')\, ds'. \end{cases} \tag{11}$$

Thus (10) tells us that to find the profile of our string at the given time T (i.e., $u(x, T)$) we need only take the graph of P on the x axis, translate it to the left by cT units, and then add it to the translation of Q on the x axis, to the right by cT units. Thinking of the graph of P and Q on the x

---

[3] The reader is reminded that the Leibniz rule for differentiating integrals with respect to parameters states that, under appropriate smoothness conditions on $x(t)$, $\beta(t)$, and $g(s, t)$, we have that

$$\frac{d}{dt}\int_{\alpha(t)}^{\beta(t)} g(s, t)\, ds = g(\beta(t), t)\beta' - g(\alpha(t), t)\alpha' + \int_{\alpha(t)}^{\beta(t)} g_t(s, t)\, ds.$$

axis as *waveforms* we see that $P(x + ct)$ is the waveform P moving uniformly in time to the left with velocity c. Similarly, $Q(x - ct)$ is just the waveform Q moving uniformly in time to the right with velocity c. Thus the solution of any problem (CP) is the sum of two *traveling waves*, one moving to the right, the other to the left, as is evident from (10).

Remark. So far we have isolated a class of solutions of the wave equation which have the form (10) and apparently involve two "arbitrary" functions. It is an interesting fact that the wave equation has no other solutions; i.e., every solution of the wave equation has the form (10) for suitable functions $P(\cdot)$ and $Q(\cdot)$. Thus (10) is the *general solution* of the wave equation. The wave equation of this section is one of the very few partial differential equations for which a general solution can be conveniently found.

2) *Huygens' Principle:* Say that the string is at rest initially except for the bounded interval J where the string is deflected and given some initial velocity; i.e., f and g in (CP) vanish identically everywhere but on J. It is of interest to determine how much time must pass for the point x_0 to "see" the "disturbance" on a set J not containing x_0. From (4) it is clear that $u(x, t)$ will vanish in any characteristic triangle $\Delta(x_0 - ct, x_0 + ct)$ whose base does not meet J. If $J = [a, b]$ and a is the nearest endpoint of J to x_0, then the earliest time that x_0 can "feel" the "disturbance" on J is $t = (a - x_0)/c$. This observation agrees with 1) in that the waveforms P and Q both travel with speed c. Thus x_0 "perceives" a "sharp beginning" for the waves generated by the disturbance on J, but observe now via (4) that if the initial velocity g is nontrivial on J, then x_0 will not in general "perceive" a "sharp end" to the waves generated by the "disturbance" on J, even when J is chosen *arbitrarily* small in length. This peculiar property of the problem (CP) is expressed by saying that "(CP) does not satisfy *Huygens' principle.*"

3) *The Semi-Group Property:* Let $u(x, t)$ be the unique solution of the Cauchy problem (CP) for given initial data f and g. Let $T > 0$ be given and let $u^*(x, t)$ be the unique solution of the Cauchy problem for the initial data $f^*(x) = u(x, T)$, $g^*(x) = u_t(x, T)$. Then we claim that

$$u^*(x, t) \equiv u(x, t + T) \tag{12}$$

for all $x \in \mathbf{R}$ and all $t > 0$. In other words the motion of the string which results when using the state of the string at $t = T$ as initial data is precisely the same as would have occurred had we used the original initial data and reset our clocks back T units when t reaches the time T. Stated again, somewhat differently, any solution $u(x, t)$ of (CP) on the strip $\{(x, t): x \in \mathbf{R}, 0 < t < T_1\}$ can be extended to a solution, $v(x, t)$ on the strip $\{(x, t): x \in \mathbf{R}, 0 < t < T_2\}$, in only one way, and moreover, $v(x, t)$ for $x \in \mathbf{R}$, $T_1 < t < T_2$ is generated by solving (CP) with initial data $f^*(x) = u(x, T_1)$, $g^*(x) = u_t(x, T_1)$ and starting the clock at $t = T_1$. We now turn to a proof of (12).

Writing via (10)

$$f^*(x) \equiv u(x, T) = P(x + cT) + Q(x - cT)$$

$$g^*(x) \equiv u_t(x, T) = c[P'(x + cT) - Q'(x - cT)].$$

Thus to construct $u^*(x, t)$ we only compute the functions $P^*(s)$ and $Q^*(s)$ via (11). It is easy to see that

$$P^*(s) = P(s + cT); \qquad Q^*(s) = Q(s - cT)$$

and hence using (10) again we obtain

$$u^*(x, t) \equiv P(x + c(t + T)) + Q(x - c(t + T)) \equiv u(x, t + T)$$

and we are done.

4) *Propagation of Singularities:* Let us consider now (CP) with data that satisfies the smoothness restrictions only piecewise. That is, suppose that for each point $x \in \mathbf{R}$, except possibly for a set Σ with no finite cluster point, the functions f and g satisfy the usual smooth requirements on some open neighborhood of x, N_x: $g \in C^1(N_x)$, $f \in C^2(N_x)$. (The exceptional set Σ can only have finitely many members in any bounded interval. Why?.) With data satisfying these piecewise smoothness restrictions, we can still use (4) to obtain a function $u(x, t)$ which satisfies the wave equation in the region formed by deleting from $\{(x, t): x \in \mathbf{R}, t > 0\}$ all characteristic lines of the form $x \pm ct = s_0$, for $s_0 \in \Sigma$. We shall also call such functions $u(x, t)$ *solutions* of (CP) even though, as we shall see, they do not satisfy the smoothness requirements imposed at the beginning of this section. Now it follows via (6) that $u(x, t)$ cannot have continuous first derivatives in any region R of the xt plane which contains a segment of a characteristic line of the form $x \pm ct = s_0$, where the function $g(s) + cf'(s)$ or the function $g(s) - cf'(s)$ is discontinuous at s_0. The interesting thing here is that discontinuities in the data cause discontinuities in the solution *for arbitrarily large times*; i.e., the solution $u(x, t)$ cannot become smooth eventually if the initial data is not smooth in the first sense defined above. Similarly, $u(x, t)$ cannot have continuous second-order derivatives in any region R in the xt plane which contains a segment of a characteristic line of the form $x \pm ct = s_0$ where the function $g'(s) + cf''(s)$ or the function $g'(s) - cf''(s)$ is discontinuous at s_0. Thus we see that discontinuities for solutions of (CP) can travel only along characteristic lines. See Figure 21.4 for a sketch of profiles of a taut string of length L released when at rest from a "plucked" initial deflection.

8. Solution of the Mixed Initial/Boundary Problem

Now we shall show how the D'Alembert solution (4) of the Cauchy problem (CP) can be used to solve the mixed initial/boundary value problem (BP) in the case where the boundary data vanishes (i.e., $\phi(t) \equiv \mu(t) \equiv 0$). An

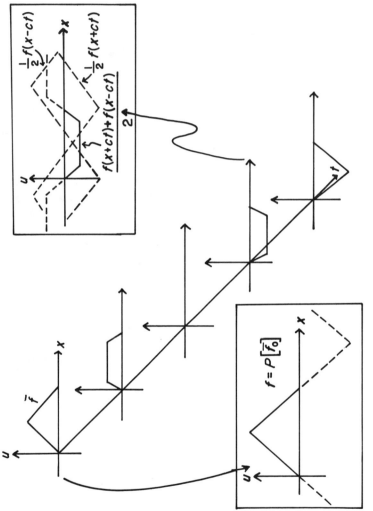

Figure 21.4. Profiles of Vibrating String, Traveling Waves

example will clarify the procedure. Let us consider the case of a taut, flexible string of length L which is fastened at both endpoints; then the motion of the string is characterized by the problem (BP) with the boundary conditions

$$u(0, t) = 0 \qquad u(L, t) = 0, \qquad \text{for all } t \geq 0.$$

Thus, in the case, (BP) takes the form

$$\text{(PDE)} \qquad u_{tt} = c^2 u_{xx}, \qquad x \in (0, L), t > 0$$

$$\text{(IC)} \qquad \begin{cases} u(x, 0) = \bar{f}(x), & x \in [0, L] \\ u_t(x, 0) = \bar{g}(x), & x \in [0, L] \end{cases} \qquad \overline{(\text{BP})}$$

$$\text{(BC)} \qquad \begin{cases} u(0, t) = 0, & t \geq 0 \\ u(L, t) = 0, & t \geq 0. \end{cases}$$

It is a remarkable fact that we can construct a suitable Cauchy problem for an infinite string whose solution will turn out to be a solution of $\overline{(\text{BP})}$ as well. To show this, the first thing we should do, obviously, is to find some conditions on the data f and g in (CP) such that the solution of (CP) satisfies as many properties of a solution of $\overline{(\text{BP})}$ as possible. With this end in mind, let us denote the solution of an appropriate Cauchy problem (CP) also by $u(x, t)$. Then it is certainly necessary that $f \in C^2(\mathbf{R})$, $g \in C^1(\mathbf{R})$, and that both coincide with $\bar{f}$ and $\bar{g}$, respectively, on $[0, L]$. A glance at (4) indicates that if f and g are odd functions about the points $x = 0$ and $x = L$, then $u(0, t) = u(L, t) = 0$ for all $t \geq 0$. Thus we may try to construct f and g from $\bar{f}$ and $\bar{g}$ in the following manner. Extend $\bar{f}$ and $\bar{g}$ into $[-L, 0]$ as odd functions about $x = 0$. Then extend these functions into $[L, 3L]$ as odd functions about $x = L$. Continuing this procedure, we arrive at two functions f and g which are evidently odd about $x = 0$ and $x = L$. (If $\bar{f}_0$ and $\bar{g}_0$ are the extensions of $\bar{f}$ and $\bar{g}$ into $[-L, 0]$ as odd functions about $x = 0$, then it is not hard to show that in fact $f = P[\bar{f}_0]$, $g = P[\bar{g}_0]$, the periodic extensions of $\bar{f}_0$, $\bar{g}_0$ into $\mathbf{R}$ with period $= 2L$.) In order to guarantee the required smoothness properties for f and g, it is necessary and sufficient that

$$\bar{f} \in C^2[0, L], \qquad \bar{g} \in C^1[0, L]$$

$$\bar{g}(0) = \bar{g}(L) = 0$$

$$\bar{f}(0) = \bar{f}(L) = \bar{f}''(0^+) = \bar{f}''(L^-) = 0.$$

The simple proof of this fact is left to the reader. Now by using (4) we can easily verify that the solution $u(x, t)$ of (CP) using the data f, g constructed above does in fact solve the problem $\overline{(\text{BP})}$.

Remark. Problems like $\overline{(\text{BP})}$ but with free endpoint conditions (or one endpoint fixed and the other free) can also be treated in the above manner. See Exercise 2. We may also seek solutions of $\overline{(\text{BP})}$ when the data $\bar{f}$ and $\bar{g}$ only satisfy the smoothness requirements piecewise. The procedure above works perfectly well in this case, and it is not hard to see that discontinuities

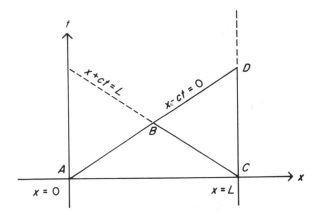

Figure 21.5. Geometry for Uniqueness Proof.

in the data propagate into the region $\{(x, t): 0 < x < L, t > 0\}$ along characteristic line segments $x \pm ct = s_0$ (where s_0 is a point of discontinuity for the data) and their "reflections" in the lines $x = 0$, $x = L$.

It can be shown that $(\overline{BP})$ cannot have more than one solution, and we shall do this now. The above method for constructing a solution to $(\overline{BP})$ appears to contain a certain arbitrariness, and hence the uniqueness result does not follow trivially from our considerations here. Uniqueness of the solution of $(\overline{BP})$ is implied by the following theorem.

Theorem 4 (Uniqueness). *Consider the region* $R = \{(x, t): 0 < x < L, t > 0\}$, *and let* $u, v \in C^2(R) \cap C^1(\bar{R})$ *be solutions of the wave equation* (PDE) *in R and such that*

$$u(x, 0) = v(x, 0), \qquad u_t(x, 0) = v_t(x, 0), \qquad 0 \le x \le L$$

and

$$B_0[u] = B_0[v], \qquad B_L[u] = B_L[v], \qquad t \ge 0$$

where B_0 *and* B_L *are any of the boundary operators described in Section* 3. *Then* $u \equiv v$ *in R.*

PROOF. To be specific we shall take $B_0[u] \equiv u(0, t)$ and $B_L[u] \equiv u(L, t)$; the other cases are treated similarly. The proof will be iterative in nature, and so we shall only give the general idea behind the proof by considering the region in Figure 21.5 instead of R.

If we put $w \equiv u - v$, then as an immediate consequence of Theorem 2 we have that $w \equiv 0$ in the characteristic triangle $\triangle ABC$. Now since $w \equiv 0$ on the line $\overline{DC}$ it follows that $w_t \equiv 0$ on $\overline{DC}$, too. Applying Theorem 1 to characteristic lines of the form $x - ct = $ constant, we conclude to $w_x \equiv 0$ on $\overline{DC}$ as well. Thus, applying Theorem 1 to characteristic lines which intersect in the triangle $\triangle BCD$ we easily conclude that $w \equiv 0$ in $\triangle BCD$.

If this process is applied, in turn, to each of the triangles in R formed by intersecting characteristics, then it is not difficult to see that $w \equiv 0$ in all of R. $\square$

9. Shaking a String to Rest

To conclude, we shall discuss a problem in control theory for a system which engineers describe as a "system with distributed parameters."

Consider a taut string of length L which is fastened at $x = 0$ and suppose that the position of the end of the string at $x = L$ can be controlled as a given continuously differentiable function of time. Then the motion of the string is the solution of the problem (where for convenience we put $c = 1$, and $L = 1$)

$$(\text{BP})_\mu \quad \begin{cases} u_{tt} = u_{xx}, & 0 < x < 1, t > 0 \\ u(x, 0) = f(x), & 0 \le x \le 1 \\ u_t(x, 0) = g(x), & 0 \le x \le 1 \\ u(0, t) = 0, & t \ge 0 \\ u(1, t) = \mu(t), & t \ge 0 \end{cases}$$

for some control function $\mu(t)$.

The control problem we wish to solve is the following: find the smallest value for $T > 0$ such that, for any given initial data f, g, there exists a control, $\mu(t)$, such that the solution $u(x, t)$ of $(\text{BP})_\mu$ has the property

$$u(x, T) \equiv 0, \qquad u_t(x, T) \equiv 0, \qquad 0 \le x \le 1. \tag{13}$$

Because solutions of $(\text{BP})_\mu$ satisfy the semi-group property (C) we see that if (13) can be achieved, then the string will remain at rest for all $t > T$, provided $\mu(t) \equiv 0$ for $t \ge T$. Thus the problem we wish to solve is *how to shake a piece of string to a standstill in the shortest possible time*.

Now the control problem as it stands is still somewhat imprecisely stated. Indeed, we have not stated the smoothness properties demanded of the solution, $u(x, t)$, of the control problem, and the class of admissible initial data has not been defined. To be specific let us demand that the solution of our control problem be such that $u \in C^2(R) \cap C^1(\bar{R})$, where $R = \{(x, t): 0 < x < 1, t > 0\}$, and let us decide later what smoothness and consistency conditions the initial data f and g must satisfy. Since $u(1, t) \equiv \mu(t)$, the smoothness condition we have imposed on $u(x, t)$ implies that μ is a continuously differentiable function.

First, let us find some necessary conditions on the control function $\mu(t)$ which guarantee the relations (13) when $T = 2$. Now from Theorem 1 we observe that the functions $\partial_t u + \partial_x u$ and $\partial_t u - \partial_x u$ must be constant along

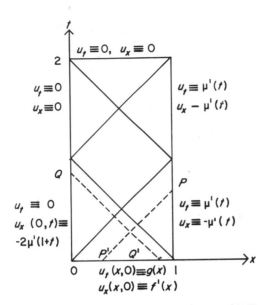

Figure 21.6. Boundary Values of Solution of $(BP)_\mu$.

characteristics in R of the form $x + t = $ constant, and $x - t = $ constant, respectively. Using this fact we easily construct the boundary values for u_t and u_x on R as shown in Figure 21.6. Now if we connect the points $P(1, t)$ and $P'(x, 0)$ with a characteristic line of the form $x - t = $ constant, then the values of μ' can be compared with the values of f and g via Theorem 1. Thus we obtain

$$2\mu'(t) = f'(1 - t) + g(1 - t), \qquad 0 \le t \le 1. \tag{14}$$

Similarly, if we connect the points $Q(0, t)$ and $Q'(x, 0)$ with a characteristic line of the form $x + t = $ constant, then we obtain via Theorem 1 that

$$-2\mu'(1 + t) = f'(t) + g(t), \qquad 0 \le t \le 1. \tag{15}$$

Thus we obtain the condition

$$\mu'(t) = H(t) \equiv \begin{cases} \frac{1}{2}\{g(1 - t) - f'(1 - t)\}, & 0 \le t \le 1 \\ -\frac{1}{2}\{g(t - 1) + f'(t - 1)\}, & 1 \le t \le 2 \end{cases} \tag{16}$$

Observing that $\mu(2) = 0$ we see that the ODE (16) characterizes $\mu(t)$ uniquely to be

$$\mu(t) = -\int_t^2 H(s)\, ds. \tag{17}$$

In Theorem 4 $(BP)_\mu$ was shown to have at most one solution for any $\mu \in C^1[0, 2]$. That a solution of $(BP)_\mu$ exists and satisfies the optimality criterion—if μ is determined via (17) and the data f and g satisfy suitable conditions—can be shown[4] using Theorem 1. Before we begin, however, let us assume at the outset that $f \in C^2[0, 1]$ and $g \in C^1[0, 1]$. Then we see from (17) that $\mu \in C^1[0, 2]$ if and only if $g(0) = 0$, and that $\mu \in C^2[0, 2]$ if and only if $g(0) = 0$ and $f''(0) = 0$. Next, let us observe from (17) that $\mu(0) = 0$, and so if $u(x, t)$ is to be of class $C^1(\bar{R})$, we must at least have that $f(1) = 0$. Similarly, we must demand that $f(0) = 0$.

Now observe via Theorem 1 that any solution of the boundary problem (BP)$_\mu$ which meets the condition $u(x, 2) = u_t(x, 2) = 0$, for $0 \le x \le 1$, must satisfy the relation

$$(\partial_t - \partial_x)u = \begin{cases} g(x - t) - f'(x - t), & 0 \le x - t \le 1 \\ -\{g(t - x) + f'(t - x)\}, & -1 \le x - t \le 0 \\ 0, & -2 \le x - t \le 0 \end{cases} \quad (18)$$

$$(\partial_t + \partial_x)u = \begin{cases} g(x + t) + f'(x + t), & 0 \le x + t \le 1 \\ 0 & 1 \le x + t \le 2 \\ 0 & 2 \le x + t \le 3. \end{cases} \quad (19)$$

The differential operators $\partial_t + \partial_x$ and $\partial_t - \partial_x$ are proportional to directional derivatives in the directions $(\vec{i} + \vec{j})/\sqrt{2}$, and $(\vec{i} - \vec{j})/\sqrt{2}$, respectively (where $\vec{i}$ and $\vec{j}$ are unit vectors in the positive t and x directions, respectively). Thus if we can show that a function $u(x, t)$ defined via (18), (19) has continuous derivatives $(\partial_t - \partial_x)u$, $(\partial_t - \partial_x)^2 u$, $(\partial_t + \partial_x)u$, and $(\partial_t + \partial_x)^2 u$, then the same can be said for u_u, u_x, u_{tt}, and u_{xx}.

Now it is trivial to find conditions on the data f and g which would insure that the functions defined by the right-hand sides of (18) and (19) are of class $C^2(R) \cap C^1(\bar{R})$. These conditions are as follows:

$$g(0) = 0, \quad f(0) = f(1) = f''(0) = 0$$
$$g(1) + f'(1) = 0 \quad (20)$$
$$g'(1) + f''(1) = 0.$$

Let us define the class of *admissible data* Ω as follows:

$\Omega = \{(f, g): f \in C^2[0, 1], g \in C^1[0, 1],$ and the identities (20) are satisfied$\}$.

If initial data (f, g) are chosen to be admissible and if (18) and (19) are solved for a function $u(x, t)$, then we have essentially shown that
1. $u \in C^2(R) \cap C^1(\bar{R})$,
2. u is a solution of (BP)$_\mu$ with μ defined via (17),
3. $u(x, 2) = u_t(x, 2) \equiv 0, 0 \le x \le 1$.

On the other hand, we can easily see that for general admissible data no control function μ exists which will bring the string to rest in less than

---

[4] The technique described here is not very practical if explicit solutions are sought, but it will serve our purposes.

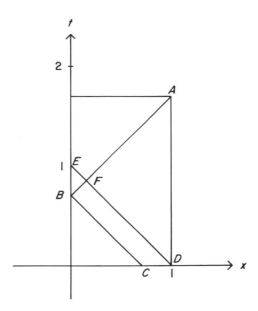

Figure 21.7. Shaking a String to Rest: $T < 2$.

two units of time. Indeed, say this were the case. Then the solution $u(x, t)$ of $(BP)_\mu$ must vanish identically above the characteristic line AB in the rectangle $\{(x, t): 0 \le x \le 1, 0 \le t \le T\}$ (see Figure 21.7). Now for any (x, t) in the triangle BFE we see via Theorem 1 that

$$u_t(x, t) + u_x(x, t) = g(x) + f'(x). \tag{21}$$

However, the left-hand side of (21) vanishes identically for (x, t) in the triangle BFE, and hence the data f and g must necessarily satisfy a consistency relation beyond admissibility. This establishes our assertion.

Summarizing our results now we have the following.

Theorem 5. $T = 2$ is the smallest value for T such that for any data $f \in C^2[0, 1]$, $g \in C^1[0, 1]$ satisfying the conditions (20), a control function μ on $[0, \infty)$ exists such that shaking problem $(BP)_\mu$ has a solution $u \in C^2(R) \cap C^1(\bar{R})$, $R = \{(x, t): 0 < x < 1, t > 0\}$, such that $u(x, t) \equiv 0$ for all $0 \le x \le 1, t \ge T$. This control function is unique and is given by

$$\mu(t) = \begin{cases} -\displaystyle\int_t^2 H(s) \, ds, & 0 \le t \le 2 \\ 0, & t \le 2, \end{cases}$$

where

$$H(s) = \begin{cases} \frac{1}{2}\{g(1 - s) - f'(1 - s)\}, & 0 \le s \le 1 \\ -\frac{1}{2}\{g(s - 1) + f'(s - 1)\}, & 1 \le s \le 2. \end{cases}$$

Remark. It can be shown that it is possible to break the string in a finite time by an appropriate choice of a *periodic* control function $\mu(t)$ of arbitrarily small amplitude. Resonance is the phenomenon involved.

Exercises

1. Prove assertion (i) of Theorem 1 in the way described below. Let k be any constant such that the line $x - ct = k$ intersects Ω. Then for some interval $s_1 < s < s_2$ the points (x, t) defined by

$$x = k - cs, \ t = s, \qquad \text{for } s_1 < s < s_2$$

   describes a segment of the line $x - ct = k$ which lies in Ω. Show via the chain rule that for the function

$$F(s) \equiv u_t(k - cs, s) - cu_x(k - cs, s)$$

   is such that $F'(s) \equiv 0$ on (s_1, s_2).

2. Solve the problem $(\overline{BP})$ where the (BC) are replaced by
   (a) $u_x(0, t) = 0, u_x(L, t) = 0, t \geq 0$.
   (b) $u(0, t) = 0, u_x(L, t) = 0, t \geq 0$.
   Hint: In a) construct from $\bar{f}, \bar{g}$ on $[0, L]$ data f, g on **R** which is even about $x = 0$ and $x = L$. What conditions on $\bar{f}, \bar{g}$ guarantee the appropriate smoothness for f and g? In b) construct from $\bar{f}, \bar{g}$ on $[0, L]$ data f, g on **R** which is odd about $x = 0$ and even about $x = L$. What conditions on $\bar{f}, \bar{g}$ guarantee the appropriate smoothness for f and g in this case?

3. Let R be a convex region in the xt plane. Show that any solution $u \in C^2(R)$ of the wave equation $u_{tt} = c^2 u_{xx}$ has the form

$$u(x, t) = F(x + ct) + G(x - ct)$$

   for some suitable, twice continuously differentiable functions, $F(\cdot)$ and $G(\cdot)$. *Hint:* Introduce new coordinates r and s in the xt plane whose level curves are the two families of characteristic lines (2) and (3). Show that the wave equation assumes the form $u_{rs} = 0$ in the new coordinates. Incidentally, this argument also shows that any solution of the wave equation in a convex region R of the xt plane can be extended to a solution in the smallest parallogram formed by segments of characteristic lines. Why? Moreover, this extension is unique. Why?

4. Modify the procedure presented in Section 9 for shaking a string to rest so that it will apply when the string has length $= L$ and the constant c in the wave equation is not necessarily unity.

5. Show that when $T > 2$ a control function $\mu(t)$ always exists such that for any admissible data, the solution of $(BP)_\mu$ has the property that $u(x, T) = u_t(x, T) = 0$, for all $0 \leq x \leq 1$. Show, however, that control functions with this property are not unique.

References

[1] P. C. Parks, "On how to shake a piece of string to a standstill," in *Recent Mathematical Developments in Control*, D. J. Bell, Ed. New York: Academic, 1973, pp. 267–287. (*Proceedings of the Univ. of Bath Conference*, Sept. 1972.)

[2] A. N. Tychonov and A. A. Samarski, *Partial Differential Equations of Mathematical Physics*, vol. II. Holden-Day, 1967.

[3] H. Weinberger, *A First Course in Partial Differential Equations*. Blaisdell.

Notes for the Instructor

Objectives. To show that under suitable conditions a taut flexible string, fastened at one end can be brought to rest, regardless of its initial deflection and initial velocity, by an appropriate shaking action applied to the other end of the string. Moreover, a minimal time is shown to exist in which the string can be brought to rest by shaking, and the optimal shaking strategy which will actually bring the string to rest within this minimal time is explicitly constructed.

The material is fairly selfcontained and would fit well into an Advanced Calculus or Applied Analysis course. A linearized model for the vibrating string is presented in some detail as are several techniques for solving the wave equation. Properties of solutions of the wave equation are examined and geometric characterizations are developed which are then used in the shaking problem for the vibrating string.

Prerequisites. Some familiarity with multivariable calculus and application of Newton's Laws to continuous media.

CHAPTER 22
Heat Transfer in Frozen Soil

Gunter H. Meyer*

1. Description of the Problem

Between the Arctic Ocean and the Brooks Range of Alaska lies the North Slope. In this cold and barren tundra oil has been found and considerable oil exploration and production activity is anticipated for the coming years. A good part of this activity will require the erection of engineering structures such as drill rigs, pipelines, work camps, and, of course, roads and airfields.

Construction on the North Slope, as elsewhere in Arctic and Antarctic regions, is complicated by the fact that the ground is in a state of permafrost. Except for an active surface layer of several feet the moisture in the ground stays permanently frozen down to a depth of 600 ft and more. During most of the year the top layer is also frozen, but during the summer thaw the ice in the soil will melt to an average depth of 2 to 3 ft. It is easy to visualize that man-made structures placed on frozen ground will sink into the ground if the ice in the underlying soil is allowed to melt. The problem is rendered even more severe if the structure emits heat, such as a buried pipeline carrying hot oil.

Over many decades of experience with construction in the Arctic both here and abroad, methods have evolved for coping with the problem of permafrost under buildings and roads. One possible avenue is the insulation of the structure from the ground in some form to prevent melting of the ice and the resulting loss of bearing strength. This insulation usually takes the form of a foundation for the structure consisting of sand, gravel, wood, and possibly man-made material arranged in various layers of different thick-

* Department of Mathematics, Georgia Institute of Technology, Atlanta, GA 30332.

ness. If building material is plentiful within easy access of the construction site the choice of design may not influence the construction cost unduly. On the other hand, many North Slope construction sites are far from any sources of raw building material. Moreover, it may at times be necessary to fly the material to the construction site, since crossing the frozen tundra with tracked vehicles, at this time the main transportation alternative, tears up the ground cover. This vegetation provides a natural insulation and stabilization of the ground; if it is disturbed the soil will melt, the water runoff will wash out gullies which disturb more of the vegetation. A non-reversible cycle will have begun which will scar the countryside permanently.

If indeed building materials have to be transported over long distances, construction cost will depend critically on the weights of the various construction components. In order to find the most economical design for the structure, a fine-tuned method is needed to evaluate whether a given design can meet the requirements placed on it. For structures in permafrost regions, the main requirement is prevention of thawing of unconsolidated moisture-rich soil so that the structure will not sink into muck. Consequently, a major component of a method of evaluation is the determination of the temperature profile in the ground.

This module will describe the formulation and application of a mathematical model for the simulation of heat flow in a layered medium which has been used by a company for a screening examination of various design alternatives for the construction of roads and airstrips on the North Slope. We shall derive in some detail the equations for the temperature field in a partially frozen medium, present an approximate (eventually numerical) method for its solution, and discuss some representative results and conclusions obtained from this model. These points were an integral part of bringing the model to bear on pressing engineering problems. Finally, to satisfy a mathematician's curiosity, we shall discuss briefly the convergence of the approximate solution to the actual solution of the heat flow equation.

Since this chapter was written, oil exploration and production have become commonplace on the North Slope. Extensive oil fields have been developed, and a huge pipeline now carries hot crude oil from the North Slope to the Gulf of Alaska. Engineering problems of considerably greater complexity than those described here have arisen and been solved. But one of the fundamental concerns of construction in the Arctic has remained unchanged: How can frozen soil be kept frozen year-round under and around man-made structures. Thus, while our model is simple compared to thermally active installations such as producing oil wells or hot pipelines, the basic physics discussed here remains valid.

Comments. Some background material on the Arctic and the problems faced by the oil industry may be found in [1]–[3]. A reader interested in the technical aspects of permafrost is referred to the *Proceedings, Permafrost International Conference* [4]. More advanced design problems and thermal models may be found in [11].

2. The Model

A mathematical model is desired for heat flow in and under such structures as roads, airstrips, and building foundations resting on a layered soil in a state of permafrost. Let us focus, for example, on the construction of an airfield. The natural ground may consist of several layers of frozen sand and silt onto which several man-made layers of sand, gravel, possibly an insulator like styrofoam, and concrete or asphalt have been placed to distribute the expected loads and to prevent the frozen ground from melting. It can be expected that during the summer thaw moisture will enter into the man-made top layers which will participate subsequently in the seasonal freezing process and, as will be shown, materially influence the insulating properties of the top layers.

It is possible to construct a full three-dimensional model for heat flow under the structure. However, the resulting equations will be quite complicated and difficult to solve. On the other hand, many years of experience have taught engineers that the thermal effects at the edge of the structure will be felt only within 4–5 ft of the edges. Since the structure may well extend over 30 or more feet in each direction, the temperature field under the major portion of the structure will depend only on depth. For this reason the decision is made to ignore edge effects and to model the temperature field as a function of time and depth only. In this setting the structure foundation becomes an "infinite" layered slab.

It must be emphasized that this approximation is based on past experience; should the results obtained from such a simplified model not be consistent with experience, the model must be improved. Remaining aware of the limitations built into the model is important in any study of this kind.

One distinguishes among three different modes of heat transmission: radiation, convection, and conduction. On the scale of the individual sand grains and pebbles, all three modes are undoubtedly present in a water-logged soil. However, experiments have shown that the overall transmission of heat in such a medium can be described adequately by a pure heat conduction model. We shall now present this model.

Let us begin by considering a single layer of moisture-free homogeneous soil, say, sand. The top of the layer is thought to be exposed to the atmosphere, and the bottom, for the time being, is held at a fixed temperature, In order to determine the temperature profile between top and bottom, a mathematical model for the prediction of heat flow in the layer can be formulated. As stated previously, we will assume that heat flow in this layer is due to conduction, which means that the following three empirical laws (or experimental facts) govern the flow of heat in the layer [5]:

(1) heat flows in the direction of decreasing temperature;
(2) the quantity of heat gained or lost by a body during a temperature change is proportional to the mass of the body and to the temperature change;

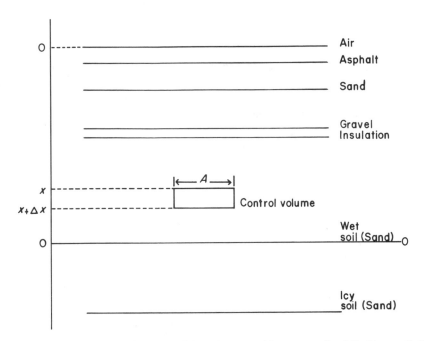

Figure 22.1. Typical Arrangement of Four Structural Layers on Partially Frozen Soil

(3) the rate of heat flow through an area is proportional to the area and to the temperature gradient normal to the area.

If u is the average temperature of a control volume, Q its heat content, and ρ its density, then with reference to Figure 22.1 the second of these laws can be expressed as

$$\Delta Q = c \Delta u \, \rho \Delta x A \tag{1}$$

The proportionality constant c is known as the specific heat of the material; $\rho \Delta x A$ is, of course, the mass of the control volume. The third law is known as Fourier's law of heat conduction. If we assume that the heat flow in a neighborhood of the point of interest in the layer is strictly vertical, then we can write for the heat loss ΔQ per time span Δt at the mean time t through an area A at depth x in the direction of increasing x,

$$\frac{\Delta Q}{\Delta t} = -kA \frac{\partial u}{\partial x}(x, t), \tag{2}$$

where the constant k of proportionality is known as the conductivity of the material. Consider now an energy balance for the control volume in Figure 22.1. The amount of heat flowing through the face at $x + \Delta x$ minus that flowing through the area at x must be equal to the amount of heat lost per time span Δt, or

$$-kA\frac{\partial u}{\partial x}(x + \Delta x, t) + kA\frac{\partial u}{\partial x}(x, t) = -c\rho A\Delta x\frac{\Delta u}{\Delta t}.$$

Assuming, as is reasonable for the system at hand, that the temperature varies smoothly with time and space, we can take the limit as Δx, $\Delta t \to 0$ and obtain Fourier's heat conduction equation for the temperature u at any point (x, t) in space and time

$$k\frac{\partial^2 u}{\partial x^2}(x, t) = c\rho\frac{\partial u}{\partial t}(x, t). \tag{3}$$

At $x = X$ the temperature is held fixed at some value u_X, i.e.,

$$u(X, t) = u_X. \tag{4}$$

The proper boundary condition at $x = 0$ is not quite as easy to formulate, since the soil and surrounding air may well be at different temperatures. We shall say more about this later. At this point we shall simply assume that the temperature of the top layer of soil is a specified function of time, say

$$u(0, t) = \alpha(t). \tag{5}$$

Finally, we need an initial temperature profile for the soil at some starting time t_0 of our observation. For convenience, we shall choose $t_0 = 0$ and write

$$u(x, 0) = u_0(x). \tag{6}$$

The actual form of u_0 may vary from problem to problem.

Equations (3)–(6) define a well-posed initial boundary value problem. Its solution for reasonable (say, continuous) data is known to exist, and experience has shown that it predicts quite well the temperature field within a homogeneous medium subject to one-dimensional heat flow. Of course, in order to use this model for engineering calculations, one needs to know the functions $\alpha(t)$, $u_0(x)$ and the physical constants k, c, and ρ for the layer. Such data either are already, or are rapidly becoming, available from geological, climatological, and soil studies.

So far, we have assumed that the layer was homogeneous and dry. Let us now go one step further and assume that the layer still is homogeneous but contains moisture in the form of water and ice. For definiteness, let us choose the layer of sand of Figure 22.1 but suppose that at a given time t_0 the sand is wet above the line 00, while frozen below it (this requires, of course, that $\alpha(t_0) > 32°F$, $u_X < 32°F$). Considering the sand as two layers is now convenient. For the upper layer equation (3) and the boundary condition (5) are valid, where k, c, and ρ are values determined for a wet sand. Below the freezing line the heat equation (3) is valid when values for a frozen sand are chosen for k, c, and ρ. Of course, the boundary condition (4) remains in effect. In order to compute the temperature profile at future times, we need to link the temperatures and heat fluxes in the wet and frozen sand at the interface 00 (see Figure 22.1), the so-called free interface. The temperature condition is simple. Both the bottom of the wet sand and the top of the frozen sand must be exactly at the freezing temperature of water (taken to be 32°F) so that

$$u_{SW}(s(t), t) = u_{SI}(s(t), t) = 32 \qquad (7)$$

where SW and SI denote the wet and icy layer of sand and where $s(t)$ denotes the location of the free interface at time t. The free interface, of course, will move in time. Suppose that in a time span Δt the interface has moved downward by a distance Δs. Then the amount of heat used up in melting a block $\Delta s \cdot A$ of frozen sand is given by

$$Q = \rho_{SI}\lambda\Delta s \cdot A$$

where ρ_{SI} is the density of the frozen sand and λ is the latent heat of fusion of the frozen sand. The heat required for melting is provided by the heat flux out of the wet sand minus the heat lost by transmission into the icy sand. Thus

$$Ak_{SI}\frac{\partial u_{SI}}{\partial x} - Ak_{SW}\frac{\partial u_{SW}}{\partial x} = \rho_{SI}\lambda\frac{\Delta s}{\Delta t}A.$$

Letting $\Delta t \to 0$, we find the second free interface condition as

$$k_{SI}\frac{\partial u_{SI}}{\partial x} - k_{SW}\frac{\partial u_{SW}}{\partial x} = \rho_{SI}\lambda s'(t). \qquad (8)$$

In order to find the temperature distribution in a partially frozen homogeneous soil, we need to solve simultaneously for the solution $\{u_{SW}, u_{SI}, s(t)\}$ of the two heat equations valid above and below $s(t)$, subject to the given boundary and interface conditions.

The discussion of our model is not yet complete. The structures we have in mind consist of several homogeneous but dissimilar layers which in turn may rest on a layered soil. For each homogeneous layer the above development holds. At the boundary between two layers we shall assume perfect thermal contact so that the temperature and the heat flux are continuous at this point. For example, if at depth $x = x_k$ an interface occurs between a sand and underlying gravel layer, then

$$\lim_{\varepsilon \to 0} [u(x_k + \varepsilon, t) - u(x_k - \varepsilon, t)] = 0$$

$$\lim_{\varepsilon \to 0} \left[k_g\frac{\partial u}{\partial x}(x_k + \varepsilon, t) - k_s\frac{\partial u}{\partial x}(x_k - \varepsilon, t) \right] = 0,$$

where the indices s and g denote sand and gravel. Note that the actual values for the conductivities k_g and k_s depend on whether the sand and gravel at location x_k are wet or frozen.

Let us summarize the equations for the mathematical model of heat conduction in a layered medium. We shall suppose that altogether M fixed dissimilar layers, exist where the ith layer extends from depth x_{i-1} to x_i. The thermal parameters are as follows.

$k_i^f(k_i^w)$ thermal conductivity of the ith layer when frozen (wet);

ρ_i density of the ith layer (assumed equal for wet and frozen soil— not a serious restriction);

$c_i^f(c_i^w)$ heat capacity of the ith layer when frozen (wet);

λ_i latent heat of fusion for the ith layer.

The object is to find the functions

$u(x, t)$ temperature at x at time t;

$s(t)$ interface between the frozen and wet ground.

As additional data we specify

$\alpha(t)$ surface temperature of the top layer, i.e., at $x = 0$;

u_X constant temperature at the bottom of the last layer, i.e., at $x = x_M$, always held below freezing;

$u_0(x)$ temperature profile in the layers at the start of the computation.

With this notation the temperature profile in the layered medium can be determined from the following free boundary problem:

$$k_i^\gamma \frac{\partial^2 u}{\partial x^2} - \rho_i c_i^\gamma \frac{\partial u}{\partial t} = 0, \tag{9a}$$

with the boundary conditions at the surface,

$$u(0, t) = \alpha(t), \tag{9b}$$

the heat flow conditions at the fixed interfaces,

$$\lim_{\varepsilon \to 0} \left[u(x_i - \varepsilon, t) - u(x_i + \varepsilon, t) \right] = 0, \tag{9c}$$

$$\lim_{\varepsilon \to 0} \left[k_i^\gamma \frac{\partial u}{\partial x}(x_i - \varepsilon, t) - k_{i+1}^\gamma u(x_i + \varepsilon, t) \right] = 0, \tag{9d}$$

and the bottom boundary condition

$$u(X, t) = u_X, \tag{9e}$$

where $\gamma = w$ if the ground is wet $(x \in (0, s(t)))$, and where $\gamma = f$ if the ground is frozen $(x \in (s(t), X))$. The position of $s(t)$ is determined from the free interface condition

$$\lim_{\varepsilon \to 0} u(s(t) + \varepsilon, t) = \lim_{\varepsilon \to 0} u(s(t) - \varepsilon, t) = 32 \tag{9f}$$

$$\lim_{\varepsilon \to 0} \left[k_j^f \frac{\partial u}{\partial x}(s(t) + \varepsilon, t) - k_j^w \frac{\partial u}{\partial x}(s(t) - \varepsilon, t) \right] = \rho_j \lambda_j \frac{ds}{dt} \tag{9g}$$

where j is the index of the layer which contains $s(t)$. Finally, an initial temperature distribution at the beginning of the computation, say at $t = 0$, is assumed

$$u(x, 0) = u_0(x). \tag{9h}$$

The point s_0 where $u_0(x) = 32$ is the initial location of free interface so that

$$s(0) = s_0. \tag{9i}$$

We shall always assume that s_0 is uniquely determined by $u_0(x)$. It is true, however, that the effect of $u_0(x)$ and s_0 on the solution $\{u(x, t), s(t)\}$ diminishes rapidly as the system evolves with time.

At first glance, the equations (9) look forbidding; however, with a little effort some regularity will become apparent. Pick a point with coordinate (depth) x under the structure. This point will fall into one particular layer of material, say, the ith layer (which may be gravel, soil, or whatever other materials are present). The temperature at time t at this point is $u(x, t)$. It is either above freezing or below. If $u(x, t) > 32°$, then the moisture at this point is present as water and thermal parameters are indexed by $\gamma = w$. If $u(x, t) < 32°$ the moisture is present as ice and the thermal parameters are indexed by $\gamma = f$. The dividing line between the wet and frozen material lies at depth $s(t)$ which falls into some layer, say, the jth layer. The movement of this line is given by (9g) and must be determined together with the temperatures $u(x, t)$ for all M layers.

The problem (9a)–(9i) is today commonly known as a two-phase Stefan problem in honor of the Austrian physicist J. Stefan (1835–1893) who used such models to study ice–water systems. Problems of this type, however, occur quite frequently in a variety of applications and not just heat transfer with change of phase. They arise in chemical reaction, biological diffusion, viscoplastic diffusion, and even stellar evolution [6], [7]. The development in the succeeding sections, particularly the solution algorithms, can be readily modified to accomodate many problems with similar mathematical structure.

3. A Solution Algorithm

In an industrial environment the formulation of the mathematical model and the choice of the method of solution, either exact or numerical, are of paramount importance. In general, little leeway, other than some standard linearizing assumptions, exists in formulating the model because the laws of physics have to be obeyed. On the other hand, if the resulting equations do not have a readily obtainable exact solution, considerable latitude exists in the choice of mathematics for solving them approximately. The full interface problem (9) with a time-varying surface temperature $\alpha(t)$ does not easily lend itself to a closed-form solution in terms of Green's and Neumann's functions, which are often used for elementary calculations, so that one

usually resorts to a numerical solution technique. The following approach
has proved efficient.

First of all, the so-called method of lines is employed to replace the heat
equation by a sequence of boundary value problems for ordinary differential
equations. Thus, suppose that we need the solution of our freezing–melting
problem over the time interval $[0, T]$. Let $N > 0$ be an integer, and define
the partition $\{0 = t_0 < t_1 < \cdots < t_N = T\}$ of the interval $[0, T]$ where
$\Delta t = T/N, t_n = n\Delta t$. Replace the partial derivative u_t at $t = t_n$ and the front
velocity $\dot{s}(t)$ by the backward difference quotients

$$u_t(x, t_n) \approx \frac{u(x, t_n) - u(x, t_{n-1})}{\Delta t} \equiv \frac{u_n(x) - u_{n-1}(x)}{\Delta t}$$

$$\dot{s}(t_n) \approx \frac{s(t_n) - s(t_{n-1})}{\Delta t} \equiv \frac{s_n - s_{n-1}}{\Delta t}$$

where $u_n(x)$ and s_n are meant to denote the solution at time $t = t_n$. Then at
each time level t_n, (9) can be approximated by the following free interface
problem for ordinary differential equations

$$k_i^\gamma(u_n)'' - \frac{\rho_i c_i^\gamma}{\Delta t}[u_n - u_{n-1}(x)] = 0 \qquad (u_n)'' \equiv \frac{d^2 u_n}{dx^2} \tag{10}$$

$$u_n(0) = \alpha(n\Delta t) \equiv \alpha_n \tag{11}$$

$$\lim_{\varepsilon \to 0}[u_n(x_i + \varepsilon) - u_n(x_i - \varepsilon)] = 0$$
$$n = 1, \cdots, N, i = 1, \cdots, M \tag{12}$$

$$\lim_{\varepsilon \to 0}[k_i^\gamma u_n'(x_i - \varepsilon) - k_{i+1}^\gamma u_n'(x_i + \varepsilon)] = 0 \tag{13}$$

$$u_n(X) = u_X \tag{14}$$

$$\lim_{\varepsilon \to 0} u_n(s_n + \varepsilon) = \lim_{\varepsilon \to 0} u_n(s_n - \varepsilon) = 32 \tag{15}$$

$$\lambda_j \rho_j \frac{s_n - s_{n-1}}{\Delta t} - \lim_{\varepsilon \to 0}[k_j^f u_n'(s_n + \varepsilon) - k_j^w u_n'(s_n - \varepsilon)] = 0 \tag{16}$$

where, again, $\gamma = w$ if $u_n \geq 32$, $\gamma = f$ if $u_n < 32$. Since $u_0(x)$ and s_0 are
given, these problems must be solved successively for $n = 1, 2$, etc., for the
temperature u_n and the interface s_n.

In the terminology of ordinary differential equations the system (10)–(16)
describes an interface problem with $M - 1$ (interior) fixed interfaces and
one free interface, the location of the freezing melting isotherm. In general,
fixed interface problems for linear equations are fairly easy to solve, but
free interface problems are not because the problem is no longer linear due
to the presence of the unknown interface. The solution algorithm proposed
here is based on a conversion to initial value problems. To make the ideas
precise, consider for the moment the simple two-point boundary value
problem

$$u' = Au + Bv + F(x) \qquad u(0) = a$$
$$v' = Cu + Dv + G(x) \qquad u(X) = b \tag{17}$$

where F and G are assumed to be continuous.

A possible way of solving (17) is to assume an initial value $v(0) = r$ and to integrate the differential equations for u and v over $[0, X]$. Since (17) is a linear inhomogeneous system, this integration can always be carried out. In fact, the solution may be written as

$$u(x) = \varphi_1(x)r + \psi_1(x)$$
$$v(x) = \varphi_2(x)r + \psi_2(x)$$

where φ_i and ψ_i are determined from the variation of constants procedure. If the second equation is used to eliminate r from the first, then it is seen that, regardless of the initial condition r, the functions u and v are related through the transformation

$$u(x) = U(x)v(x) + w(x) \tag{18}$$

which is known as the Riccati transformation. The functions U and w can be computed directly without going through the variation of constants procedure. Since $u(0) = a$ regardless of what we choose for v it follows that $U(0) = 0$ and $w(0) = a$. To satisfy $u(X) = b$, we must choose

$$b = U(X)v(X) + w(X)$$

or

$$v(X) = \frac{b - w(X)}{U(X)}. \tag{19}$$

The defining equations for U and w can be obtained from the variation of constants approach described in great detail in [8]. Here we shall merely verify the validity of the appropriate equations.

Let U and w be solutions of

$$U' = B + AU - DU - CU^2, \quad U(0) = 0 \tag{20a}$$
$$w' = [A - CU(x)]w - U(x)G(x) + F(x), \quad w(0) = a. \tag{20b}$$

These are two initial value problems which can be solved at least in a neighborhood of the initial point $x = 0$. Suppose that $\{U, w\}$ exist over $[0, X]$. Let $\hat{v}$ be given by (19), i.e.,

$$\hat{v} = \frac{b - w(X)}{U(X)}$$

and let v be the solution over $[0, X]$ of the linear equation

$$v' = [CU(x) + D]v + Cw(x) + G(x), \qquad v(X) = \hat{v}. \tag{21}$$

Then the pair

$$\{u(x) \equiv U(x)v(x) + w(x), v(x)\} \qquad (22)$$

is a solution of (17). Indeed, we see that

$$u(0) = U(0)v(0) + w(0) = w(0) = a \qquad u(X) = U(X)\hat{v} + w(X) = b$$

and that $v' = Cu + Dv + G(x)$; moreover,

$$u' = U'v + Uv' + w' = [B + AU - DU - CU^2]v$$
$$+ U[CU + D]v + UCw + UG + Aw - UCw - UG + F$$
$$= A(Uv + w) + Bv + F$$
$$= Au + Bv + F(x).$$

Equations (20)–(22) are commonly known as the invariant imbedding equations for the boundary value problem (17). Suppose next that a fixed interface is present at $x = L$. Let us write the problem as

$$u_i' = A_i u_i + B_i v_i + F_i(x),$$
$$\qquad\qquad\qquad\qquad\qquad\qquad i = 1, 2 \qquad (23)$$
$$v_i' = C_i u_i + D_i v_i + G_i(x),$$

$$u_1(0) = a \qquad u_2(X) = b \qquad u_1(L) = u_2(L) \qquad k_1 v_1(L) = k_2 v_2(L)$$

$$(24)$$

where the subscript 1 refers to the solution $\{u_1, v_1\}$ over $[0, L]$, and where 2 designates the solution over $[L, X]$.

We have seen that over each subinterval the solutions can be represented by (22), where U_i and w_i are given by the differential equations (20a) and (20b) and where $\hat{v}_2$ is given by (19), all functions being properly subscripted with 1 for the interval $[0, L]$ and 2 for $[L, X]$. Missing still are the initial values for $U_2(L)$, $w_2(L)$, and $v_1(L)$. From the fixed interface condition we obtain

$$U_1(L)v_1(L) + w_1(L) = U_2(L)v_2(L) + w_2(L)$$
$$k_1 v_1(L) = k_2 v_2(L).$$

We do not yet know $v_1(L)$; hence these two equations must be solved for $U_2(L)$ and $w_2(L)$ such that the above relations hold for arbitrary $v_1(L)$.

Notice that if we set

$$U_2(L) = U_1(L)\frac{k_2}{k_1} \qquad w_2(L) = w_1(L) \qquad v_1(L) = \frac{k_2}{k_1}v_2(L) \qquad w_2(L) = w_1(L)$$

$$(25)$$

then the functions

$$\{u_1(x) \equiv U_1(x)v_1(x) + w_1(x), v_1(x)\} \text{ and } \{u_2(x) \equiv U_2(x)v_2(x) + w_2(x)\}$$

solve the fixed interface problem.

Finally, let us assume that the boundary $x = X$ is not fixed but free, i.e., undetermined, but that also a condition on $v(X)$ is specified. To be specific, let us suppose that we wish to solve (23) subject to

$$u_1(0) = a \qquad u_1(L) = u_2(L) \qquad k_1 v_1(L) = k_2 v_2(L)$$

$$u_2(s) = f(s) \qquad v_2(s) = g(s)$$

(26)

where s denotes the (unknown) location of the free boundary and g is a given function over $[L, \infty]$. As outlined previously, over $[0, L]$ and $[L, s]$ we have available the representation

$$u_1(x) = U_1(x)v_1(x) + w_1(x)$$

$$u_2(x) = U_2(x)v_2(x) + w_2(x)$$

where U_i and w_i, $i = 1, 2$, are known functions, while v_2 and consequently v_1 can be found only after s is known. However, the boundary conditions at s require that s be chosen such that

$$f(s) = U_2(s)g(s) + w_2(s).$$

Thus consider the functional $\varphi(x) \equiv U_2(x)g(x) + w_2(x) - f(x)$. Since U_2 and w_2 are assumed to be known (or computed) over $[L, \infty]$, we simply need to find a root s of $\varphi(x) = 0$. If such an s can be found, then we integrate (21) for $i = 2$ subject to $v(s) = g(s)$ backward over $[L, s]$ and continue with v_1 subject to $v_1(L) = k_2 v_2(L)/k_1$ over $[0, L]$ in order to obtain the complete solution $\{u_i, v_i\}$, $i = 1, 2$. For easy reference we shall call the integration of U_i and w_i the "forward sweep" and that of v_i the "backward sweep." Thus the free boundary problem is solvable by completing the forward sweep, by determining the free boundary, and by carrying out the backward sweep. This approach will now be adapted to the Stefan problem for a layered soil.

Starting with the initial data $u(x, 0) = u_0(x)$ and $s(0) = s_0$, we shall advance the solution from time to time. Thus let us suggest that the temperature profile $u_{n-1}(x)$ and the freezing melting isotherm s_{n-1} are known at time $t_{n-1} = (n - 1)\Delta t$ and that we are to find u_n and s_n. For ease of notation, let us suppress the subscript n denoting time and instead introduce the subscript i denoting the temperature in the ith layer (at time t_n). In addition, it will be helpful on occasion to specifically indicate whether the temperature is above or below freezing. We shall do so with the superscript w for wet and f for frozen ground. For example, with this notation the interface condition (15) (at $t = t_n$) can be written as

$$u^f(s) = u^w(s) = 32.$$

(The reader is cautioned to distinguish between the function $w_i(x)$ arising in the Riccati transformation and the superscript w denoting the wet phase.)

In order to cast the problem (10)–(16) into a form similar to (17), we shall choose instead of (10), the equivalent first-order system

$$u_i' = v_i$$

(27)

$$v_i' = \frac{c_i^\gamma \rho_i}{k_i^\gamma \Delta t}[u_i - u_{n-1}(x)] \equiv \eta_i^\gamma [u_i - u_{n-1}(x)] \tag{28}$$

where $\eta_i^\gamma = c_i^\gamma \rho_i / k_i^\gamma \Delta t$ and where $i = 1, \cdots, M$.

The fixed boundary and interface conditions are

$$u_1(0) = \alpha(n\Delta t),$$

$$u_i(x_i) = u_{i+1}(x_i),$$

$$k_i^\gamma v_i(x_i) = k_{i+1}^\gamma v_i(x_{i+1}), \qquad i = 1, \cdots, M - 1 \tag{29}$$

$$u_M(X) = u_X.$$

In addition, we have the free interface conditions

$$u_j^f(s) = u_j^w(s) = 32 \tag{32}$$

$$k_j^f v_j^f(s) - k_j^w v_j^w(s) = \lambda_j \rho_j \frac{s - s_{n-1}}{\Delta t} \tag{33}$$

where j is the index of the layer to which s belongs at $t = t_n$.

The solution algorithm outlined above is readily adapted to this problem. Suppose for the moment that $\alpha(n\Delta t) > 32$ so that a water phase is present. Then we carry out the forward sweep. We know that u_i^w and v_i^w are related through the Riccati transformation

$$u_i^w = U_i(x)v_i^w + w_i(x) \tag{34}$$

where U_i and w_i are found successively from the initial value problems

$$U_i' = 1 - \eta_i^w U_i^2 \qquad U_1(0) = 0 \qquad U_{i+1}(x_i) = \frac{k_{i+1}^w U_i(x_i)}{k_i^w} \tag{35}$$

$$w_i' = -\eta_i^w U_i(x)[w_i - u_{n-1}(x)]$$

$$w_1(0) = \alpha(n\Delta t) \qquad w_i(x_i) = w_{i+1}(x_i). \tag{36}$$

Similarly, the solutions u_i^f and v_i^f for the ice phase are related through a Riccati transformation which we shall write as

$$u_i^f = R_i(x)v_i^f + z_i, \tag{37}$$

where R_i and z_i are determined from the forward sweep

$$R_i' = 1 - \eta_i^f R_i^2 \qquad R_M(X) = 0 \qquad R_{i-1}(x_{i-1}) = \frac{k_{i-1}^f R_i(x_{i-1})}{k_i^f} \tag{38}$$

$$z_i' = -\eta_i^f R_i(x)[z_i - u_{n-1}(x)]$$

$$z_M(X) = u_X \qquad z_{i-1}(x_{i-1}) = z_i(x_{i-1}). \tag{39}$$

(Note that this forward sweep carries us through the layers from the lower horizon X to the surface so that these equations are integrated in the direction of decreasing x.) It now remains to find the free interface. If at an arbitrary point x in the jth layer the temperature of both phases is equal to 32, then the corresponding gradients v_i are found from (34) and (37) as

$$v_j^w(x) = \frac{32 - w_j(x)}{U_j(x)} \qquad v_j^f = \frac{32 - z_j(x)}{R_j(x)}.$$

If this representation is substituted into the interface condition (23), we find that s must be a root of the functional (at the nth time level)

$$\varphi(x) \equiv k_j^f \left(\frac{32 - z_j(x)}{R_j(x)} \right) - k_j^w \left(\frac{32 - w_j(x)}{U_j(x)} \right) - \lambda_j \rho_j \frac{x - s_{n-1}}{\Delta t} = 0. \qquad (40)$$

Once such a root has been found we can complete the backward sweep by integrating (backward over $[0, s]$)

$$v_i^{w'} = \eta_i^w U_i(x)v_i + \eta_i^w [w_i(x) - u_{n-1}(x)] \qquad v_i^w(s) = \frac{32 - w_i(s)}{U(s)} \qquad (41)$$

and (forward over $[s, x_m]$)

$$v_i^{f'} = \eta_i^f R_i(x)v_i^f + \eta_i^f [z_i(x) - u_{n-1}(x)] \qquad v_i^f(s) = \frac{32 - z_i(s)}{R_i(s)}. \qquad (42)$$

The complete temperature profile in the layered soil is given by

$$u_n(x) = U_i(x)v_i^w(x) + w_i(x), \qquad x \in [x_{i-1}, x_i], \qquad x < s$$
$$u_n(x) = R_i(x)v_i^f(x) + z_i(x), \qquad x \in [x_{i-1}, x_i], \qquad x > s.$$

If $\alpha(n\Delta t) < 32$ then no water phase is present. In this case, we can dispense with the forward sweep involving U_i and w_i and with the functional (38); instead we carry out the forward sweep for the ice phase by computing R_i and z_i. At the surface $x = 0$ the temperature is $u^f(0) = \alpha(n\Delta t)$. Hence in view of the Riccati transformation (37) the gradient $v_1^f(0)$ must be chosen such that

$$\alpha(n\Delta t) = R_1(0)v_1^f(0) + z_1(0)$$

or

$$v_1^f(0) = \frac{\alpha(n\Delta t) - z_1(0)}{R_1(0)}.$$

This value for v_1^f and $s = 0$ are used to complete the backward sweep by integrating (42).

In summary, we have a well-defined (analytical) solution algorithm for the free interface problem (1) of heat transfer with change of phase in a layered soil. We have still to show that this algorithm, which was presented only formally, is actually applicable. Thus we need to show that given the solution $\{u_{n-1}(x), s_{n-1}\}$ at the previous time level we can actually compute $\{u_n, s_n\}$ with this algorithm.

In order to demonstrate that the above sweep method can be used to compute a solution $\{u_n, s_n\}$, three properties of the above equations must be established, namely, that the Riccati equations (35) and (38) have bounded solutions and that the functional equation (40) has a root. If this is indeed the case, then all other equations have solutions since they are linear ordinary differential equations. The existence question for the Riccati equations is readily resolved. We see from (35) that $U_1(0) = 0$ and $U_1'(0) = 1$ and $U_1'(\hat{x}) = 1$ whenever $U_1(\hat{x}) = 0$. Hence $U_1 > 0$ on $[0, x_1]$. Moreover, $U_1' < 0$ if $U > \sqrt{1/\eta_1^w}$ which assures that $U_1 \leqq 1/\sqrt{\eta_1^w}$ on $[0, x_1]$. A similar analysis can be applied to U_2 and successively to U_j, $j = 3, \cdots, M$. A similar argument shows that $-\infty < R_j < 0$ on $[0, X]$.

The next problem to be tackled is the existence of a zero s for the function (38). Since this function does not come into play if the surface temperature is below freezing, let us suppose that $\alpha(n\Delta t) > 32$. It follows from (38) and (40) that $\lim_{x \to 0+} U_1(x) = 0+$ and $\lim_{x \to 0+} \Phi(x) = -\infty$, and similarly that $\lim_{x \to X-} \Phi(x) = +\infty$, regardless of where s_{n-1} may be located. The question thus would resolve itself if Φ were continuous. Because z_j and w_j are continuous at fixed interfaces and $k_j^w U_{j+1}(x_j) = k_{j+1}^w U_j(x_j)$ and $k_j^f R_{j-1}(x_{j-1}) = k_{j-1}^f R_j(x_{j-1})$, we see that the first two terms of (40) are continuous. The last term, however, exhibits jumps at the interfaces because, in general, $\lambda_j \rho_j \neq \lambda_{j+1} \rho_{j+1}$. From a conceptual point of view, this behavior does not cause any difficulties. We shall simply trace $\Phi(x)$ from $x = 0$ until it first turns nonnegative; this point is well-defined if Φ is taken to be continuous from the left at the interfaces.

In summary, then, the above algorithm constitutes a feasible mathematical method to compute the solution $\{u_n, s_n\}$ for the free interface problem (4), since at each time level all equations have solutions.

Two important aspects of this problem remain to be investigated. The first of these is of overriding concern in an industrial environment. How does one obtain actual (i.e., numerical) answers with the above method? After all, the method is stated in terms of initial value problems for ordinary differential equations. Hence methods for their integration must be considered. In addition, it is necessary to find the zero of a nonlinear, possibly discontinuous function. This problem also needs some comments. We shall next discuss in some detail how to obtain numerical answers with the above method. The second aspect concerns convergence of the computed answers to the solution of the continuous problem. This point is generally of interest to the mathematician and will be discussed in the last section; nonmathematicians usually decide on convergence on the basis of the computed results as illustrated below.

The numerical solution of the forward and backward sweep equations can be obtained quite simply. Since in replacing the continuous Stefan problem with the "by lines approximation" a simple backward difference quotient was employed, the truncation error of the approximation is $0(\Delta t)$. Hence it seems more than adequate to apply a second-order integrator for the special variable. Moreover, it is generally observed that implicit integration techniques exhibit better stability properties than explicit methods. The drawback of implicit methods is due to the fact that an equation must be solved for the unknown. If this equation is nonlinear this problem can be complicated. Fortunately, this difficulty does not occur if the trapezoidal rule is applied to carry out the sweeps. Let us trace out the application of the trapezoidal rule for the equations at hand. For definiteness, let us suppose we need to find $U_1(x)$ and $w_1(x)$ over $[0, x_1]$. We define a partition $\{0 = x^0 < x^1 < \cdots < x^m = x_1\}$ of $[0, x_1]$ and set $\Delta x_j = x^j - x^{j-1}, j = 1, \cdots, M$. Then U_1 and w_1 at the mesh points $\{x^j\}$ are found recursively from the trapezoidal formula

$$U_1(x^j) = U_1(x^{j-1}) + \frac{\Delta x_j}{2}[2 - \eta_1^w(U_1^2(x^j) + U_1^2(x^{j-1}))], \qquad U_1(x^0) = 0$$

$$w_1(x^j) = w_1(x^{j-1}) + \frac{\Delta x_j}{2}[-\eta_1^w U_1(x^j)(w_1(x^j) + u_{n-1}(x^j))$$

$$-\eta_1^w U_1(x^{j-1})(w_1(x^{j-1}) + u_{n-1}(x^{j-1}))],$$

$$w_1(x^0) = \alpha(n\Delta t).$$

The first of these equations is quadratic and can be solved in closed form for the unknown $U_1(x^j)$; once $U_1(x^j)$ is known, the second equation is linear in $w_1(x^j)$ and can be solved without difficulty. Similar expressions are used for R_j and z_j. In this manner, U_i, w_i, R_j, and z_j can be computed at discrete mesh points distributed over the total interval $[0, X]$.

Once U_i, R_j, w_i, and z_j are known at the mesh points over the entire interval $[0, X]$ the value of Φ is known at the mesh points. In order to find the free interface where $\Phi(x) = 0$, we may simply evaluate the functional at each mesh point and place s by linear interpolation between the first two mesh points $\{x^k, x^{k+1}\}$ for which $\Phi(x^k) \cdot \Phi(x^{k+1}) \leq 0$. In general, s will not coincide with a mesh point of the partition. In this case several options exist. Either s is added to the partition and $U_i(s)$, $w_i(s)$, $R_j(s)$, and $z_j(s)$ are computed with the trapezoidal rule, or s is moved to the nearest mesh point. The latter course is quite simple and proved adequate. Finally, once the location of s is fixed, the trapezoidal rule is used to integrate v_i^w and v_j^f, which then are used to define the complete temperature profile $u_n(x)$ over $[0, X]$ at time $t = n\Delta t$.

Comment. The forward/backward sweep method is commonly called the method of invariant imbedding. A detailed discussion of this initial value solution algorithm may be found in [8].

4. An Application

The above model and numerical solution technique were used for the evaluation of various design alternatives for an airstrip on permafrost. The model was considered applicable because thermal edge effects do not appear to extend beyond 4–5 ft from the edge of the runway so that heat flow under the load bearing portion of the structure is essentially one-dimensional.

Because some uncertainty exists in the data necessary to carry out the computation, all parameters had to vary over considerable ranges which necessitated a lot of computation. Fortunately, the facilities available and the speed of the computer algorithm allowed the effective use of computer graphics. Rather than on reams of paper the computed results were displayed immediately after computation on a television screen. Two illustrations, redrawn from photographs of the television screen show representative results for a structure of four man-made layers on a permafrost soil for two different temperature inputs.

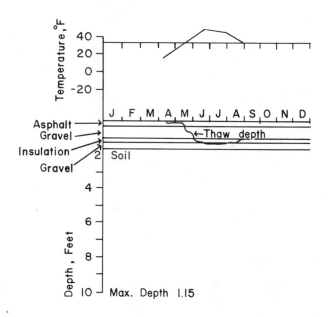

LYR	MATERIAL	THICK	DEPTH		MONTH	TEMPERATURE
1	Asphalt	0.5	0.5		JAN	−9.9
2	Gravel	0.5	1.0		FEB	−13.3
3	Insulation	0.2	1.2		MAR	−7.7
4	Gravel	0.5	1.7		APR	8.6
5	Soil	200.0	201.7		MAY	25.9
					JUN	39.1
	CONDUCTIVITY				JUL	47.5
LYR	ICE	WATER	SAT.	POR.	AUG	45.2
1	20.00	20.00	0.00	0.20	SEP	36.0
2	75.00	50.00	1.00	0.35	OCT	22.6
3	1.20	1.20	0.00	0.20	NOV	6.5
4	75.00	50.00	1.00	0.35	DEC	−3.7
5	60.00	40.00	1.00	0.50		

Figure 22.2. Thaw Depth Under Asphalt-Topped Gravel-Insulation Sandwich

Extensive experiments had shown that the results are seasonally periodic
and that no long-range effects accumulate over large time spans (up to 40
years). Therefore, only one cycle is exhibited. The computation starts in
April and continues until the surface temperature reaches the freezing point.
The main quantity of interest is, of course, the maximum thaw depth; since
it always had begun decreasing at this time the computation was terminated
at this point (between the early and middle part of September).

Figure 22.2 shows a plot of the surface temperature $\alpha(t)$ versus time (the
time axis is horizontal and labeled with J(anuary), F(ebruary), etc.) and a

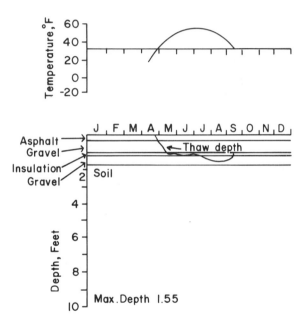

LYR	MATERIAL	THICK	DEPTH	MONTH	TEMPERATURE
1	Asphalt	0.5	0.5	JAN	−9.9
2	Gravel	0.5	1.0	FEB	−13.3
3	Insulation	0.2	1.2	MAR	−7.7
4	Gravel	0.5	1.7	APR	8.6
5	Soil	200.0	201.7	MAY	36.0
				JUN	50.0

		CONDUCTIVITY		JUL	57.5
LYR	ICE	WATER	SAT.	POR.	AUG	55.0
1	20.00	20.00	0.00	0.20	SEP	46.0
2	75.00	50.00	1.00	0.35	NOV	6.5
3	1.20	1.20	0.00	0.20	DEC	−3.7
4	75.00	50.00	1.00	0.35		
5	60.00	40.00	1.00	0.50		

Figure 22.3. Thaw Depth Under Asphalt-Topped Gravel-Insulation Sandwich with
Abnormally High Surface Temperature

plot of the computed thaw depth $s(t)$ through a structure of asphalt, gravel, an insulator like styrofoam, and gravel resting on a soil in a state of permafrost. As in seen, for the thermal parameters displayed on the right of the illustration the interface $s(t)$ does not enter the soil.

Figure 22.3 shows the same system when the summer temperatures are increased by 10°F. The interface is now seen to enter the moisture-bearing gravel underneath the insulator; however, the permafrost still escapes the summer thaw. Because of the speed of the algorithm, many other temperature inputs $\alpha(t)$ were tried. This was important because of the uncertainty of the soil surface temperature. It has been found that the annual median soil temperature is several degrees higher than the corresponding air ground temperature. However, daily records are needed for this model which are not yet known. An attempt has also been made to include convective Newtonian cooling on the surface in the form of

$$k\frac{\partial u}{\partial x}(0, t) = h(t)(u(0, t) - \alpha(t))$$

where the heat transfer coefficient h is adjusted according to snow cover and wind velocity. Again, reliable data are lacking, so that the computed results are taken to indicate trends rather than absolute values.

In carrying out these computations the influence of the time and space steps on the computed answers was examined. It is generally believed than an implicit approximation of the type discussed here is stable; hence it is common to choose the mesh sizes sufficiently small so that the computer answers change little on further mesh refinements, and sufficiently large so that the computation is rapid. Practical experience will usually point to the optimum mesh sizes.

Comments. The reader interested in further details on the parameter study is referred to [9] and to the references given there. Additional data on soil and permafrost may be found in [4], [10].

5. Convergence

From an engineering and management point of view, the model developed represents a fairly complete description of heat transfer in a layered soil. It was demonstrated that this model has a solution $\{u_n, s_n\}$ at each time level and that the solution can be obtained numerically by quite elementary techniques. For a mathematician the problem does not end here, since no assurance has been given that the approximate solutions do indeed come close, in some sense, to the continuous solution of the Stefan problem (9). It is still required to prove convergence of $\{u_n, s_n\}$ to $\{u(x, n\Delta t), s(n\Delta t)\}$ as $\Delta t \to 0$. With our present understanding, this convergence proof cannot be carried through without imposing several hypotheses on the model which may not necessarily be justified by the physics of heat transfer with change of phase and, in fact, may not be satisfied by our model. Thus the convergence theorem stated must be taken only as an indication that the method of lines approximation can be made to work, at least in certain special cases.

In order to keep the problem tractable we shall make the simplification that the model consists of only one layer with prescribed surface temperature $\alpha(t)$ and bottom temperature U_X. Let us index all thermal quantities above the interface $s(t)$ with 1, those below $s(t)$ by 2. Then the continuous and by-lines equations at the nth time level are

$$k_i u_{i_{xx}} - c_i \rho_i u_{i_t} = 0; \qquad k_i u_i'' - \frac{c_i \rho_i}{\Delta t}[u_i - u_{n-1}(x)] = 0, \qquad i = 1, 2$$

$$u_1(t) = \alpha(t) \qquad u_2(X) = U_X \qquad u_1(0) = \alpha(n\Delta t) \qquad u_2(X) = U_X$$

$$u(x_1, 0) = u_0(x) \qquad u(s(0), 0) = 0$$

with the free interface condition

$$0 = k_1 u_{1x} - k_2 u_{2x} + \rho_1 \lambda \frac{ds}{dt}; \qquad k_1 u_1' - k_2 u_2' + \rho_1 \lambda \frac{s - s_{n-1}}{\Delta t} = 0$$

$$u_1(s(t), t) = u_2(s(t), t) = u_1(s) = u_2(s) = 0.$$

(For ease of exposition we have chosen $u = 0$ as the phase transition temperature.) It is known that this problem has a unique classical solution. Let us state the final convergence theorem.

Theorem. *If $u_1(x, t)$, $u_2(x, t)$, and $s(t)$ are the solutions of the continuous free interface problem, and if $U_1^N(x, t)$, $U_2^N(x, t)$, and $s^N(t)$ are approximate solutions defined by setting*

$$U_i^N(x, t) = u_i(x)$$

$$s^N(t) = \frac{t_n - t}{\Delta t} s_{n-1} + \frac{t - t_{n-1}}{\Delta t} s_n, \, t \in (t_{n-1}, t_n),$$

then $U_1^N(x, t) \to u_1(x, t)$, $U_2^N(x, t) \to u_2(x, t)$ and $s^N(t) \to s(t)$ (in a weak sense) as $\Delta t = T/N \to 0$ for some fixed final time T.

Considerable ground work and a certain sophistication are required for the proof of this theorem, which is omitted. For details the reader is referred to [12].

Exercises

1. Derive Fourier's heat conduction equation from Fourier's law of conduction for variable conductivity, i.e., $k = k(x, t)$. Can you think of a thermal model where a variable k may occur?

2. Derive the heat equation in three space dimensions. Use either the divergence theorem for arbitrary control volumes or consider a cube as control volume.

3. Suppose water is in contact with a warm wall at $x = 0$ and at $s(t)$ with a slab of ice held exactly at 32°F at all times. Find a free boundary problem describing the temperature in the water. Is this model realistic for an ice and water system?

4. Verify by back-substitution the correctness of the invariant imbedding equations for an interface problem.

5. Find the closed-form solution of the Riccati equation (35).

6. Find the asymptote to the solution of $U' = 1 + 4U - U^2$, $U(0) = 0$.

7. Discuss the difficulties in applying the trapezoidal rule to $u' = \sin(u + t)$, $u(0) = 0$.

8. Set up the invariant imbedding equation for the problem $u_{xx} - u_t = 0$, $u(x, 0) = u_0(x)$, $(\partial u/\partial x)(0, t) = \alpha(t)$, $u(1, t) = g(t)$.

9. Write out the invariant imbedding equations for the free boundary value problem of Exercise 3.

10. Set up the invariant imbedding equations when each second-order problem is written as

$$u_i' = \frac{1}{k_i} v_i$$

$$v_i' = \frac{c_i \rho_i}{\Delta t} [u_i - u_{n-1}(x)]$$

instead of 23.

11. Discuss how the program in Section 4 can be incorporated into a complete model for the most economical construction of an airstrip on permafrost.

12. Integrate $u_{xx} - u_t = 0$, $u(0, t) = t$, $u(1, t) = 1$, $u(x, 0) = x$ numerically with
   a) a forward finite difference scheme,
   b) the method of lines.
   In both cases examine the behavior of the computed solutions as $\Delta t, \Delta x \to 0$.

13. Modify the model and solution technique for a layered medium in order to include convective surface cooling of the form

$$k\frac{\partial u}{\partial x}(0, t) = h(t)(u(0, t) - \alpha(t)).$$

14. Formulate the Stefan problem for the freezing of a large lake, write the computer program, and examine the effect of the various thermal and numerical parameters.

15. Discuss why one cannot ignore the change of phase in the layers and solve the system as a simple one-phase heat flow problem.

References

[1] W. S. Ellis, "Will oil and tundra mix?" *National Geographic*, vol. 140, p. 485, 1971.

[2] R. D. Guthrie *et al.*, "North to the tundra," *National Geographic*, vol. 141, p. 294, 1972.

[3] B. Keating, "North for oil," *National Geographic*, vol. 137, p. 294, 1970.

[4] *Proceedings Permafrost International Conf.*, 1963, National Academy of Sciences, National Research Council Publication 1287, Washington, DC.

[5] L. R. Ingersoll, O. J. Zobel and A. C. Ingersoll, *Heat Conduction with Engineering and Geological Applications*, McGraw-Hill, New York, 1948.

[6] J. Ockendon and W. Hodgkins, Eds., *Moving Boundary Problems in Heat Flow and Diffusion*. Oxford: Clarendon, 1975.

[7] L. Rubinstein, *The Stefan Problem*, Transl. Math. Monog. 27, A. Solomon, tr. Providence, RI: Amer. Math. Soc., 1971.

[8] G. H. Meyer, *Initial Value Methods for Boundary Value Problems*. New York: Academic, 1973.

[9] G. H. Meyer, N. N. Keller and E. J. Couch, "Thermal model for roads, airstrips, and building foundations in permafrost regions," *J. Canadian Petroleum Technol.*, Apr.–June 1972.

[10] M. S. Kersten, *Thermal Properties of Soils*, Bull. 28, Eng. Exp. Sta., University of Minnesota, 1949.

[11] J. A. Wheeler, "Permafrost thermal design for the Trans-Alaska Pipeline," in *Moving Boundary Problems*, D. G. Wilson, A. D. Solomon, and P. T. Boggs, Eds. New York: Academic, 1978.

[12] G. H. Meyer, "A numerical method for two-phase Stefan problems," *SIAM J. Num. Anal.*, vol. 8, pp. 555–568, 1971.

Notes for the Instructor

Objectives. Section 1 contains a description of the problem to be solved and the motivation for the work. Section 2 presents a selfcontained formulation of the mathematical model based on elementary heat transfer principles. Section 3 contains a solution algorithm for the mathematical model. It illustrates the method of lines for partial differential equations and an initial value solution technique for two point boundary value problems for ordinary differential equations. Section 4 contains some representative results from an industrial parameter study. Section 5 presents a convergence theorem and a discussion of convergence, but no proofs.

Prerequisites. For Sections 2 and 3 only calculus and some mathematical maturity are needed. No particular knowledge of partial differential equations is required, but previous exposure to ordinary differential equations is helpful.

Time. The material may be covered in three or four lectures.

CHAPTER 23
Network Analysis of Steam Generator Flow

T. A. Porsching*

1. Origin of the Problem

In this first section we want to examine the role of the steam generator in the overall operation of a nuclear power plant. In this way we hope to put into perspective the mathematical problem which will eventually evolve and, at the same time, to emphasize its importance.

Figure 23.1 shows the main flow paths in a nuclear power plant. Using this

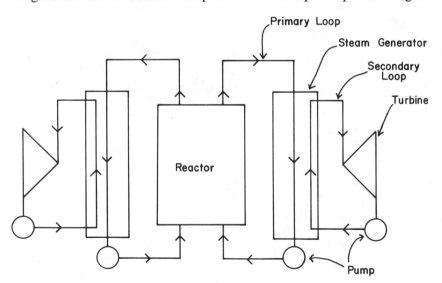

Figure 23.1. Schematic of Nuclear Power Plant.

* Department of Mathematics and Statistics, University of Pittsburgh, Pittsburgh, PA 15260.

figure, it is possible to trace the way in which the thermal energy generated by the fissioning of the nuclear fuel is converted into the mechanical energy of the turbine. Notice that independent fluid circuits exist called primary and secondary loops. The fluid in the primary loops is pumped through the reactor core where it undergoes a temperature rise of typically 40–70°F. From the core the fluid passes through the hot leg by piping and into the tubes of the steam generator. Here, its acquired heat is transferred through the tube walls to the shell side of the steam generator. The cooler (∼ 500°F) fluid leaving the steam generator is then pumped through the cold legs and returned to the reactor core.

Water in the secondary loop enters the shell side of the steam generator and, passing along the outside of the hot tubes, is converted into steam. The thermal energy of this steam is then expended in the form of mechanical work which drives the turbines. Once through the turbines, the condensed steam is returned to the steam generator.

Figure 23.2 shows a cross section of a steam generator. To obtain some idea of the data associated with this device, we refer to that reported in [1, ch. 23] for the steam generators of the Oconee Nuclear Station located in South Carolina. These two units are approximately 73 ft long and 12 ft in diameter. Each contains about 15,500 tubes and weighs about 600 tons.

Since the tubes are over 60 ft long and only about 3/4-in in diameter, supporting them at intermediate points along their length is obviously necessary. The detail in Figure 23.2 shows a portion of one of these tube support plates. Notice the three areas where the holes have been broached to provide flow passages for the secondary fluid.

> "The principal maintenance problem associated with these heat-exchanger-type steam generators is tube leakage, which can be caused by chemical or mechanical action or a combination of the two" [1].

This is an important concern regarding nuclear steam generators because the tubes contain the radioactive reactor coolant. Although the exact mechanism which causes tube pitting is not fully understood, it is strongly correlated with the chemistry and flow distribution of the shell side fluid. For example, evidence exists that, periodically, parts of the tubes are not blanketed with the steam–water mixture of the shell side flow. This condition is called "dryout" and is believed to be a key factor in causing tube damage. Also, in regions of low flow, particles may precipitate out of the fluid and form caustic deposits on the tube surfaces. Indeed, it has been argued [19], [20] that particle deposition on a surface is governed by an equation of the form $N = KC$, where N (mass/time × area) is the particle deposition flux, C (mass/volume) is the average particle concentration, and K (length/time) is the deposition coefficient. This last quantity is a function of particle size and local fluid velocity. From Figure 23.3, which shows a typical plot of K, we see that K depends strongly on this velocity. Utilizing N, it is a simple

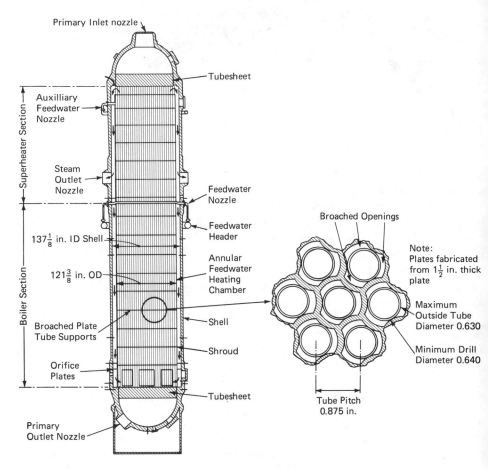

Figure 23.2. Cross Section of Nuclear Once-Through Steam Generator (Reproduced from [1])

matter to calculate the buildup of particles of given size on a surface, for if we assume that N is constant, then Nt represents the mass of these particles which settle onto a unit area of surface in t units of time. Hence if the surface is assumed to be clean at $t = 0$, then h, the thickness of the deposit at time t, is $h = Nt/\rho = KCt/\rho$, where ρ is the density of the particulate matter. Since C and ρ are generally available, the ability to determine deposit buildup depends essentially on the ability to determine the fluid velocity.

In view of the above considerations obtaining a realistic approximation of the shell side flow distribution is important. Due to the complicated three-dimensional geometry of the shell side, the presence of the two-phase steam–water mixture, and the nonlinearities inherent in the conservation laws which govern the flow, solving this problem in its fullest generality

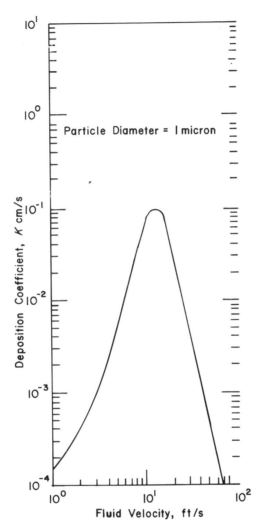

Figure 23.3. Behavior of Deposition Coefficient with Fluid Velocity (from [20]).

does not appear to be possible. However, by utilizing certain simplifying assumptions and adopting a particular point of view, we can obtain a tractable mathematical model which, nevertheless, provides an acceptable numerical solution of the flow problem.

In the following sections we shall develop a simplified version of this model and present a numerical method for its solution. Although the material presented here is intended for classroom instruction, we wish to emphasize that the ideas lie at the heart of a more elaborate method which can be used to analyze flows in steam generators.

2. Some Elements of Network Theory

As we have previously mentioned, the analysis of shell side steam generator flow is an extremely complex problem. Our approach will be to reduce the continuous problem to a discrete one in the sense that we shall consider flows along certain preselected flow paths. In effect, we shall lump the flow in a region of the steam generator into a single average flow for that region and then formulate laws which these lumped flows should obey. The interconnected system of flow paths on which the discrete flows occur is termed a *network*. The literature on networks is already quite large and continues to grow. In this section we shall deal only with those few notions which are necessary for the treatment of our problem. For further reading on the subjects of networks, graphs[1] and other discrete flow problems, see [2]–[5], [21].

We abstractly define our network η as a couple (V, S). Here V is a finite set of unordered elements called *nodes*, and S is a set of ordered pairs of elements of V called *links*. In modeling our flow problem the nodes will represent junctions where the flow changes direction, and the links will define the flow paths. We assume that there are n elements in V and m elements in S.

Since the nodes are isomorphic to the first n positive integers, we let $V = \{1, \cdots, n\}$. On the other hand, the jth link s_j of η is denoted by $(P(j), Q(j))$ where $P(j), Q(j) \in V$. Thus we can write $S = \{s_j \equiv (P(j), Q(j)) | j = 1, \cdots, m\}$. The nodes $P(j)$ and $Q(j)$ are termed, respectively, the *initial* and *terminal* nodes of link s_j and constitute the *extremities* of the link. At the same time, link s_j is said to be *incident* upon nodes $P(j)$ and $Q(j)$.

These algebraic definitions are geometrically motivated. By drawing the nodes as numbered circles and the links as arcs on which numbered arrow heads have been placed pointing from the initial to the terminal node, we obtain a complete description of the network. For example, in Figure 23.4 we have depicted the network defined by the sets

$$V = \{1, 2, 3\},$$

$$S = \{(3, 1), (1, 2), (3, 2), (3, 1), (1, 1)\}.$$

Note the presence of the self-loop $s_5 = (1, 1)$ and the parallel but distinct links $s_1 = (3, 1)$ and $s_4 = (3, 1)$. Since selfloops do not generally represent physically meaningful flow paths, we shall exclude them from further consideration.

A sequence of network links $\{s_{i_1}, s_{i_2}, \cdots, s_{i_k}\}$ is a *chain* if for $j = 2, \cdots, k - 1$, link s_{i_j} has one extremity in common with $s_{i_{j-1}}$ and the other with

---

[1] Although the distinction between networks and graphs is fuzzy, we will use the term "network" to emphasize the idea of a system which carries flow. "Graph," on the other hand, suggests a skeleton of connections without any particular flow connotation.

Figure 23.4. A Simple Network.

$s_{i_{j+1}}$. Nodes p and q are *connected* if a chain $\{s_{i_1}, \cdots, s_{i_k}\}$ exists such that they are extremities of links s_{i_1} and s_{i_k}, respectively. The network itself is connected if every pair of nodes in it is connected. Since liquid flows in an inherently continuous manner[2] we shall confine our attention to connected networks.

In modeling our flow problem we shall make use of the fundamental law of conservation of mass. The network analog of this law is known as *Kirchhoff's* node law and is mathematically formulated as follows. With each node i of η we associate two sets of links,

$$\omega^+(i) = \{(i, k)|k \in V, (i, k) \in S\},$$

and

$$\omega^-(i) = \{(k, i)|k \in V, (k, i) \in S\}.$$

Loosely speaking, $\omega^+(i)$ is the set of links incident upon node i which point away from the node, while $\omega^-(i)$ contains those which point toward the node. With each link s_i of η we associate a real number w_i. Then the m-dimensional vector[3] $w = (w_1, \cdots, w_m)^T$ is a *flow on* η if

$$\sum_{s_j \in \omega^+(i)} w_j - \sum_{s_j \in \omega^-(i)} w_j = 0, \qquad i = 1, \cdots, n. \tag{1}$$

Equation (1) expresses Kirchhoff's node law which states that the net flow into a node is equal to the net flow out of it. If w is a flow, then w_j is the flow on link s_j. Actually, it would be more precise to call the w_j link flow *rates* since their units turn out to be mass per unit time, e.g., pounds/ second. Notice that the link flows may assume both signs. We adopt the usual convention that $w_j > 0$ means that the *actual* direction of flow on link s_j is from the initial node to its terminal node, while $w_j < 0$ means that the actual direction of flow is from terminal to initial node.

Equation (1) can be written in a convenient, compact way which involves introducing the (node–link) incidence matrix. This is the $n \times m$ matrix

---

[2] There is at least one notable exception to this. See [6].

[3] We shall always use superscript T to denote transpose. Thus w is a column vector.

$A = [a_{ij}]$ where

$$a_{ij} = \begin{cases} +1, & \text{if } i = P(j) \\ -1, & \text{if } i = Q(j) \\ 0, & \text{otherwise.} \end{cases}$$

For example,

$$A = \begin{bmatrix} -1 & 1 & 0 & -1 \\ 0 & -1 & -1 & 0 \\ 1 & 0 & 1 & 1 \end{bmatrix} \tag{2}$$

is the incidence matrix corresponding to the network in Figure 23.4 when the selfloop s_5 is removed. Utilizing this matrix, (1) clearly becomes

$$Aw = 0. \tag{3}$$

Since any link is incident upon its initial and terminal nodes and only those nodes, every column of A contains, except for zeros, exactly one $+1$ and one -1. Therefore, the rows of A sum to the zero vector, and its row rank is consequently less than or equal to $n - 1$. In fact *it is exactly* $n - 1$. To prove this we discard row n of A. Then we note that the links which are incident upon the node set $V_1 = \{n\}$ determine columns containing exactly one nonzero entry. The rows in which these nonzero entries lie cannot occur in any vanishing nontrivial linear combination of the rows of A. Thus we may discard them from A, but these same rows define a set of nodes, say V_2, and we can repeat the above argument on links which are incident upon V_2 and which have their other extremities in the complement of V_2. Because of the assumed connectedness of η, this process will eventually exhaust the first $n - 1$ rows of A showing that they are independent.

Since the row and column ranks of a matrix are equal (see, for example, [7, p. 42]), we immediately have that the dimension of the null space of A is $m - (n - 1)$. Hence the most general flow on η depends on $m - n + 1$ arbitrary parameters. Clearly, we need to develop further conditions to single out a particular flow from among the multitude of solutions of (3).

In modeling the flow problem we must include a means of treating *boundary conditions*. These are the conditions which hold at those places where the fluid enters and leaves the region of interest, for example, the regions adjacent to the tube sheets in Figure 23.2. In a network, boundary conditions are accommodated by designating certain nodes as *boundary nodes* and prescribing the appropriate conditions at these nodes. Since boundary nodes correspond to points where the fluid enters and leaves the network, Kirchhoff's node law cannot be expected to hold at them. However, as a consequence of imposing this law at all the other nodes of the network, we can show that the sum of the boundary flows is zero. In other words, the total flow into the network is equal to the total flow out of it. To prove this statement, let us assume $n - v > 0$ boundary nodes exist, numbered $v + 1, \cdots, n$. Then we can partition the incidence matrix A as $\begin{bmatrix} A^0 \\ \partial A \end{bmatrix}$, where A^0 consists of the first v rows of A and represents the non-boundary or interior nodes and where ∂A consists of the last $n - v$ rows of A and represents the boundary nodes. If we define u^0 and ∂u, respectively, as v- and $n - v$-dimensional vectors of ones, then we have that since A is an incidence matrix,

$$(u^{0T}, \partial u^T) \begin{bmatrix} A^0 \\ \partial A \end{bmatrix} = u^{0T}A^0 + \partial u^T \partial A = 0.$$

Therefore, if w is a flow on η,

$$u^{0T}(A^0 w) + \partial u^T(\partial A w) = 0. \tag{4}$$

On the other hand, Kirchhoff's node law applies at the interior nodes so that

$$A^0 w = 0. \tag{5}$$

Then certainly $u^{0T}(A^0 w) = 0$, and it follows from (4) that $\partial u^T(\partial A w) = 0$, which is what we wanted to prove.

The network shown in Figure 23.5 may be used to model part of the flow region in a steam generator. In this network the boundary nodes are nodes

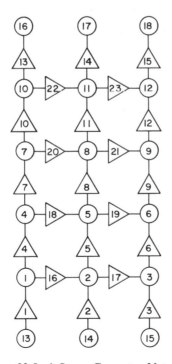

Figure 23.5. A Steam Generator Network.

13–18. The modeling procedure by which the physical flow region can be reduced to such a network will be discussed in the next section.

As previously noted, Kirchhoff's node law does not generally define a unique flow on a network.

The additional conditions which are required to do this come in the form of relations between the flows and a new set of variables $\{p_i\}$ associated with the nodes and termed *node states*. In our flow problem these states can be interpreted as the static *pressures* which exist at points in the flow region. We will assume that the flow-node state relation which holds for each link s_j is of the form

$$w_j = f_j(p_{P(j)} - p_{Q(j)}). \tag{6}$$

Here $f_j(t)$ is a function of the variable t and $\Delta p_j \equiv p_{P(j)} - p_{Q(j)}$ is the state difference or pressure drop across link s_j. Equation (6) is called a *link characteristic* and states that the flow on link j is a known function of *only* the pressure drop across the link.

We are now ready to consider the basic network flow problem. Suppose that we are given a link characteristic (6) for each link of a network η which has $n - v$ boundary nodes $v + 1, \cdots, n$ at which the pressures $p_{v+1}, \cdots, p_n$ are known. Then we seek to determine pressures $p_1, \cdots, p_v$ and flows $w_1, \cdots, w_m$ which satisfy the characteristics (6) for $j = 1, \cdots, m$ and Kirchhoff's node law (1) for $i = 1, \cdots, v$ (i.e., at the interior nodes of η). We observe that if we simply substitute the equations (6) into the first v equations (1), then we have precisely v equations and v unkowns. Thus, if a unique set of pressures exists which satisfies this system, then (6) determines a corresponding unique set of flows, and the problem is completely solved. Clearly, the equations to be satisfied by the pressures are

$$\sum_{s_j \in \omega^+(i)} f_j(p_{P(j)} - p_{Q(j)}) - \sum_{s_j \in \omega^-(i)} f_j(p_{P(j)} - p_{Q(j)}) = 0, \qquad i = 1, \cdots, v. \tag{7}$$

We term (7) the *network equation*.

By utilizing the incidence matrix once again, we can write the network equations in a concise form. Let $p = (p_1, \cdots, p_n)^T$ and $\Delta p = (\Delta p_1, \cdots, \Delta p_m)^T$ denote n- and m-dimensional vectors of pressures and pressure drops. From the manner in which the incidence matrix $A = [a_{ij}]$ has been defined, it is easy to see that

$$\Delta p_j = p_{P(j)} - p_{Q(j)} = \sum_{i=1}^{n} a_{ij} p_i, \qquad j = 1, \cdots, m.$$

Hence $\Delta p = A^T p$, where the matrix A^T is the transpose of A. Letting $f(\Delta p) = (f_1(\Delta p_1), \cdots, f_m(\Delta p_m))^T$, it follows from (5) and (6) that

$$A^0 w = A^0 f(\Delta p) = A^0 f(A^T p) = 0.$$

Therefore, another form of (7) is

$$A^0 f(A^T p) = 0. \tag{8}$$

Let us write the first three network equations for the network of Figure 23.5. Since[4]

$$\Delta p_1 = p^*_{13} - p_1, \Delta p_2 = p^*_{14} - p_2, \Delta p_3 = p^*_{15} - p_3,$$

$$\Delta p_4 = p_1 - p_4, \Delta p_5 = p_2 - p_5, \Delta p_6 = p_3 - p_6,$$

$$\Delta p_{16} = p_1 - p_2, \Delta p_{17} = p_2 - p_3,$$

and since Kirchhoff's node law for nodes 1, 2, 3 reads, respectively,

$$w_4 + w_{16} - w_1 = 0,$$

$$w_5 + w_{17} - w_2 - w_{16} = 0,$$

$$w_6 - w_3 - w_{17} = 0,$$

we obtain

$$f_4(p_1 - p_4) + f_{16}(p_1 - p_2) - f_1(p^*_{13} - p_1) = 0,$$

$$f_5(p_2 - p_5) + f_{17}(p_2 - p_3) - f_2(p^*_{14} - p_2) - f_{16}(p_1 - p_2) = 0,$$

$$f_6(p_3 - p_6) - f_3(p^*_{15} - p_3) - f_{17}(p_2 - p_3) = 0.$$

Notice that in the first equation, which corresponds to Kirchhoff's law for node 1, the unknown quantity p_1 appears in each term and is multiplied by $+1$ or -1 as the corresponding flow is added or subtracted. Similar remarks can be made about the remaining two equations. Indeed, this is a property of the network equations in general. In Section 4 we shall exploit this important property in constructing an algorithm to solve the equations numerically.

At this point we can ask two fundamental mathematical questions about the network equations.

1) Do they have a unique solution (indeed, do they have any solution at all)?
2) If the answer to 1) is yes, how can we compute this solution?

In the remainder of this section we shall try to answer question 1), postponing further consideration of 2) until Section 4.

First of all, unless we say more about the functions $f_j(t)$, the answer to question 1) is "NO!" For example, suppose that we examine the network equations corresponding to the simple one link network shown in Figure 23.6. Let us assume that node 2 is the boundary node and that the pressure there, p^*_2, is zero. Then only one equation exists, and it reads

$$f_1(p_1) = 0. \tag{9}$$

If we are given that $f_1(t) \equiv 1$, then (9) has no solution. On the other hand, if we are given $f_1(t) = \sin t$, then (9) has an infinite number of solutions.

---

[4] To emphasize that nodes 13–15 are boundary nodes where the pressures are known, we have "starred" these quantities.

Figure 23.6. A One Link Network.

Fortunately, experience has shown that the functions $f_j(t)$ occurring in practice do not resemble either of the above functions. In fact, it is reasonable to assume that, as we increase the pressure drop (or driving force) across a given link, the flow in that link also increases. This leads us to assume that *for each j, $f_j(t)$ is a strictly increasing function of t*. This is still not enough to guarantee an affirmative answer to 1) since the functions

$$f_1(t) = \begin{cases} t, & \text{if } t < 0 \\ 1 + t, & \text{if } t \geq 0 \end{cases} \tag{10}$$

and

$$f_1(t) = 1 + e^t \tag{11}$$

are both strictly increasing but neither allows a solution of (9). The difficulty with (10) is the discontinuity at $t = 0$, while the problem with (11) is that its range is restricted to positive numbers. We can remove both of these difficulties by assuming that $f_j(t)$ is a continuous function of t and has an unrestricted range.[5] This leads us to define an *admissible characteristic* as one for which $f_j(t)$ is defined on $-\infty < t < \infty$, is continuous and strictly increasing there, and satisfies $\lim_{t \to \pm\infty} f_j(t) = \pm\infty$. Note that if $f_1(t)$ defines an admissible characteristic, then (9) has exactly one solution. In fact, we have the following theorem.

Theorem 1. *If the links of a network η have admissible characteristics and if η has at least one boundary node, then the network equations (7) or (8) have a unique solution.*

PROOF. We shall not prove this theorem in its full generality since such a proof involves notions which we are not prepared to introduce. A proof can be found, for example, in [8]. We will prove it, however, in the special case that the functions $f_j(t)$ are *linear*, i.e., of the form

$$f_j(t) = d_j t + c_j, \tag{12}$$

where d_j and c_j are constants. These define admissible characteristics if and only if $d_j > 0$.

   Let $D = \text{diag}(d_1, \cdots, d_m)$, i.e., the diagonal matrix whose ith diagonal elements is d_i. Also, let $c = (c_1, \cdots, c_m)^T$ and assume that the last $n - v > 0$

---

[5] To assume that the flow in any link is a continuous function of the pressure drop across that link is quite natural. However, the assumption that the flow becomes infinite with its pressure drop is somewhat artificial since the link is more likely to saturate at a finite flow.

nodes are the boundary nodes. If we go back to the form of the network equations given by (8), we see that the ith equation is

$$\sum_{j=1}^{m} a_{ij}d_j(A^Tp)_j + \sum_{j=1}^{m} a_{ij}c_j = 0, \qquad i = 1, \cdots, v$$

or

$$A^0 D(A^Tp) + A^0 c = 0. \tag{13}$$

By writing

$$p^0 = (p_1, \cdots, p_v)^T$$

$$\partial p = (p^*_{v+1}, \cdots, p^*_m)^T$$

$$p = \begin{bmatrix} p^0 \\ \partial p \end{bmatrix}$$

$$A = \begin{bmatrix} A^0 \\ \partial A \end{bmatrix},$$

we can split off the boundary dependent part of (13). In fact, we see that (13) may now be written as

$$A^0 D[A^{0^T} | (\partial A)^T] \begin{bmatrix} p^0 \\ \partial p \end{bmatrix} + A^0 c = 0$$

or

$$A^0 D[A^{0^T} p^0 + (\partial A)^T \partial p] + A^0 c = 0$$

or finally,

$$(A^0 D A^{0^T})p^0 = -(A^0 c + A^0 D(\partial A)^T \partial p). \tag{14}$$

Equation (14) represents a nonhomogeneous system of v linear equations in the v unknown pressures of the vector p^0 (note that the right-hand side of (14) is known). By the fundamental theorem on the solvability of linear equations, (14) has a unique solution if and only if the $v \times v$ coefficient matrix $A^0 D A^{0^T}$ is nonsingular.

We show that $A^0 D A^{0^T}$ is nonsingular by assuming the contrary. Then a vector $x = (x_1, \cdots, x_v)^T \neq 0$ exists such that $A^0 D A^{0^T} x = 0$. Thus $x^T A^0 D A^{0^T} x = 0$. If we let $z = (z_1, \cdots, z_m)^T = A^{0^T} x$, this last equation becomes

$$z^T D z = 0. \tag{15}$$

Since D is a diagonal matrix, we can write (15) in longhand as

$$\sum_{j=1}^{m} d_j z_j^2 = 0, \tag{16}$$

but each $d_j > 0$, so that the only way in which (16) can hold is for $z_j = 0$,

$j = 1, \cdots, m$. Hence $z = 0$. Now according to the definition of z, it is a linear combination of the columns of A^{0^T}, i.e., the rows of A^0. In fact if we denote the ith row of A^0 by a_i, then we have

$$z = \sum_{j=1}^{v} a_i x_i = 0. \tag{17}$$

However, we have seen that the rank of A is $n - 1$. Therefore, the vectors $a_i, i = 1, \cdots, v \leq n - 1$ are linearly independent, and (17) implies that $x_i = 0, i = i, \cdots, v$, which is a contradiction. This establishes the non-singularity of $A^0 D A^{0^T}$ and completes the proof. □

We remark once more that, although the proof applies only to the linear case (12), the theorem is true for a much wider class of functions. For example, if $b_j > 0$, $d_j > 0$, and c_j are constants, then

$$f_j(t) = b_j t^3 + d_j t + c_j$$

defines an admissible *nonlinear* characteristic. The characteristics we shall introduce in the next section to model the steam generator flow problem will be nonlinear.

3. Modeling the Flow Problem

We are now ready to show how to model the steam generator flow problem so that the network equations of Section 2 apply. Although the development will be quite general, it will be helpful to apply the ideas to a specific example. Suppose then that we wish to determine the flow distribution in a portion of the shell side of the steam generator pictured in Figure 23.2. This region, which is shown in Figure 23.7, is symmetric about one of the tube support plates and extends to the adjacent plates. For simplicity, we have not shown the tubes, but one must remember that they occupy the region and are vertically oriented. Furthermore, although the actual flow region is three-dimensional, the region in Figure 23.7 is a two-dimensional "slice" taken at the symmetry plane of the steam generator as shown in the plan view of the figure.

Now let us divide this region into a finite number of subregions or cells by inserting fictitious boundaries. We have elected to use 12 such cells, and these are numbered by associating nodes with them in the manner of Figure 23.7. We have indicated the fictitious boundaries by dashed lines.

Since movement of fluid between two adjacent cells must occur across their common boundary, we can account for the flow communication between them by lumping the distributed flow across the whole boundary into a single quantity which represents the *total* flow across the boundary.

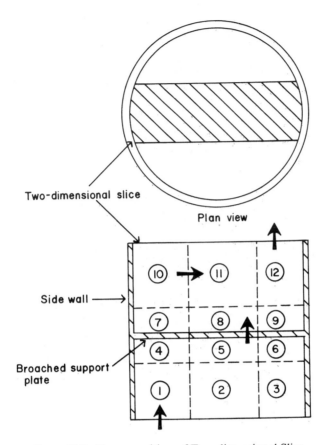

Figure 23.7. Decomposition of Two-dimensional Slice.

It is natural to signify geometrically this flow communication by inserting a link between adjacent cells. Thus in Figure 23.7 we connect the nodes of the 12 cells by inserting 17 links. To recognize that the fluid enters and leaves the region across the horizontal boundaries of cells 1, 2, 3, and 10, 11, 12, we connect each of these nodes to a boundary node. The geometric realization of this procedure is the network of Figure 23.5.

Consider cell i. According to the principle of conservation of mass, if no fluid is being created or destroyed in the cell and if the density of the fluid in the cell is *constant*, then the total mass efflux across the boundaries of the cell is equal to the total mass influx; but the movement across boundaries can now be thought of in terms of flows on the links of an associated network. Thus, if we let w_j denote the mass flow, say, in pounds per second, across the boundary which is penetrated by link s_j, and if $w_j > 0$ corresponds to fluid movement from the initial to the terminal node of s_j, then for cell i we have

$$\text{total mass efflux} = \sum_{\substack{s_j \in \omega^+(i) \\ w_j > 0}} w_j - \sum_{\substack{s_j \in \omega^-(i) \\ w_j < 0}} w_j$$

$$\text{total mass influx} = \sum_{\substack{s_j \in \omega^-(i) \\ w_j > 0}} w_j - \sum_{\substack{s_j \in \omega^+(i) \\ w_j < 0}} w_j.$$

Therefore,

$$\sum_{\substack{s_j \in \omega^+(i) \\ w_j > 0}} w_j + \sum_{\substack{s_j \in \omega^+(i) \\ w_j < 0}} w_j = \sum_{\substack{s_j \in \omega^-(i) \\ w_j > 0}} w_j + \sum_{\substack{s_j \in \omega^-(i) \\ w_j < 0}} w_j$$

or

$$\sum_{s_j \in \omega^+(i)} w_j - \sum_{s_j \in \omega^-(i)} w_j = 0, \qquad (18)$$

which is Kirchhoff's node law for node i.

The static pressure will in general assume different values at the different nodes of the cells. The pressure difference between two nodes of adjacent cells—which may be regarded as the pressure drop across the link connecting the two cells—is due to a number of factors. We shall consider only two of these:

1) the effect of the fluid's weight.
2) the frictional resistence of the tube surfaces and other impervious boundaries on the moving fluid.

If we refer to Figure 23.8, then the weight of the fluid in link s_j causes the pressure at node $P(j)$ to exceed that at node $Q(j)$ by an amount given by

$$(\Delta p_j)_{\text{el}} = \rho L_j g \cos \theta_j. \qquad (19)$$

Here ρ is the (mass) density of the fluid, g is the acceleration due to gravity, L_j is the length of the link and θ_j is the angle at which the link is inclined to the vertical. Notice that we have used the subscript el to denote this pressure drop since it is due to the difference in *elevation* of the link's extremities.

As the fluid moves over the links, it passes around the tubes, along the shroud, and through the broached openings in the tube support plates.

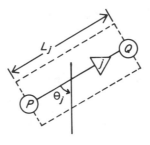

Figure 23.8. Elevation Pressure Drop.

Each of these exerts a frictionally resistive force on the fluid which is balanced by a pressure difference across the link's extremities. Extensive experimentation has established empirically the form of this pressure drop as (see [9, ch. 11] for more details)

$$(\Delta p_j)_{\text{fr}} = \frac{F_j}{\rho A_j^2} |w_j| w_j, \tag{20}$$

where A_j is the cross sectional area[6] of the link and F_j is a proportionality constant called the friction factor.[7]

Note that if we adopt a constant set of units, for example, if we measure mass in pounds (lb), force in poundals (pdl), time in seconds (s), and length in feet (ft), then the units of the right-hand side of (19) are

$$\frac{\text{lb}}{\text{ft}^3} \text{ft} \frac{\text{ft}}{\text{s}^2} = \left[\text{lb} \frac{\text{ft}}{\text{s}^2} \right] \frac{1}{\text{ft}^2} = \frac{\text{pdl}}{\text{ft}^2}$$

and those of (20) are (F_j being dimensionless)

$$\frac{\text{ft}^3}{\text{lb}} \frac{1}{\text{ft}^4} \frac{\text{lb}^2}{\text{s}^2} = \left[\text{lb} \frac{\text{ft}}{\text{s}^2} \right] \frac{1}{\text{ft}^2} = \frac{\text{pdl}}{\text{ft}^2}.$$

Thus in both cases we obtain the units of pressure drop as required.

To obtain the total pressure drop across link s_j, we simply add the contributions due to friction and elevation. That is

$$\Delta p_j = (\Delta p_j)_{\text{fr}} + (\Delta p_j)_{\text{el}} = \frac{F_j}{\rho A_j^2} |w_j| w_j + \rho L_j g \cos \theta_j. \tag{21}$$

Since the friction factor and the density are positive quantities, this equation defines the pressure drop as a strictly increasing, nonlinear function of the link flow. A graph of this function is shown in Figure 23.9. This figure also shows the inverse function obtained by solving (21) for w_j. It is easy to see that this inverse is given explicitly by

$$w_j = f_j(\Delta p_j) \equiv A_j \sqrt{\frac{\rho}{F_j} |\Delta p_j - \rho L_j g \cos \theta_j|} \, \text{sgn}(\Delta p_j - \rho L_j g \cos \theta_j), \tag{22}$$

[6] Since the links are conceptual devices, to speak of their cross sectional areas is somewhat ambiguous (the same is true of their lengths, but it is natural to take these to be the distances between the nodes). One way to assign cross sectional areas is to divide the volumes of the cells by the number of links incident upon them, assign these subvolumes to the appropriate links, and then *define* the cross sectional area to be the ratio of a link's volume to its length. For example, if s_{16} connects nodes 1 and 2 in Figure 23.7, then

$$A_{16} = \frac{1}{L_{16}} \left(\frac{\text{volume of cell 1}}{3} + \frac{\text{volume of cell 2}}{4} \right).$$

[7] The definition of friction factor varies from author to author. Our definition has been motivated by a desire to produce a simple but realistic form of the frictional pressure drop $(\Delta p_j)_{\text{fr}}$.

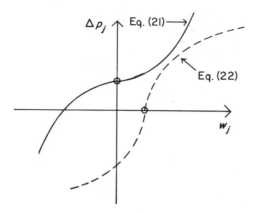

Figure 23.9. Link Characteristic and Inverse.

where sgn (x) is the so-called sign function and is defined by

$$\mathrm{sgn}\,(x) = \begin{cases} 1, & \text{if } x > 0 \\ 0, & \text{if } x = 0 \\ -1, & \text{if } x < 0. \end{cases}$$

Now we observe that (22) is precisely of the form (6) and so defines a link characteristic. Moreover, since (18) is Kirchhoff's node law, we see that we have reduced the flow problem to that of solving the network equations. Not only does (22) define a link characteristic, but—even better—it defines an admissible characteristic. Therefore, by Theorem 1, if we have at least one boundary node in our network, then the flow problem has a unique solution.

We have arrived at this fortunate mathematical state of affairs by simplifying certain aspects of the actual problem. One of the most crucial simplifying assumptions we have tacitly made is to *neglect the influence of heat*. Recall that in the early discussion of the steam generator, we mentioned that the flowing quantity of interest is a two-phase mixture of vapor and water. Now the modeling of a two-phase flow process is in general quite complicated and utilizes some assumptions which cannot be derived from first principles (that is, they are empirical correlations which have evolved through experimentation). Possibly the simplest approach to the problem is to assume that w_j represents the mass flow rate of the two-phase mixture and that the fraction of this which is vapor—this fraction is termed the *quality*—is a simple function of the thermal energy of the mixture. Among other things this assumption implies that the two phases move together at the same velocity, a condition which is known to be violated for certain types of slow motion.

One measure of the thermal energy of the mixture is the *enthalpy H*. Under the above assumption, if we know the mass flow and enthalpy of the mixture, then we know (via the quality) the mass flow of the vapor present in the mixture. The introduction of the new variable H requires the addition of another equation to the model. This comes from the principle of Conservation of Thermal Energy, and for the network model we have been considering it may be written

$$H_i\left(\sum_{\substack{s_j\in\omega^+(i)\\w_j>0}} w_j - \sum_{\substack{s_j\in\omega^-(i)\\w_j<0}} w_j\right) = \sum_{\substack{s_j\in\omega^+(i)\\w_j>0}} (w_j H_{P(j)} + \varphi_j) - \sum_{\substack{s_j\in\omega^+(i)\\w_j<0}} (w_j H_{Q(j)} - \varphi_j),$$

$$i = 1, \cdots, v. \tag{23}$$

Here we have denoted the external rate of heat addition[8] to link j by φ_j. Note that like the pressure, the enthalpy is a node variable.

If we measure heat in British thermal units (BTU's), then the units of enthalpy can be taken as BTU/lb. If the units of φ_j are BTU/s, then both sides of (23) have units of BTU/s. Thus (23) equates heat rates at node i. The left side is the rate at which heat is being removed from the node by fluid efflux, and the right side is the rate of heat addition to the node by the incoming fluid.

We notice the presence of the flow rates w_j in (23). If we presume that we know these, and if we are given the φ_j (that is, if we know the heat addition from the tubes), then (23) constitutes a set of linear equations for the enthalpies. On the other hand, because the enthalpies do not appear in the network equations, we can solve them *without reference to* (23). This is a consequence of our assumption that the density ρ is constant. In the more general case, the density at any point in the flow region is a function of the pressure and enthalpy at the point, say

$$\rho = R(p, H). \tag{24}$$

Equation (24) is known as the *equation of state*. In steam generator calculations, the pressure variation on the shell side is sufficiently small to justify regarding ρ as being independent of p. In other words ρ may be evaluated at a constant "system pressure" p^* so that (24) becomes[9]

$$\rho = R(p^*, H). \tag{25}$$

If H were constant, then indeed that density would also be constant. However, the assumption of constant H is difficult to justify, especially with the formation of vapor. If we assume that ρ is given by (25) and introduce this into the characteristic (22), then the network equations contain the node enthalpies and *must be solved simultaneously with* (23). This is a much more difficult problem than the one involving only the network equations and goes well beyond the instructional intent of this chapter. Therefore, we shall avoid it by assuming that the density is constant.

---

[8] In our steam generator model this is, of course, the heat which is conducted through the tube walls from the primary loop.

[9] See footnote to Table 2.

We shall devote the remainder of this section to an appraisal of link characteristic (22) by examining its equivalent form (21). This requires the introduction of certain *partial differential equations* but does not affect either the network equations or the subsequent development. Therefore, the continuity of the presentation will not be interrupted if the rest of the section is omitted.

The application of Newton's second law to an infinitesimal element of fluid results in a set of partial differential equations known as the Navier–Stokes equations. A derivation of these equations can be found in almost any book on hydrodynamics, for example [9] or [10]. If the motion is steady[10] and two-dimensional, as in the case of the steam generator slice considered earlier, then these equations are

$$\rho\left(u\frac{\partial u}{\partial x} + v\frac{\partial u}{\partial y}\right) = -\frac{\partial p}{\partial x} + \mu\left(\frac{\partial^2 u}{\partial x^2} + \frac{\partial^2 u}{\partial y^2}\right) + F_1, \tag{26}$$

$$\rho\left(u\frac{\partial v}{\partial x} + v\frac{\partial v}{\partial y}\right) = -\frac{\partial p}{\partial y} + \mu\left(\frac{\partial^2 v}{\partial x^2} + \frac{\partial^2 v}{\partial y^2}\right) + F_2. \tag{27}$$

Equations (26) and (27) are written in terms of the rectangular coordinates (x, y). The quantities u and v denote the fluid velocity components in the x and y directions, respectively, p is the pressure, and ρ is the (constant) density. The constant μ is called the viscosity and reflects the fact that elements of a moving fluid exert shear forces on each other. The terms F_1 and F_2 represent the x and y components of forces[11] such as gravitational pull and frictional drag at boundaries. For our applications we can take F_1 to be (see [11])

$$F_1 = -\frac{\hat{f}}{2D}\rho|u|u - \rho g \cos\theta, \tag{28}$$

where $\hat{f}$ is another friction factor (cf. (20)) and D is a quantity having the units of length known as the hydraulic or equivalent diameter.[12] The angle θ is that between the x axis and the direction of the gravitational vector.

If we integrate (26) along a link s_j which is parallel to the x axis, we obtain

$$\Delta p_j = p_P - p_Q = -\int_P^Q F_1\,dx + I, \tag{29}$$

where

$$I = \int_P^Q \left[\rho\left(u\frac{\partial u}{\partial x} + v\frac{\partial u}{\partial y}\right) - \mu\left(\frac{\partial^2 u}{\partial x^2} + \frac{\partial^2 u}{\partial y^2}\right)\right]dx.$$

By the mean value theorem for integrals

---

[10] In steady motion all partial derivatives with respect to time vanish.

[11] The dimensions of each term in (26) and (27) are actually force per unit volume.

[12] Except in the case of flow in circular pipes, where D coincides with the pipe diameter, the physical significance of this quantity is ambiguous (see [9, ch. 11] for further discussion).

$$-\int_{P}^{Q} F_1 \, dx = \frac{f}{2D} L_j \rho |\bar{u}| \bar{u} + \rho g L_j \cos \theta, \tag{30}$$

where $\bar{u}$ is the velocity at some point on the link.

Therefore, if we define the link mass flow rate to be $w_j = \rho A_j \bar{u}$ and let $F_j = \hat{f} L_j / 2D$, then (30) becomes

$$-\int_{P}^{Q} F_1 \, dx = \frac{F_j}{\rho A_j^2} |w_j| w_j + \rho g L_j \cos \theta.$$

Consequently, it follows from (29) that

$$\Delta p_j = \frac{F_j}{\rho A_j^2} |w_j| w_j + \rho g L_j \cos \theta + I. \tag{31}$$

Comparing this equation with (21), we see that they differ only by the integral I. If we accept the Navier–Stokes equations as providing an accurate description of the fluid's motion, then I is the error introduced by using the characteristic (22).

Let us examine the terms in I. Since

$$u = \frac{dx}{dt}, \quad v = \frac{dy}{dt},$$

we have from the chain rule

$$\frac{d^2x}{dt^2} = \frac{du}{dt} = \frac{\partial u}{\partial x}\frac{dx}{dt} + \frac{\partial u}{\partial y}\frac{dy}{dt} = u\frac{\partial u}{\partial x} + v\frac{\partial u}{\partial y},$$

but $a = d^2x/dt^2$ is the acceleration of a fluid element in the x direction. Thus

$$\int_{P}^{Q} \rho \left(u\frac{\partial u}{\partial x} + v\frac{\partial u}{\partial y} \right) dx = \int_{P}^{Q} \rho a \, dx \tag{32}$$

represents the integral of the inertial forces due to the motion of the fluid along the link.

As we have already mentioned, the remaining part of I arises from the shear effects due to the relative motion of fluid elements. If we call (32) the inertial effect, then $-\int_{P}^{Q} \mu(\partial^2 u/\partial x^2 + \partial^2 u/\partial y^2) \, dx$ may be called the viscous effect. Since our characteristic (22) assumes that both of these are zero, we expect the error in (22) to be significant in situations when either the viscous or inertial effects are comparable to those produced by the friction and elevation.

4. Numerical Solution of the Network Equations

Having reduced the steam generator flow problem to one of solving the network equations, it remains to formulate an algorithm for their solution. We have seen in Section 2 that these equations possess a unique solution. However, because of the nature of the characteristic (22), they are necessarily nonlinear. Therefore, a closed-form solution, even in the sense of Gauss elimination, is out of the question, and so we employ an iterative method.

The iterative method that we shall consider is called the *nonlinear Gauss–Siedel (NGS) method* and is a natural extension of the well-known method for the linear case.[50] We shall formally define it for a general system of v equations containing v unknowns.

Suppose that $F_i(x_1, \cdots, x_v)$, $i = 1, \cdots, v$, are continuous real valued functions of the real variables x_i, $i = 1, \cdots, v$. Then

$$F_i(x_1, \cdots, x_v) = 0, \qquad i = 1, \cdots, v \tag{33}$$

represents a system of v equations, and any set of real numbers $x_1^*, \cdots, x_v^*$ which satisfies (33) is called a solution of this system. The *NGS* method defines a sequence of vector iterates $x^k = (x_1^k, \cdots, x_v^k)$, $k = 0, 1, \cdots$ in the following manner.

Step 1. Choose x^0.

Step 2. Suppose that $x^k(k \geq 0)$ and x_q^{k+1}, $q = 1, \cdots, i - 1$ ($1 \leq i \leq v$) have been determined.

Step 3. To determine x_i^{k+1}, find s such that

$$F_i(x_1^{k+1}, \cdots, x_{i-1}^{k+1}, s, x_{i+1}^k, \cdots, x_v^k) = 0, \tag{34}$$

and then let $x_i^{k+1} = s$.

Notice that when $i = 1$ in step 2, the condition on x_q^{k+1}, $q = 1, \cdots, i - 1$ is vacuous, and x_1^{k+1} is determined in step 3 as the solution s of

$$F_1(s, x_2^k, \cdots, x_v^k) = 0.$$

Clearly, the NGS method is well-defined, that is, the iterates x^k are uniquely determined, if and only if each *scalar* equation (34) has a unique solution s. When this is true, the elements of the NGS sequence are determined in the order illustrated by the tableau for the case $v = 3$. Each line of the tableau corresponds to the particular solution of (34) which determines the circled variable.

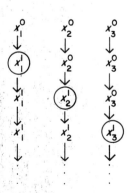

---

[13] The linear method has also been called the "single step method," the "method of successive displacements," and the "Liebmann method." For more details, see [12].

A central question about any iterative method concerns the convergence of its iterates; do numbers x_i^*, $i = 1, \cdots, v$, exist such that

$$\lim_{k \to \infty} x_i^k = x_i^*, \qquad i = 1, \cdots, v?$$

With regard to the NGS method, the iterates obviously depend not only on the functions F_i but also on the initial iterate x^0. Thus they may converge for some choices of x^0 but not for others. In the event that they converge for *any* choice of x^0, we say that the method is *globally convergent*. This is an extremely agreeable property for an iterative method to have, since it means that we do not have to be particularly concerned about how we start the method to obtain convergence.

Since the functions F_i are continuous, it follows from (34) that the NGS iterates converge to a solution of (33) providing that they converge at all. Therefore, the question of computing a solution to (33) by the NGS method hinges on the convergence of its iterates.

That the network equations are a special case of (33) may be seen by going back to the form (8) and letting

$$F_i(p_1, \cdots, p_v) = \sum_{j=1}^{m} a_{ij} f_j \left(\sum_{q=1}^{n} a_{qj} p_q \right), \qquad i = 1, \cdots, v. \tag{35}$$

At first glance this equation appears to be inconsistent since the left side involves only the first v pressures whereas the right side involves all $n > v$ pressures. However, it must be remembered that nodes $v + 1, \cdots, n$ are boundary nodes where the corresponding pressures are presumed known. So, in fact, the right side involves only v unknowns.

The following theorem completely answers the question of how the NGS method behaves relative to network equations. It guarantees if not the best of all possible worlds, at least a well-ordered state of affairs.

Theorem 2. *If the links of a network η have admissible characteristics, and if η has at least one boundary node, then the NGS iterates are well-defined and globally convergent to the unique solution of the network equations.*

The proof of this theorem, although not particularly difficult, is too long to include here. A proof may be found in [13].

Theorem 2 asserts that we can use the NGS method to calculate an approximate solution of the network equations and that this solution can be made to agree with the exact solution to any degree of accuracy provided that we solve (34) exactly and iterate long enough. Of course, since (34) is a nonlinear scalar equation in the unknown s, solving it exactly will not be possible in general. In fact, a subsidiary numerical method is usually employed to obtain an approximate solution. This means that in practice we cannot realize the hypotheses of Theorem 2. Nevertheless, we expect that if (34) is solved accurately, then the resulting approximate NGS iterates will converge to an accurate solution of the network equations.

A second point to be made regarding the practical implementation of the NGS method concerns the question of when to terminate the iterations. A completely satisfactory answer to this question is not known.

Usually, the decision to terminate is based on some measure of the difference of successive iterates. For example, if we define the norm of x, $\|x\|$, to be $(\sum_{i=1}^{v} x_i^2)^{1/2}$, then we might stop the iteration at the first value of k such that

$$\|x^k - x^{k-1}\| < \varepsilon \|x^k\|, \tag{36}$$

where $\varepsilon > 0$ is some preselected convergence criterion. Note that (36) in no way guarantees that the norm of the error $\|x^k - x^*\|$ is less than ε.

A more sophisticated stopping criterion can be formulated by taking into account the *rate of convergence* of the method. Several ways to define this rate exist (see, for example, [14 ch. 9]), but in practice it usually must be estimated as the iteration proceeds. A stopping criterion which employs the convergence rate is contained in [15].

The idea of a convergence rate serves another useful purpose—it frequently allows us to compare different iterative methods. For instance, we obtain the nonlinear successive overrelaxation method from the NGS

Table 1

j	L_j(ft)	A_j(ft^2)	F_j	θ_j(deg)
1	0.5	0.28	0.106×10^{-3}	0.
2	0.5	0.21	0.136×10^{-3}	0.
3	0.5	0.28	0.106×10^{-3}	0.
4	1.25	0.124	0.53×10^{-3}	0.
5	1.25	0.093	0.7×10^{-3}	0.
6	1.25	0.124	0.53×10^{-3}	0.
7	0.16	0.01	0.6×10^{-3}	0.
8	0.16	0.01	0.6×10^{-3}	0.
9	0.16	0.01	0.6×10^{-3}	0.
10	1.25	0.124	0.53×10^{-3}	0.
11	1.25	0.093	0.7×10^{-3}	0.
12	1.25	0.124	0.53×10^{-3}	0.
13	0.5	0.28	0.106×10^{-3}	0.
14	0.5	0.21	0.136×10^{-3}	0.
15	0.5	0.28	0.106×10^{-3}	0.
16	0.5	0.42	0.112×10^{-2}	90.
17	0.5	0.42	0.112×10^{-2}	90.
18	0.5	0.047	0.0105	90.
19	0.5	0.047	0.0105	90.
20	0.5	0.047	0.0105	90.
21	0.5	0.047	0.0105	90.
22	0.5	0.42	0.112×10^{-2}	90.
23	0.5	0.42	0.112×10^{-2}	90.

method simply by letting $x_i^{k+1} = \omega s + (1 - \omega)x_i^k$ in step 3. Here, $1 \leq \omega < 2$ is called the relaxation parameter. The NGS method is recovered by letting $\omega = 1$. However, choices of ω significantly greater than one frequently give iterates whose convergence is dramatically faster than that of the NGS iterates (see the sample problem in the next section). What is even more remarkable is that, for a large class of problems, this is quantitatively predicted by the convergence rates. For more information on the successive overrelaxation method, the reader is referred to [12], [16] for the linear case and [13], [15] for the nonlinear case.

We conclude this section with a final word about the numerical solution of (34) in the case where the underlying system (33) represents the network equations. From (35), we see that the ith network equation is

$$\sum_{j=1}^{m} a_{ij}f_j\left(\sum_{q=1}^{n} a_{qj}p_q \right) = 0, \tag{37}$$

from which it follows that the ith equation (34) is

$$\sum_{j=1}^{m} a_{ij}f_j\left(\sum_{1 \leq q < i} a_{qj}p_q^{k+1} + a_{ij}s + \sum_{i < q \leq v} a_{qj}p_q^k + \sum_{v < q \leq n} a_{qj}p_q^* \right) = 0. \tag{38}$$

We recall that a_{ij} is the element in the ith row and jth column of the incidence matrix and the functions $f_j(t)$ define the admissible link characteristics.[14] For the steam generator flow problem they are given explicitly by (22).

Now each term in (38) may be written as $a_{ij}f_j(a_{ij}s + C_{ij}^k)$ where $a_{ij} = 0, \pm 1$ and C_{ij}^k is a constant. Since the characteristics are admissible, it follows that *each nonvanishing term* of the sum (38) is a strictly increasing function of s which changes sign, and so the same is true of the sum. Therefore, exactly one value of s solves (38), and we compute this numerically by such methods as bisection, *regula falsi*, etc.[15]

5. Illustrative Problem

In this section we pursue the investigation of the flow problem associated with the network of Figure 23.5. This problem has already been discussed in Section 3 and gives rise to a set of 12 network equations of the form (37). These are completely described by the incidence matrix, the data which defines the link characteristics (22), and the boundary conditions.

The elements of the incidence matrix may be obtained directly from Figure 23.5 in an obvious manner. For example, the only nonzero elements of the first row are $a_{11} = -1$, $a_{14} = 1$, and $a_{1,16} = 1$.

---

[14] To test his understanding of the nature of (38), the reader should convince himself that for the network of Figure 23.5, the equation which applies when $i = 2$ is $-f_2(p_{14}^* - s) + f_5(s - p_5^k) - f_{16}(p_1^{k+1} - s) + f_{17}(s - p_3^k) = 0$.

[15] See [17], [18] for a description and analysis of these basic methods.

Table 2

Node	Pressure[†] (pdl/ft^2)	Node	Pressure (pdl/ft^2)	Node	Pressure (pdl/ft^2)
1	6810.6	2	6810.6	3	6810.6
4	4826.6	5	4820.2	6	4826.6
7	2773.4	8	2779.8	9	2773.4
10	789.41	11	789.41	12	789.41

[†] To convert poundals per square foot to the more familiar pounds per square inch multiply by 0.216×10^{-3}. Thus the pressure at node 1 is 1.47 lb/in^2. Note that the total pressure variation over the domain of the problem is less than 1.35 lb/in^2. This is consistent with our earlier assumption of a constant system pressure (see (25)).

Table 3

Link	Flow (lb/s)	Link	Flow (lb/s)	Link	Flow (lb/s)
1	136.3	2	90.8	3	136.3
4	129.3	5	104.7	6	129.3
7	121.3	8	120.8	9	121.3
10	129.3	11	104.7	12	129.3
13	136.3	14	90.8	15	136.3
16	6.94	17	−6.95	18	8.08
19	−8.08	20	−8.08	21	8.08
22	−6.95	23	6.93		

To define the characteristics (22), we take $\rho = 49$ lb/ft^3, $g = 32.2$ ft/s^2, and the remaining data from Table 1. The density we have chosen is approximately the density of water about to boil at a system pressure of 600 psia. As boundary conditions we assume that $p_{13} = p_{14} = p_{15} = 7600$ pdl/ft^2 and $p_{16} = p_{17} = p_{18} = 0.0$ pdl/ft^2.

The problem defined by these conditions was solved on a computer by the NGS method described in Section 4. The initial pressures were taken to be $p_i^\circ = 330$ pdl/ft^2, $i = 1, \cdots, 12$. After 36 iterations the quantity $[\sum_{i=1}^{12} (p_i^{36} - p_i^{35})^2]^{1/2}$ was less than 10^{-4}, and the problem was declared to be converged. The pressures and flows thus obtained are given in Tables 2 and 3. The problem was then resolved by the nonlinear successive overrelaxation method with $\omega = 1.33$. Essentially the same solution was obtained in 19 iterations, illustrating the advantage that can be gained by iterating with $\omega > 1$.

References

[1]   *Steam, Its Generation and Use.* New York: Babcock and Wilcox, 1972.
[2]   R. B. Potts and R. M. Oliver, *Flows in Transportation Networks.* New York: Academic, 1972.
[3]   L. R. Ford and D. R. Fulkerson, *Flows in Networks.* Princeton, NJ: Princeton Univ. Press, 1962.

[4] C. Berge and A. Ghouila-Houri, *Programming, Games and Transportation Networks.* New York: Wiley, 1965.

[5] R. G. Busacker and T. L. Saaty, *Finite Graphs and Applications.* New York: McGraw-Hill, 1965.

[6] *The Holy Bible*, Exodus 14:21.

[7] C. C. MacDuffee, *Vectors and Matrices*, Carus Mathematical Monograph Number 7, Mathematical Association of America, 1943.

[8] G. Birkhoff, "A variational principle for nonlinear networks," *Q. Appli. Math.*, vol. 21, pp. 160–162, 1963.

[9] J. G. Knudsen and D. L. Katz, *Fluid Dynamics and Heat Transfer*, New York: McGraw-Hill, 1958.

[10] R. Von Mises and K. O. Friedrichs, "Fluid dynamics," in *Applied Mathematical Sciences*, vol. 5. New York: Springer-Verlag, 1971.

[11] J. E. Meyer, "Hydrodynamic models for the treatment of reactor thermal transients," *Nucl. Sci. and Eng.* vol. 10, pp. 269–277, 1961.

[12] R. S. Varga, *Matrix Iterative Analysis*, Prentice-Hall, Englewood Cliffs, New Jersey, 1962.

[13] W. Rheinboldt, "On M-functions and their application to nonlinear Gauss–Seidel iterations and network flows," *J. Math. Anal. Appl.*, vol. 32, pp. 274–307, 1970.

[14] J. Ortega and W. Rheinboldt, *Iterative Solution of Nonlinear Equations in Several Variables.* New York: Academic, 1970.

[15] L. A. Hageman and T. A. Porsching, "Aspects of nonlinear block successive overrelaxation," *SIAM J. Numerical Anal.*, 1975.

[16] D. Young, *Iterative Solution of Large Linear Systems.* New York: Academic, 1971.

[17] A. Ostrowski, *Solution of Equations and Systems of Equations*, 2nd ed. New York: Academic, 1966.

[18] J. Traub, *Iterative Methods for the Solution of Equations.* Englewood Cliffs, NJ: Prentice-Hall, 1964.

[19] S. K. Beal, "Deposition of particles in turbulent flow on channel or pipe walls," *Nucl. Sci. Eng.*, vol. 40, pp. 1–11, 1970.

[20] ——, "Prediction of heat exchanger fouling rates—A fundamental approach," preprint of paper presented at AICHE Meeting, Nov. 1972.

[21] W. A. Blackwell, *Mathematical Modeling of Physical Networks.* New York: Macmillan, 1968.

Notes for the Instructor

Prerequistes. The contents of this chapter should be accessible to a student who has had courses in the elementary calculus and linear algebra. Some familiarity with the basic conservation laws of physics (i.e., mass, energy, and momentum) is desirable, but not essential. The same is true of computer programming, since a student should be extremely gratified by utilizing a computer to administer the *coup-de-grace* to *a* problem on which he has spent considerable time.

Time. The unit can be covered in three or fewer lectures—less if much of the material is left to the student.

Remarks. Section 1: The opening section attempts to motivate the whole study by relating it to an actual problem encountered in the operation of a nuclear power plant. The instructor should emphasize that this problem is a real concern to the nuclear power industry and involves plant components costing millions of dollars. He should also show the students schematics of the plant and steam generator and in general terms describe the heat transfer and hydraulic processes which take place. More details of plant operation may be found in [1] and "Systems Summary of a Westinghouse Pressurized Nuclear Power Plant" by G. Masche, which can be obtained by writing to Westinghouse Electric Corporation.

After Section 1 has been discussed, the class will have been presented with a general, nontechnical statement of the problem. At this point the instructor may wish to consider other methods of attack in addition to the network approach. For example, if the mathematical sophistication of the class is sufficient to allow consideration of partial differential equations, then the problem can be restated in terms of finding an appropriate solution of the classical equations of fluid dynamics—the momentum (Navier–Stokes) equations and the continuity equation. This is the most straightforward method of attack, but it should be quickly realized that, because of the geometry of the flow region and the complex nature of the equations themselves, such an attack is not likely to succeed. As an alternative to a complete solution of the equations, one might settle for a numerical solution by finite differences. This opens the door on a whole new subject—computational fluid dynamics. In this connection, *Computational Fluid Dynamics*, by P. J. Roach, Hermosa Publishers, 1972, may be consulted. However, the direct finite difference approach leads to an enormous computational problem whose solvability is in turn open to question.

Section 2: Section 2 lays the ground work for the development of a model which is based on simple network concepts. This section is autonomous and could also be used to supplement courses in graph theory or the numerical solution of systems of nonlinear equations.

The physical significance of Kirchhoff's node law as a discrete conservation law should be emphasized. It is, of course, the same law that is satisfied by the current in electrical networks.

It would be instructive for the class to construct the incidence matrices of several simple graphs. The proof that an incidence matrix with n rows has rank n-1 is not an empty academic exercise. This fact is used later to complete the proof of an existence and uniqueness theorem for network flows. The determination of the rank could be assigned as a class exercise. This is also true of the demonstration that the sum of the boundary flows is zero.

In electrical network terminology, a link characteristic is the familiar Ohm's law. The instructor can motivate the form of (6) by reminding the class that many flows in physics have been observed to depend on the difference of values of other variables; for example, heat on a temperature difference, electrical current on a voltage difference, and air flow on a difference in atmospheric pressure.

It is extremely important to have the class understand the nature of the network equations (7). These have been written out in detail for a specific network. The instructor might carry this example further by substituting some simple functions for the link characteristics.

To save time, the proof of Theorem 1 in the case of linear link characteristics may be omitted. However, it does constitute a useful application of matrix algebra.

Section 3: This section develops the network model of the steam generator flow problem. The procedure is essentially the "control volume" approach which is quite common in fluid dynamics.

Some other factors which contribute to the pressure drop in addition to elevation and friction are sudden expansions or contractions of the flow region and sharp bends or corners in the flow region. (See [9] for more details.)

Of the empirical quantities introduced in this section, the friction factor is by far the most elusive. Relative to this, [9] may be consulted for more details. Alternately, the instructor could invite an engineering oriented colleague to give a guest lecture on the nature and determination of friction factors and other correlations.

To give the student some appreciation of the complexity of the real problem and the magnitude of the assumptions required to formulate it in terms of a network model, the instructor should discuss the nature of the thermal effects. On the other hand, the appraisal of the link characteristics, although it provides some understanding of the pressure drop effects not taken into account, may be omitted if the class is not familiar with partial differential equations.

Section 4: This section describes an iterative method which may be used to solve the network equations numerically. Like Section 2, it is autonomous and constitutes a small exercise in numerical analysis. The only notions employed are those of limit, continuity, function composition, and the simple bisection method for solving single equations. The bisection method or an equivalent is likely to be a part of any numerical analysis subroutine package supplied by computer manufacturers.

The ideas of convergence rate and successive overrelaxation are not essential for the solution of the network problem. They are intended to provide further avenues of exploration for the curious.

Section 5: Herein is presented a contrived but nontrivial illustrative problem and its numerical solution. Ideally, the class should write a computer program to solve this problem or one that they themselves have developed. If time or computer is not available to allow such an ambitious undertaking, the class should at least hand calculate a few pressure iterates at a given node, assuming that the pressures at the remaining nodes are those of Table 2. In this way they will see that the iterates do indeed tend to the appropriate quantity in Table 2.

After the flow problem has been solved, the instructor may wish to refer back to the simple particle deposition model introduced in Section 1 and have the class compute some particle deposition fluxes for a variety of particle concentrations. Furthermore, if it is assumed that the steam generator must be shut down for cleaning when the average deposit thickness exceeds a preselected amount (say 0.25 ft), then calculations could be done to determine when this occurs.[16] This is but one way in which the answers of the flow problem can be used to provide relevant information about the actual steam generator.

---

[16] For example, one could assume that only particles 1 micron in diameter are involved, that their density is 5 lb/ft^3, and that their average concentration is 10^{-4} lb/ft^3. The calculation could then be performed using the data of Table 3 and Figure 23.3. Note that the units on the ordinate of this figure are cm/s and not ft/s.